PHOSPHOLIPIDS
Biochemical, Pharmaceutical, and Analytical Considerations

PHOSPHOLIPIDS
Biochemical, Pharmaceutical, and Analytical Considerations

Edited by

Israel Hanin
Loyola University of Chicago
Stritch School of Medicine
Maywood, Illinois

and
Giancarlo Pepeu
University of Florence
Florence, Italy

PLENUM PRESS • NEW YORK AND LONDON

Library of Congress Cataloging-in-Publication Data

International Colloquium on Lecithin (5th : 1989 : Cannes, France)
   Phospholipids : biochemical, pharmaceutical, and analytical
considerations / edited by Israel Hanin and Giancarlo Pepeu.
      p.   cm.
   "Proceedings of the Fifth International Colloquium on Lecithin
... held April 10-12, 1989, in Cannes, France"--CIP t.p. verso.
   Includes bibliographical references and index.
   ISBN 0-306-43698-1
   1. Phospholipids--Congresses.   I. Hanin, Israel.   II. Pepeu.
Giancarlo.   III. Title.
   QP752.P53I58   1989
   612'.01577--dc20
                                                     90-48077
                                                         CIP
                                                          AC

QP
752
.P53
I58
1989

Proceedings of the Fifth International Colloquium on Lecithin;
Phospholipids: Biochemical, Pharmaceutical, and Analytical Considerations,
held April 10-12, 1989, in Cannes, France

ISBN 0-306-43698-1

© 1990 Plenum Press, New York
A Division of Plenum Publishing Corporation
233 Spring Street, New York, N.Y. 10013

Printed in the United States of America

PREFACE

This book is the result of the Proceedings of the 5th International Colloquium on Lecithin, which took place in Cannes, France, on April 10-12, 1989.

It follows what is becoming now a tradition of excellent International Conferences dealing with the focal subject of Lecithin and all of its applications and uses in every day life. Lecithin and its phospholipid components are used extensively in many fields, including human and animal nutrition, medicine, cosmetics and a variety of industrial applications. There is, therefore, considerable interest in the study of its chemical nature, of the biological properties of its phospholipid components, and in the identification of new potential uses.

In 1980, on the initiative of Lucas Meyer GmbH (Hamburg) the first Colloquium took place in Rome, Italy, and 4 more were to follow in 1982 (Brighton, England), 1984 (Vienna, Austria), 1986 (Chicago, USA), and 1989 in Cannes, France.

The scope of subjects covered during these five Colloquia has broadened from one to the next, reflecting the focus of scientific and technological endeavors in this field, at the time of each Colloquium. In this, the 5th Colloquium the focus was on a broader aspect of the field in comparison with the four earlier Colloquia, going beyond lecithin to the properties, biochemistry and biological actions of the individual phospholipids. This Colloquium, as is evident from the Table of Contents, dealt with general aspects of phospholipid technology, terminology and analysis. In addition, it also discussed critically the pathogenetic role of natural lipids such as the platelet aggregating factor and lysophosphatidylcholine, and the potential therapeutic usefulness of phosphatidylcholine and phosphatidylserine. It also focused on liposomes, both as drug delivery vehicles and as therapeutic agents.

Participants in the Colloquium again consisted of an interdisciplinary, International group of scientists from universities, clinics, research centers, official organs and industry. They represent a true cross-section of investigators interested in the complex field of phospholipids, as it relates to a variety of applications.

It is hoped that this book, which contains all the lectures and posters presented at the Colloquium, will provide a meaningful contribution to our knowledge regarding the biological importance of phospholipids, and will offer a useful reference source for future research in the field.

The Colloquium, which was also very successful in combining the scientific sessions with the elegant and enjoyable atmosphere of Spring in Cannes, was sponsored and financed again by the Lucas Meyer GmbH (Hamburg), for which the editors are most grateful.

Israel Hanin, Ph.D.
Chicago, Illinois

Giancarlo Pepeu, M.D.
Florence, Italy

CONTENTS

PHOSPHOLIPIDS - NATURAL, SEMISYNTHETIC, SYNTHETIC

F. Paltauf and A. Hermetter

Department of Biochemistry and Food Chemistry
Graz University of Technology
Graz, Austria

INTRODUCTION

Nature was remarkably inventive in creating the variety of polar lipids which form the matrix of biological membranes. The rationale for the variability of membrane lipids is not clear, it might simply be that these amphiphilic structures have in common the capability to arrange as bilayers in an aqueous environment. However, lipids are not only the plaster which holds the membrane together; there is ample evidence that distinct phospholipid classes or species serve additional tasks which make them indispensable for the functioning of membrane-linked pocesses. For example, phosphatidylinositols are involved in signal transduction and are therefore essential for the viability of eukaryotic cells. Most likely adaptation of polar lipid structures to specific requirements has occurred during evolution. The cell envelope of thermoacidophilic archaebacteria consists of chemically stable tetraether glyceroglyco- (or phospho-) lipids[19] that might be essential for these organisms to survive at the extremes of high temperature and low pH. The pulmonary surfactant coating the mammalian alveolus contains dipalmitoylphosphatidylcholine (DPPC) as a major constituent. Together with other phospholipids and specific proteins DPPC reduces the surface tension at the air-liquid interface[26]. Reversible changes in the membrane phospholipid pattern in response to environmental stress, e.g., temperature or solvents, have repeatedly been observed (see respective chapters in ref.17).

The same chemical and physicochemical properties that determine the role of phospholipids as structural components of biological membranes are relevant for their usefulness in numerous practical applications. It is the propensity of phospholipids to form bilayer structures or micelles, to stabilize dispersions or to coat surfaces, that is exploited in the manufacture of food, the production of cosmetics, the formation of liposomes employed as drug carriers, the stabilization of fat emulsions for intravenous administration, and for manifold industrial applications.

Phospholipids
Edited by I. Hanin and G. Pepeu
Plenum Press, New York, 1990

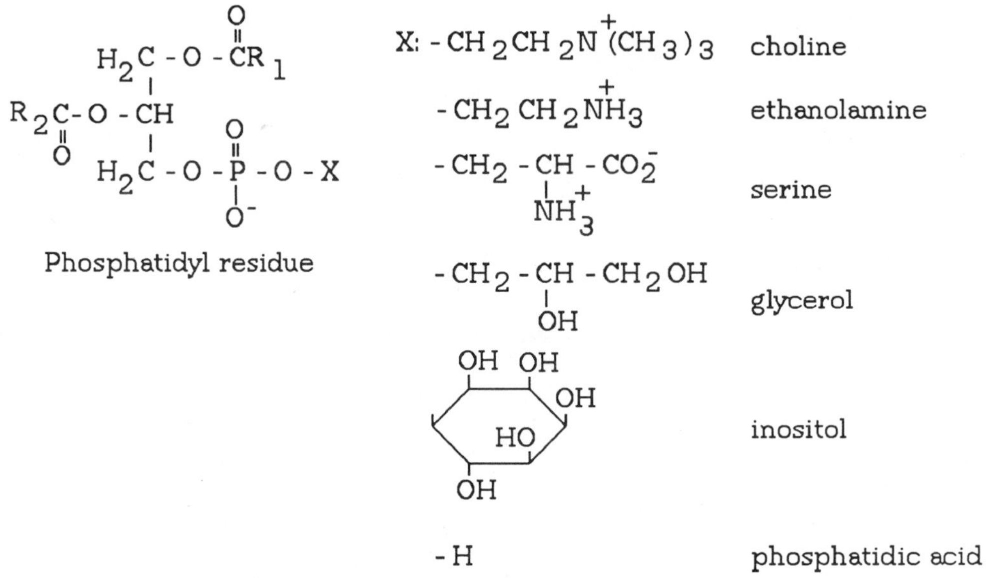

$$H_2C - O - \overset{\overset{\textstyle O}{\|}}{C} R_1$$
$$R_2 \overset{\|}{\underset{O}{C}} - O - CH$$
$$H_2C - O - \overset{\overset{\textstyle O}{\|}}{\underset{\underset{\textstyle O^-}{|}}{P}} - O - X$$

Phosphatidyl residue

X: $- CH_2 CH_2 \overset{+}{N}(CH_3)_3$ choline

$- CH_2 CH_2 \overset{+}{N}H_3$ ethanolamine

$- CH_2 - \underset{\underset{\textstyle \overset{+}{N}H_3}{|}}{CH} - CO_2^-$ serine

$- CH_2 - \underset{\underset{\textstyle OH}{|}}{CH} - CH_2 OH$ glycerol

inositol

$- H$ phosphatidic acid

Fig.1. Structure of glycerophospholipids.
R = long - chain alkyl.

In this chapter currently used methods for the preparation of phospholipids will be described, including isolation from natural sources, chemical modification of natural phospholipids, semisynthetic procedures and total chemical synthesis.

GLYCEROPHOSPHOLIPIDS

The chemical structure of glycerophospholipids allows modification in the head group, the hydrophobic side chains and the type of bonding between the aliphatic moieties and the glycerol backbone. Variation in the head group leads to different glycerophospholipid classes, such as phosphatidylcholine, - ethanolamine, - serine, - inositol, - glycerol etc. (see Figure 1). The composition of the apolar moieties characterizes phospholipid species, e.g., palmitoyloleoyl phosphatidylcholine, stearoyl-arachidonoyl phosphatidylethanolamine. The type of bonding (ester or ether) between the aliphatic chains and glycerol determines the phospholipid subclass (see Figure 2). Ether lipids can be either alkenyl (or vinyl) ether lipids, usually termed plasmalogens; these are found in practically all phospholipids from animal sources[15], and in some (mainly

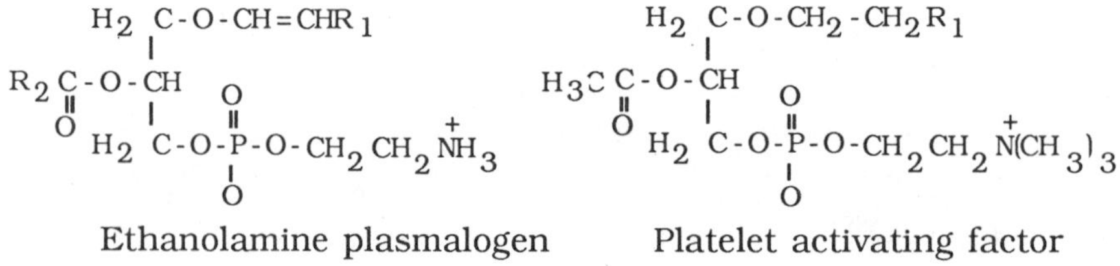

$$H_2C - O - CH = CHR_1$$
$$R_2 \overset{\|}{\underset{O}{C}} - O - CH$$
$$H_2C - O - \overset{\overset{\textstyle O}{\|}}{\underset{\underset{\textstyle O}{|}}{P}} - O - CH_2 CH_2 \overset{+}{N}H_3$$

Ethanolamine plasmalogen

$$H_2C - O - CH_2 - CH_2 R_1$$
$$H_3C \overset{\|}{\underset{O}{C}} - O - CH$$
$$H_2C - O - \overset{\overset{\textstyle O}{\|}}{\underset{\underset{\textstyle O}{|}}{P}} - O - CH_2 CH_2 \overset{+}{N}(CH_3)_3$$

Platelet activating factor

Fig.2. Structure of ether lipids

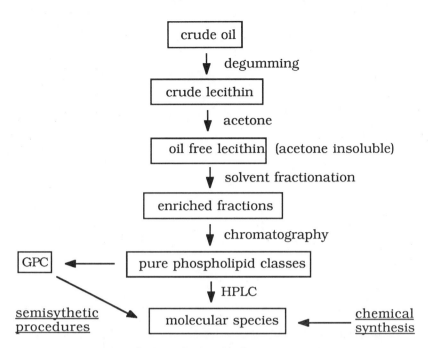

Scheme 1. Preparation of phospholipids from natural sources.
HPLC = high - pressure liquid chromatography.
GPC = sn - glycero - 3 - phosphocholine; it is obtained by
hydrolytic deacylation of natural phosphatidylcholine.

anaerobic) bacteria[10], but not in plants[20]. Their biological role as
membrane components is largely unknown. Another type are alkylacyl-
glycerolipids which are the metabolic precursors of plasmalogens[22], but
have biological significance by themselves, e.g., the so called platelet
activating factor, which is the first phospholipid known to exhibit a wide
spectrum of physiological activities[2]. In contrast to ether lipids shown in
Fig.2 archaebacterial di- and tetraether lipids contain only ether - linked
aliphatic moieties (and no acylester groups)[16,19]. Phospholipids from
natural sources will always be complex mixtures of phospholipid classes,
subclasses and species.

The main source of natural glycerophospholipids is soybean oil.
Other crude seed oils are quantitatively less important. Animal sources
are egg yolk or brain. Another prospective source of phospholipids are
microorganisms, such as bacteria, algae, fungi or yeasts[27].

The process of phospholipid preparation is schematically
depicted in Scheme 1. Plant phospholipids can be regarded as a
by-product in the manufacture of seed oil; therefore crude lecithin
is rather low priced. The more pure the fractions, the more
expensive they are, but phospholipids from natural sources will
always be less costly than the corresponding phospholipids
prepared by semisynthetic or synthetic methods. Isolation of
molecular species from natural sources can to some extent, be
achieved by HPLC[25], but available methodology does not allow
preparation on a large scale.

TABLE 1

PHOSPHOLIPID COMPOSITION (%)
OF SOYBEAN AND EGG YOLK LECITHIN[23].

Phospholipid class	Soybean	Egg yolk
Phosphatidylcholine	31,7	78,8
Phosphatidylethanolamine	20,8	17,1
Phosphatidylserine	3,0	trace
Phosphatidylinositol	17,5	0,6
Phosphatidic acid	2	trace
Plasmalogen	-	1
Sphingomyelin	-	2,5
Other phospholipids	10,2	-
Phytoglycolipids	14,8	-

TABLE 2

FATTY ACID COMPOSITION (%) OF PHOSPHATIDYLCHOLINE
FROM SOYBEAN AND EGG YOLK [a].

Fatty acid	Soybean	Egg yolk
$C_{16:0}$	14	31
$C_{16:1}$	-	1
$C_{18:0}$	4	13
$C_{18:1}$	10	30
$C_{18:2}$	64	15
$C_{18:3}$	7	0,5
$C_{20:4}$	-	4
$C_{22:6}$	-	3

[a] Data were kindly provided by Dr. M Schneider, Lucas Meyer, Hamburg.

The pattern of glycerophospholipid classes and species varies greatly depending on the source. Table 1 shows the phospholipid compositions of soya and egg yolk phospholipids. In Table 2 the fatty acid compositions of soybean and egg yolk phosphatidyl-choline are shown.

PHYSICAL AND CHEMICAL PROPERTIES OF GLYCEROPHOSPHOLIPIDS

The physical and chemical properties of glycerophospholipids are determined by the kind of head group and by the chain length and degree of unsaturation of the constituent aliphatic moieties. Therefore, modification of the chemical structure of phospholipids will affect their physical properties, which, in turn, will determine their practical usefulness, as was mentioned above. One method to obtain phospholipids with desired properties is to select the right source. For example, egg lecithin contains relatively more choline glycerophospholipids and less polyunsaturated species than soya lecithin (see Tables 1 and 2). In order to obtain tailored natural phospholipids one might attempt to manipulate the phospholipid pattern of plants or microorganisms, either by breeding mutants or by applying methods of genetic engineering[21].

Phospholipids contain a number of functional groups that make them chemically reactive. Chemical modification leads to phospholipids with altered physicochemical properties; such products are available on an industrial scale.

Treatment of crude phospholipids with acetic anhydride[24] leads to the acetylation of amino groups of phosphatidylethanolamine and phosphatidylserine, and of hydroxy groups of phosphatidylinositol. N - Acetyl phosphatidylethanolamine is less polar at the nitrogen and has a net negative charge (whereas phosphatidylethanolamine is zwitterionic over a wide pH range). Therefore, N - acetyl phosphatidylethanolamine has improved oil in water emulsifying properties.

Oxidation with hydrogen peroxide[9] leads to products with polar hydroxy groups in the otherwise apolar side chain region. The hydrophilic character of the phospholipids is thus increased, improving the dispersion of phospholipids in water. Oxygenation can also be achieved by the use of enzymes, such as lipoxygenases[31].

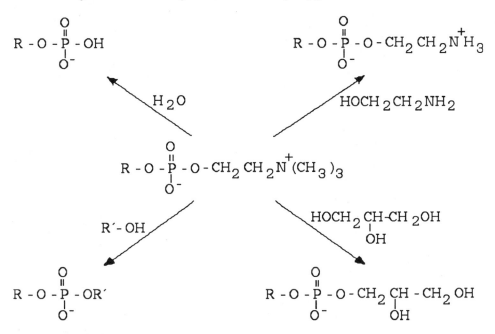

Fig.3. Phospholipase D - catalyzed transphosphatidylation.
R can be 1,2 - diacylglycerol; 1,2 - dialkylglycerol; 1(2),2(1) alkylacyl-glycerol; 1 - O (1´- alkenyl) 2 - acylglycerol; long - chain alkyl.
R´= primary alcohol not exceeding 6 carbon atoms.

5

The acylester bonds of phospholipids can be hydrolyzed either chemically or enzymatically. Complete acid or base - catalyzed hydrolysis yields fatty acids and water soluble glycerophosphoric acid diesters, e.g., glycerophosphocholine, which is of central importance in the semisynthetic route to defined glycerophospholipids. Controlled hydrolysis by phospholipase A_2 leads to lysophospholipids which greatly enhance the emulsifying properties of lecithins[31]. Phospholipases A_2 from natural sources are rather expensive, but cloning of the respective structural genes and expression in microorganisms should make these enzymes accessible on a large scale. Lysophosphatidylcholine forms micelles or mixed micelles together with other lipids, which might be used as drug delivery systems.

The double bonds of natural phospholipids can be hydrogenated using Ni or Pd as catalysts. Saturated phospholipids have higher melting points and greater oxidative stability. They form stable emulsions, which are used for manufacturing pharmaceutical or cosmetic ointments.

Phospholipase D not only hydrolyzes phospholipids to phosphatidic acid, but also catalyzes transphosphatidylation[3], as shown in Figure 3. This reaction, when carried out with crude phospholipid preparations in the presence of glycerol, leads to phosphatidylglycerol - rich phospholipids[18]. Phosphatidylglycerol is a potent emulsifier even in the presence of Ca ions. When carried out with defined phosphatidylcholine species, phospholipase D - catalyzed transphosphatidylation is a relatively simple and straightforward method for the semisynthetic preparation of various phospholipid classes and species.

Chemically defined phospholipids are required for the preparation of liposomes used in the parenteral or enteral applicaton of drugs, as emulsifiers with specific properties, for the preparation of biocompatible coatings or vesicles consisting of polymerizable phospholipids, as

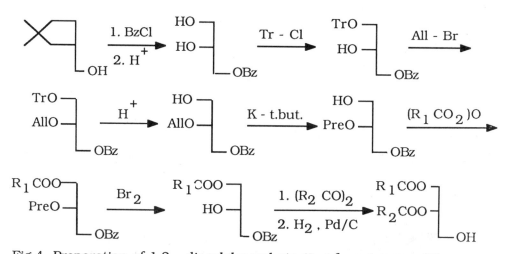

Fig.4. Preparation of 1,2 - diacylglycerol starting from isopropylidene glycerol.
Bz, benzyl; All, allyl; Tr, triphenylmethyl; Pre, propenyl; K - t.but. potassium tert. butylate.

pharmacologically active compounds, in biochemical or biophysical research, etc. Phospholipids with defined chemical structure are accessible by chemical synthesis or semisynthetic procedures. Methods are available for the total chemical synthesis of essentially any phospholipid or phospholipid analog[6], but the reaction sequences comprise numerous steps, and the final products are therefore rather expensive.

SYNTHESIS OF PHOSPHOLIPIDS

Phospholipid synthesis involves the formation of ester or ether bonds linking the aliphatic moieties to the glycerol backbone, and the attachement of the polar head group. The most common educts for the glycerol part of phospholipids are ispropylidene glycerols or sugar alcohols, such as D - mannitol. Racemic isopropylidene glycerol is rather inexpensive in contrast to the enantiomers, which are required when optically active phospholipids are to be synthesized.

In Figure 4 examples are given to illustrate the reaction sequences leading from isopropylidene glycerol to diacylglycerols[7]. There are multiple variants to this reaction scheme, using different arrangements of

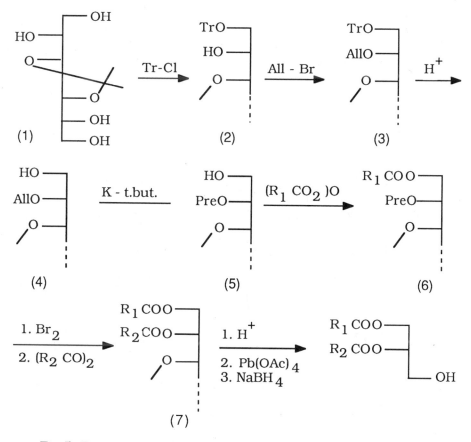

Fig.5. Preparation of 1,2 - diacylglycerols staring from 3,4 - isopropylidene-D - mannitol.
For abbreviations, see Fig.4.

7

protecting groups that can selectively be removed and replaced with the desired substituent. The benzyl group is resistant against hydrolysis and can be removed only by catalytic hydrogenation. The trityl group can be introduced specifically to primary hydroxy groups and is removed by acid treatment. Under strong basic conditions (potassium - t. - butylate) the allyl group is converted to the propenyl group[4] which can be cleaved off in the presence of bromine or by mild acid treatment.

Similar protecting groups as used with isopropylidene glycerol can be used when D - mannitol is the starting material[8] (see Figure 5). D - Mannitol is first converted to the 3,4 - isopropylidene derivative. This is then converted to 1,6 - ditrityl - 2,5 - diacyl - 3,4 - isopropylidene D - mannitol. The steps leading to (7) are similar to those described for the synthetic route starting from isopropylidene glycerol (see above). Then, the isopropylidene group is removed by acid treatment and the resulting diol is oxidatively cleaved with lead tetraacetate. The aldehyde thus formed is reduced with sodium borohydride to yield the desired

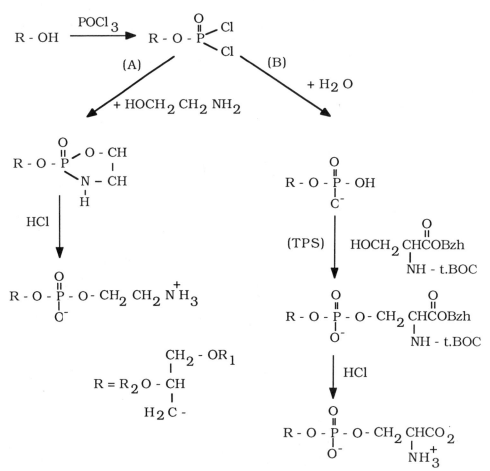

Fig.6. Synthesis of phospholipids starting from 1,2 - diradylglycerols. TPS, triisopropylbenzenesulfonyl chloride; Bzh, benzhydryl; t.BOC, tert.- butyloxycarbonyl; R_1 and R_2, long - chain acyl or alkyl

mixed - acid diglyceride. At this step acyl migration is inevitable, leading to at least 10 % of the 1,3 - isomer[8]. By analogous routes 1,2 - di - O - alkylglycerols can be prepared.

The phosphodiester head group can be attached by a variety of methods (for a recent review, see ref.28). Two possible routes are depicted in Figure 6. Route A involves a cyclic oxazaphospholane intermediate[5], route B employs phosphatidic acid as an intermediate, which can be coupled with different alcohols, such as choline[1], N - protected ethanolamine, protected glycerol[30] or, as shown here, protected serine[13]. In addition to triisopropylbenzenesulphonyl chloride other coupling reagents can be used, such as dicyclohexylcarbodiimide or trichloroacetonitrile.

The examples given are representative insofar as the routes described can be used for the preparation of most of the common phospholipid classes, subclasses and species. It should be noted, however, that alternative methods exist which may be more convenient for the preparation of specific glycerophospholipids.

Partial synthesis of glycerophospholipids starts from preformed glycerophosphodiesters and therefore requires much less reaction steps than total synthesis. Methods for the semisynthetic preparation of phospholipids have been reviewed at the last meeting in Chicago[12] and shall not be repeated here in detail.

The most convenient starting material for the semisynthetic preparation of phospholipids is sn - glycero - 3 - phosphocholine (GPC) which is obtained by deacylation of natural phosphatidylcholine. Acylation of GPC with an activated acyl derivative, e.g., acylimidazolide or acyl anhydride yields in a one - step process[11,29] phosphatidylcholines with

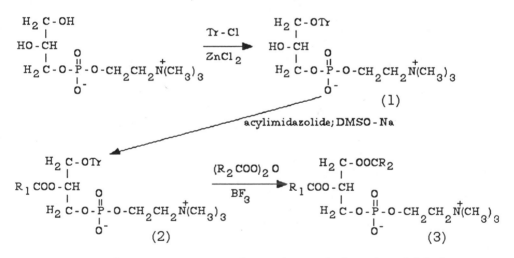

Fig.7. Semisynthetic preparation of mixed - acid phosphatidylcholines. Tr, triphenylmethyl; DMSO, dimethylsulfoxide. For the acylation of (2) to (3) acyltriazolides can be used instead of fatty acid anhydrides.

two identical acyl moieties and with the same steric configuration as naturally occurring glycerophospholipids. We have recently developed a method for the partial synthesis of mixed - acid phosphatidylcholines[14] which is described in Figure 7. GPC is converted to 1 - O - trityl GPC, which is then acylated to tritylacyl GPC. In a one - pot reaction the trityl group is replaced with an acyl group. This reaction can be carried out with different activated acyl derivatives, such as acyl anhydrides or acyltriazolides. In the presence of borontrifluoride - etherate as a catalyst, the reaction proceeds within a short time (30 min) at low temperature (around 0°C), and the final product obtained in high yield contains less than 2 - 3 % of positional isomers. Pure mixed - acid phosphatidyl-cholines are thus accessible in a three - step procedure which can be carried out on any scale. This reaction is considerably shorter and therefore less expensive than the total synthesis of the corresponding phospholipids. Modification of the polar head group using phospholipase D - catalyzed transphosphatidylation (see Figure 3.), or condensation of phosphatidylcholine - derived phosphatidic acid with any desired alcohol (e.g., serine or unnatural alcohols) gives access to the majority of phospholipids that might be required for practical or scientific purposes.

CONCLUSION

Whenever applicable, the semisynthetic route to defined phospholipids is the method of choice. It provides easy access to defined mixed - acid phospholipids, and will promote their use in diverse fields of application.

REFERENCES

1. Aneja, R., and Chadha, J. S., 1971, A total synthesis of phosphatidylcholines, Biochim. Biophys. Acta 248:455-457.

2. Benveniste, J., and Vargaftig, B. B., 1983, Platelet activating factor: An ether lipid with biological activity, in: "Ether Lipids. Biochemical and Biomedical Aspects", H. K. Mangold and F. Paltauf, eds., Academic Press, New York, pp. 355-376.

3. Comfurius, P., and Zwaal, R. F. A., 1977, The enzymatic synthesis of phosphatidylserine and purification by CM - cellulose column chromatography, Biochim. Biophys. Acta 488:36-42.

4. Cunningham, J., and Gigg, R., 1965, The preparation of 1 - O - alk - 1' - enyl ethers of glycerol, J. Chem. Soc. 2968-2975.

5. Eibl, H., 1978, Phospholipid synthesis: Oxazaphospholanes and dioxaphospholanes as intermediates, Proc. Natl. Acad. Sci. USA 75:4074-4077.

6. Eibl, H.,1980, Synthesis of glycerophospholipids, Chem. Phys. Lipids 26:405-429.

7. Eibl, H., and Woolley, P.,1986, Synthesis of enantiomerically pure glyceryl esters and ethers. I. Methods employing the precursor 1,2 - isopropylidene - sn - glycerol, Chem. Phys. Lipids 41:53-63.

8. Eibl, H., andWoolley, P., 1988, Synthesis of enantiomerically pure glyceryl esters and ethers. II. Methods employing the precursor 3,4 - isopropylidene - D - mannitol, Chem. Phys. Lipids 47:47-53.

9. El-Tarras, M. F., Abdel Moety, E. M., Ahmad, A. K. S., and Amer, M. M., 1976, Studies on rancidity of oils and fats. On the autoxidation of phospholipids, Oleagineux 31:229 ; Chem. Abstr. 85:122107.

10. Goldfine, H., and Hagen, P. - O., 1972, Bacterial plasmalogens, in: "Ether Lipids. Chemistry and Biology," F. Snyder, ed., Academic Press, New York, pp. 329-350.

11. Hermetter, A., and Paltauf, F., 1981, A facile procedure for the synthesis of saturated phosphatidylcholines, Chem. Phys. Lipids 28:111-115.

12. Hermetter, A., and Paltauf, F., 1987, Partial synthesis of glycerophospholipids, in: "Lecithin. Technological, Biological and Therapeutic Aspects," I. Hanin and G. B. Ansell, eds., Plenum Press, New York, pp. 37-45.

13. Hermetter, A., Paltauf, F., and Hauser, H., 1982, Synthesis of diacyl and alkylacyl glycerophosphoserines, Chem. Phys. Lipids 30:35-45.

14. Hermetter, A., Stütz, H., Franzmair, R., and Paltauf, F., 1989, 1 - O - Trityl - sn - glycero - 3 - phosphocholine: a new intermediate for the facile preparation of mixed - acid 1,2 - diacyl-glycerophosphocholines, Chem. Phys. Lipids, 50: 57-62.

15. Horrocks, L. A. , 1972, Content composition and metabolism of mammalian and avian lipids that contain ether groups, in: "Ether Lipids. Chemistry and Biology," F. Snyder, ed., Academic Press, New York, pp. 177-272.

16. Kates, M., 1972, Ether - linked lipids in extremely halophilic bacteria, in: "Ether Lipids. Chemistry and Biology," F. Snyder, ed., Academic Press, New York, pp. 351-398.

17. Kates, M., and Kuksis, A., eds., "Membrane Fluidity. Biophysical Techniques and Cellular Regulation", 1980, The Humana Press Inc., Clifton.

18. Kudo, S., 1988, Biosurfactants as food additives, in: "Proceedings World Conference on Biotechnology for the Fats and Oils Industry," T. H. Applewhite, ed., Amer. Oil Chem. Soc., pp. 195-201.

19. Langworthy, T. A., 1983, Dialkyldiglyceroltetraethers, in: "Ether Lipids. Biochemical and Biomedical Aspects," H. K. Mangold and F. Paltauf, eds., Academic Press, New York, pp. 161-175.

20. Mangold, H. K., 1972, The search for alkoxylipids in plants, in: "Ether Lipids. Chemistry and Biology," F. Snyder, ed., Academic Press, New York, pp. 399-405.

21. Nielsen, N. C., and Wilcox, J. R., 1988, Biotechnology for soybean improvement, in: "Proceedings. World Conference on Biotechnology for the Fats and Oils Industry," T. H. Applewhite, ed., Amer. Oil Chem. Soc., pp. 58-64.

22. Paltauf, F. , 1983, Biosynthesis of 1 - O - (1'alkenyl) glycerolipids (plasmalogens), in: "Ether Lipids. Biochemical and Biomedical Aspects," H. K. Mangold and F. Paltauf, eds., Academic Press, New York, pp. 107-128.

23. Pardun, H., 1982, Progress in productions and processing of vegetable lecithins, Fette Seifen, Anstrichm. 84:1-11.

24. Pardun, H., 1982, An empiric method to determine the emulsifiability of vegetable lecithins in O/W - systems, Fette Seifen, Anstrichm. 84:291-299.

25. Patton, G. M., Fasulo, J. M., and Robins, S. J., 1982, Separation of phospholipids and individual molecular species of phospholipids by high - performance liquid chromatography, J. Lipid Res. 23:190-196.

26. Possmayer, F., Metcalfe, I. L., and Enhorning, G., 1980, The pulmonary surfactant, in: "Membrane Fluidity. Biophysical Techniques and Cellular Regulation",M. Kates and A. Kuksis, eds., The Humana Press Inc., Clifton, pp. 57-67.

27. Ratledge, C., 1987, Microorganisms as sources of phospholipids, in: Lecithin. Technological, Biological and Therapeutic Aspects," I. Hanin and G. B. Ansell, eds., Plenum Press, New York, pp. 17-35.

28. Stepanov, A. E., and Shvets, V. I., 1980, Formation of phosphoester bonds in phosphoglyceride synthesis, Chem. Phys. Lipids 41:1-51.

29. Warner, T. G., and Benson, A., 1977, An improved method for the preparation of unsaturated phosphatidylcholines: Acylation of sn - glycero - 3 - phosphocholine in the presence of sodium methylsulfinylmethide, J.Lipid Res. 18:548-552.

30. Woolley, P., and Eibl, H., 1988, Synthesis of enantiomerically pure phospholipids including phosphatidylserine and phosphatidylglycerol, Chem. Phys.Lipids 47:55-62.

31. Yamane, T., 1988, Enzyme technology for the lipids industry: an engineering overview, in: "Proceedings. World Conference on Biotechnology for the Fats and Oils Industry," T. H. Applewhite, ed., Amer. Oil Chem. Soc. , pp. 17-22.

OVERVIEW OF ANALYTICAL METHODS FOR PHOSPHOLIPID STUDIES

Serge Laganiere

Department of Pharmacology
Loyola University Chicago
Stritch School of Medicine
2160 South First Avenue
Maywood, Illinois 60153

1.0 INTRODUCTION

It is now firmly recognized that phospholipids (PL) are important
biologically active mediators of signal transduction within the cell
and also between cells. These findings have completely modified the
earlier notion that phospholipids were merely involved structurally in
the hydrophobic core of cellular membranes and responsible for
excitability of neural and muscular tissue, or served as eicosanoid
precursors of prostaglandins and leukotrienes.

In recent years, lipidologists have unravelled many new roles for
the phospholipids and their derivatives. Several examples follow:

Phosphoinositide (PtdIns) turnover produces metabolites which
become intracellular mediators in mobilizing Ca^{++} (13) and stimulating
protein kinase C (94). In the central nervous system, several
neurotransmitters have been shown to stimulate phosphatidylinositol
(PI) turnover (41). Therefore, many receptor-stimulated regulatory
events are dependent on phospholipids as dynamic sources of second
messenger (61).

ABBREVIATIONS: according to the recommendations of the IUPAC-IUB
Commission on nomenclature of lipids, the following abbreviations will
be used throughout the text: Ptd$_2$Gro, cardiolipin (CL); PtdH,
phosphatidic acid (PA); PtdCho, phosphatidylcholine (PC); PtdEth,
phosphatidylethanolamine (PE); PtdGro, phosphatidylglycerol (PG);
PtdIns, phosphatidylinositol (PI); PtdSer, phosphatidylserine (PS);
CerPCho, sphingomyelin (SPM); LysoPtdCho, lysophosphatidylcholine
(LPC). Other abbreviations: PL, phospholipid(s); DAG, diacylglycerol;
PAF, platelet activating factor; AGEPC, 1-O-alkyl-2-acetyl-sn-glycero-
3-phosphocholine; HETE, hydroxyeicosatetraenoic acid; HPETE,
hydroxyperoxyeicosatetraenoic acid; FFA, free fatty acid; DEAE,
diethylaminoethylcellulose; TEAE, triethylaminoethylcellulose; TLC,
thin layer chromatography; HPTLC, high performance thin layer
chromatography; OPTLC, overpressured thin layer chromatography; HPLC,
high performance liquid chromatography; RP-HPLC, reverse-phase HPLC.

Phospholipids
Edited by I. Hanin and G. Pepeu
Plenum Press, New York, 1990

The 1-O-alkyl-2-acetyl-sn-glycero-phosphocholine, coined platelet activating factor (PAF), has been characterized as an extremely biologically active phospholipid in inflammatory and allergic reactions (11,22,51). Receptor-mediated contraction of smooth muscle cells, induction of local leukocyte infiltration (115) and increase in cell permeability are also mediated by PAF.

Phospholipids have been implicated in cancer research as antitumor compounds (118,136). The antineoplastic activity of alkyl-lysophospholipids on mouse tumors is mainly due to activation of cytotoxic macrophages by the lyso compound (5,12). Newer developments are ongoing with phospholipid derivatives, as shown elsewhere in this book.

Recently, attention has been focused on lysophosphatidylcholine (LPC) as a causal agent of atherosclerosis, as LPC can stimulate division of smooth muscle cells by an LPC-dependent increase in Ca^{++} uptake (130). Moreover, low concentrations of LPC stimulate protein kinase C in the presence of phosphatidylserine (PS) and Ca^{++} (95). LPC may also play a role in atherogenesis as a potent chemoattractant factor for human monocytes (103).

Finally, brain PL's are rich in highly unsaturated fatty acid moieties that can become a preferential target for oxygen-derived free-radicals, induce formation of lipid hydroperoxides and alter cell membrane stability.

As every lipidologist will agree, no single approach can yet be used for the analysis of phospholipids. Separation, detection, collection and quantitation of phospholipids still require multi-step operations (85). This chapter will focus on procedures that are currently available for the analysis of glycerophospholipids and some of their biologically active derivatives.

2.0 QUANTITATIVE DETERMINATION OF PHOSPHOLIPIDS

2.1 LIPID EXTRACTION

Most methods use chloroform and methanol to extract lipids from whole tissue or tissue homogenates (16,43).

Routinely, tissue lipids are extracted with a motor-driven glass-glass homogenizer with 15-20 volumes of chloroform-methanol (2:1;v/v) per gram of tissue. After sedimentation at 1500 x g for 5-10 min, the tissue residue may be re-extracted with the same solvent and the 2 organic layers are pooled. Non-lipid materials contaminate the lipid fraction and should, in most instances, be removed.

The organic phase is subsequently washed with 0.2 volume of an aqueous solution of 0.5% $CaCl_2$ or 0.75% KCl (43). Phase separation is completed overnight at 4 °C or by centrifugation (1500 x g for 10 min) in the cold. This step suffices to remove most of the gangliosides and ceramides from the organic layer but does not eliminate the proteolipid protein. Inevitably, the phase separation will yield losses of lysoglycerolipids and prostaglandins into the aqueous phase. Complete removal of non-lipid components is a difficult task that can be alleviated if further separation of the lipid mixture is conducted.

Recently, hexane-isopropanol (3:2;v/v) has been advocated (53) as a less toxic universal solvent than chloroform-methanol for lipid extraction. Only one single phase extraction is generated with this solvent system and the non-lipid material is very easily pelleted. Kolarovic and Fournier (82) evaluated the extent and completedness of the lipid extraction from biological tissue by a variety of one-phase and two-phase systems using high performance thin layer chromatography (HPTLC) combined with scanning photodensitometry. They showed that the total chromogenic response of non-acidic PL's was lowered by 10-35% when various two-phase separation systems were used as compared to one-phase systems.

Due to the ionic nature of polyphosphoinositides, their extraction benefits from using acidic procedures. Addition of 0.25-0.5% concentrated hydrochloric acid to a Folch extraction enhances total recovery of most polyphosphoinositides (56). However, free fatty acids (FFA) and diacylglycerols (DAG) are also increased as a result of this modification, possibly indicating acid breakdown (65).

Several cautionary procedures should be followed when lipid chemistry is planned. The polyunsaturated fatty acyls of the phospholipids, particularly those from neural origin, are susceptible to oxidation. For lipid work, a cold room is best. High quality solvents may be conditioned by bubbling N_2 through them and should always be used at 4°C. Antioxidants such as BHA or BHT may also be added (0.01%) during lipid extraction. The lipids, dissolved in an apolar solvent, should be stored under nitrogen or argon in a glass vessel at -20 °C or less.

In animal studies, decapitation always results in enzymatic release of FFA, diglycerides and hydrolysis of polyphosphoinositides in nervous tissue (3). Head-focused microwave irradiation inactivates enzymatic breakdown of lipids in approx. 2 sec and prevents artefactual values (19).

2.2 COLUMN SEPARATION

Once lipid extraction has been completed, the rich mixture of neutral lipids (cholesterol, free and esterified, FFA, mono, di and triglycerides), polar lipids such as phospholipids, and other complex lipids can be conveniently chromatographed by passage through a column of alumina or silicic acid (Unisil, Clarkson Chemical Co, Williamsport, Pa., or Bio-Sil BH, Bio-Rad Laboratories). The fractionation achieved by these methods (106) is not as specific as with other chromatographic separations (see below), but is still widely employed, particularly when large amounts of lipid need to be separated. The extract, redissolved in chloroform, is deposited on the column and sequentially eluted with solvents of increasing polarity. Twenty-five bed volumes of a low polarity solvent such as chloroform will elute the neutral lipid fraction on silicic acid. Phospholipids are eluted as the methanol concentration is increased in the eluent. Saunders and Horrocks (109) have added one elution step with methyl formate to obtain a fraction containing prostaglandins, lipoxygenase products and glycolipids.

The modified celluloses diethylamino (DEAE) and triethylamino (TEAE) ethylcelluloses provide better separation than that achieved with alumina or silicic acid (107). Neutral lipids, nonacidic and acidic phospholipids are the three generic classes eluted by such ion-exchange chromatography. Moreover, risks of oxidation or hydrolysis are substantially less than during silicic acid chromatography.

Recently, disposable silica cartridges were shown to facilitate separation of PL from non-phosphorus containing lipids. Juaneda and Rocquelin (70) used Sep-Pack silica cartridges (Waters-Millipore) and reported a complete partition of neutral and PL within minutes, from samples ranging between 10 and 100 mg of lipids. Alternatively, disposable aminopropyl bonded phase columns (Bond Elute, Analytichem International, Harbor City, CA), can be used to rapidly isolate and purify polar and neutral individual lipid classes from 10 mg of total lipid, with better than 95% recovery (76).

The resulting PL fractions were first analyzed by sequential chemical degradation (30). However, these pioneer works have generated impetus for development of more direct and powerful techniques such as planar and adsorption chromatography, that allow one to purify intact PL.

2.3 PLANAR CHROMATOGRAPHY

Planar chromatography (thin layer chromatography, TLC; HPTLC) has many attractive advantages over other procedures: simple and inexpensive equipment is required; ease of use for the novice chromatographer; and capacity of analysing several samples simultaneously.

A large body of methods for TLC separation of glycerolipids is available in the literature. Classically, two-dimensional systems using 20x20 cm silica gel precoated plates have been adopted to separate the acidic phospholipids (PtdH, PtdIns, PtdSer) which are difficult to resolve completely in one dimension.

Nowadays, adsorbents of different types and thickness are available to augment almost every application. Modified silica coatings have been favored to achieve one-dimensional separation (48, 58). Impregnation of silica gel with borate increases the resolution capacities of the gel, and this approach was successfully applied to improve PL class separation on both TLC (40) and HPTLC (81). Korte and Casey (83) reported use of an inert preabsorbent zone (Celite) that facilitates high resolution of the compounds to be separated.

More sophisticated two-dimensional systems are available for separating the neutral lipids from the phospholipids, such as multiple development involving sequential elution in the same dimension (98), or including in situ enzymatic treatment (38).

The major drawbacks of two-dimensional TLC reside in the multiple steps that prevent direct comparison of two samples in any single event (71). Moreover, since extensive drying is required between steps of the procedure, unwanted oxidation and hydrolysis may be promoted.

In recent years, development of HPTLC plates has greatly facilitated lipid analysis. Introducing high efficiency silica adsorbent with smaller particle size (10 microns) and narrow pore size distribution combined with thin adsorbent layer (150 microns) has resulted in faster separations (139). Since smaller sample loads are used (1 to 10 ug of total lipid) over shorter migration distances of 5-7 cm, migration time of less than 10 min can be obtained.

Several applications for polar lipid separation were recently described (50,82,89,110,126). For improving resolution of phospholipids, an HPTLC system using gradient chamber saturation was designed (81). In this model, a soybean lipid extract was first

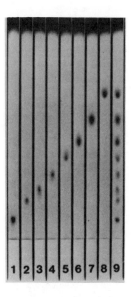

FIGURE 1. One-dimensional TLC of phospholipid standards on boric
acid-impregnated plates. TLC plates (Whatman precoated LK5
or LK5D) with a preadsorbent zone were dipped into a
solution of 1:2% boric acid in absolute ethanol-water
(1:1;v/v). After air drying and activation at 100°C for 1
hr, the undipped preadsorbent zone was spotted with the
phospholipids (3 ug) dissolved in chloroform-methanol
(2:1;v/v) and the plates were developed with chloroform-
methanol-water-ammonium hydroxide (120:75:6:2;v/v/v/v). 1,
PI; 2, LPC; 3, PS; 4, SPM; 5, PG; 6, PC; 7, PE; 8, CL; and
9, a mix of 1-8. (Reproduced, with permission, from ref.40).

separated in two acidic and non-acidic fractions on a DEAE-Sephadex A-
25 column and each fraction spotted on two different lanes to avoid
overlap. The separation revealed at least 17 identified components in
a single run, including FFA, lyso-compounds and glycerides. Yao and
Rastetter (138) chose multiple development to avoid column separation.
By using up to 4 solvent systems, they assembled a very powerful
microanalysis of complex tissue lipids, obtaining 20 different lipid
subclasses, including separation of lysoPtdCho, PtdCho, PtdSer,
PtdIns, Ptd$_2$Gro and PtdEth with a solvent system consisting of methyl
acetate-1-propanol-chloroform-methanol-0.25% KCl (25:25:25:10:9;
v/v/v/v/v). Another approach (112) could resolve completely PtdCho,
CerPCho, PtdIns, PtdSer, PtdEth and PtdGro from gastric aspirate in
less than 25 min by using small chromatographic chambers (21 x 5 x 11
cm) to develop the 10 x 10 cm HPTLC plates.

Overpressured thin-layer chromatography (OPTLC) is the newest
development in planar chromatography. Based on the work of Tyihak and
co-investigators in Hungary (127), this technique is designed to avoid
the theoretical limitations of traditional TLC such as variation in
the saturation of the vapor phase over the adsorbent in the chamber.
In OPTLC chambers, the solvent is forced-pumped through the sorbent,
which is tightly covered with a flexible membrane under an external
pressure, thus eliminating solvent vapors over the sorbent. As the
flow resistance of the eluent on the plate increases with distance,
the working pressure of the pump must also increase gradually.

The development is thus achieved in a closed system with controlled parameters, allowing one to alter the flow-rate, and the pressure of the system. Two-dimensional development may be performed on plates of large dimension (20 x 40 cm) using very low volumes of eluent (less than 5 ml for a 20 x 20 cm plate) at unmatched speed of elution (up to a front migration of about 1 cm/min) (133). OPTLC thus is capable of providing high performance liquid chromatography (HPLC)-type separability when compared to TLC, and allows for temperature programming as in gas chromatography. Very little application for lipid analysis has been conducted so far with OPTLC except for a technique which separated fourteen lipid fractions from serum lipids, including the PL, on the same plate, by unidimensional development with two solvent systems consisting of chloroform-methanol-water (65:25:4;v/v/v), followed by hexane-acetone (100:10;v/v) (102).

With tissue rich in plasmalogens such as brain or muscle, separation of the ether-vinyl species from the diacylglycerolipids can be achieved by acid cleavage of the labile O-vinyl bond (63). The plates are first developed with chloroform-methanol-15M NH_4OH (65:25:4;v/v/v) and dried. They are then exposed to concentrated HCl fumes for 10 min, dried again and developed in the 2nd dimension with a similar but more basic solvent system varied to 100:50:12(v/v/v). The migration of the resulting 2-monoacyl phosphoglycerides is easily distinguished from the intact diacylglycerophopholipid. However, the alkylacylphosphoglycerides are not resolved by this approach. A double-development system to separate the plasmalogens and the alkylacyl from the diacyl subclasses is available (105).

More recently, a two dimensional TLC procedure involving in situ chemical reactions was elaborated to resolve the three subclasses (10). A more sophisticated approach consisting of acidic and alkaline hydrolysis, under specific conditions, coupled to reactions with phospholipase A_2, C and D activity was employed to differentiate the acyl, alkyl and alk-1-enyl moieties of various phospholipids in situ (2).

Separation of the subclasses is also conveniently achieved by HPLC, as described below. Irrespective of the procedure employed for separation, quantitation is best obtained by fatty acid methyl ester derivatization and gas liquid chromatographic (GLC) analysis.

2.3.1 DETECTION

Detection of the phospholipids is a prerequisite for quantitation and collection of the isolated species. For high sensitivity, TLC plates are generally sprayed with 6-p-toluidino-2-naphthalenesulfonic acid (69), rhodamine 6G or 2,7'-dichlorofluorescein (93) and a UV lamp is used to detect the fluorogenic complexes. Phospholipids can be scraped off and extracted from the silica gel for radioactivity counting or lipid phosphorus determination. However, this task is tenuous and time-consuming. Moreover, traces of silica can easily quench the scintillation signal (134) or contaminate the lipid phosphorus determination. Additionally, when small amounts of lipids are chromatographed, the sensitivity of numerous determinations is insufficient (4,7).

Therefore, HPTLC bands are conveniently detected by the charring reaction with copper acetate-phosphoric acid (39) or sulfuric acid-dichromate at 150-180°C for 5-10 min. Under these conditions, all classes of lipids are positively detected. This approach will be further discussed under 'quantitation'.

2.3.2 QUANTITATION

Microanalysis requires a sensitive procedure for lipid phosphorus determination and, generally, the detection limit of phosphorus levels varies between 0.5 and 1.0 ug of phosphorus (4,7).

We have modified a previously published method (97) into a rapid and sensitive assay in order to achieve a lower limit of 0.05 ug of phosphorus with Malachite Green. Concentrated sulfuric acid (0.1 ml) is added to the dried sample in a 16 x 125 mm disposable test tube. The tube is heated over a gas-oxygen hand torch (Bethlehem Apparatus, Hellertown, Pa) until complete charring of the sample. Twenty ul of 6% hydrogen peroxide are added to the cooled tube and boiled again until complete discoloration. Water is next added up to 1.0 ml, followed by 1 ml of 2N NaOH. Fifty ul of 16.5% sodium sulfite (w/v) are added and samples and standards are stirred vigorously. One milliliter of the assay solution is aliquoted in a 12 x 75 mm borosilicate tube and vortexed with 50 ul of 2.9×10^{-2} M ammonium heptamolybdate. A blue complex is obtained by mixing 100 ul of 0.19% (w/v) Malachite Green oxalate (Aldrich Chemicals, Milwaukee, Wis) dissolved in aqueous 1% polyvinyl alcohol (average MW 14,000, Aldrich Chemicals), as described by Petitou et al. (97). The reaction is allowed to proceed for 15 min before reading at 640 nm. Ten to 120 ul of a standard solution which contains 20.8 mg/l of KH_2PO_4 are used to achieve a linear detection between 0.05 and 0.6 ug of phosphorus.

Scanning densitometry is used for direct quantitative photodensitometric evaluation (reflectance mode) of planar chromatograms in situ. The advantage of this more costly method is that greater sensitivity can be obtained, mostly by using fluorometric methods (55). Typical settings are the following: slit width, 0.1-0.4 mm; bandwidth, 10 mm; and scanning rate, 0.5 mm/s. The resulting signal can be interfaced to a computing integrator.

Gustavsson (50) developed a standardized procedure, using molybdenum blue, which provides quantitation of phospholipids independently of the number of double bonds. Nile red, (9-diethylamino-5H-benzo[α]-phenoxazine-5-one), is a more versatile, general purpose fluorometric reagent for quantitation of a wide variety of lipids (44). Chromatograms are dipped in a Nile red solution (8 ug/ml of methanol-water 80:20;v/v). Background fluorescence is later destroyed by dipping in a dilute solution of bleach in order to attain a lower detection of 1-4 ng of phospholipid phosphorus. Nile red staining is minimally affected by the degree of unsaturation of lipids, and will stain fully saturated chains.

2.4 HIGH PERFORMANCE LIQUID CHROMATOGRAPHY FOR SEPARATION OF PL

Phospholipid separation by HPLC has inherent advantages which are: 1) nonvolatile, thermally sensitive lipids, such as PL, can be separated at room temperature; 2) risks of oxidation are minimized by use of degassed solvents in a closed system; 3) methods can be varied specifically with a variety of solvents (delivery of which can be automated and programmed); column packing available includes bonded phases, ion exchangers and gel permeation materials; and detection modes include U.V., fluorescence, mass spectrometry and light scattering detector; and 4) each component of a complex mixture can be separated in a single run, collected and extracted from the HPLC eluent for further analysis by complementary techniques such as mass spectrometry and liquid scintillation spectrometry (1,116).

The use of HPLC in lipidology has become increasingly popular over the past 10 years (85,116), although direct monitoring of lipids separated by HPLC is not inexpensive when compared with planar chromatography. Variable wavelength monitors adapted with microcells of low volumes (8-10 ul) are used at 200-215 nm to detect the U.V. absorption by the sigma transition of the double bond electrons in the fatty acyl chain as well as functional groups such as phosphate, carbonyl and carboxyl substituents (71). Because these groups vary among the phospholipid classes, U.V.-dependent quantitation is not possible and fractions must be collected for quantitative measurement by external procedures. The description of an automated phosphorus analyzer for post-column quantitation of phospholipids can be found (75) as well as a post-column reaction system for the determination of organophosphorus compounds in the absence of organic solvents in the mobile phase (101). However, these methods are not widely used.

2.4.1 PHOSPHOLIPID CLASSES

Isocratic and gradient-driven separation of individual phospholipid classes from biological sources is primarily achieved on a silica column, and several systems have been described (see list in Table 1). These various approaches are attractive since the total lipid extract is directly injected onto the column. However, no system is free from limited applicability. For instance, due to the U.V. monitors employed, 2-lysophospholipids and sphingomyelin are hardly detected. As seen in Table 1, the presence of phosphate buffer in the mobile phase will subsequently prevent quantitation by lipid phosphorus determination of the phospholipid content of the collected fractions. Additionally, mobile phases containing strong mineral acid invariably hydrolyze the plasmalogens of the lipid extract.

We therefore modified the solvent system defined in (96), in order to obtain a one-pump separation of brain PL classes at 211 nm. A typical chromatogram is reproduced in Figure 2.

The first eluent consisted of a mixture of hexane-isopropanol-water-ethanol-glacial acetic acid (367-490-50-50-0.6;v/v/v/v/v eluent A) and was used to equilibrate the column (Ultrasphere-Si, 5 um) at room temperature. Three minutes prior to injecting the total hippocampal lipid extract (15 ul), which was resuspended in hexane-isopropanol (3:2;v/v), flow rate was fixed at 0.3 ml/min while switching to 100% eluent B, which consisted of hexane-isopropanol-ethanol-water-glacial acetic acid (367-490-80-50-0.68;v/v/v/v/v). Ten minutes after the injection of the lipid sample, the pump rate was increased to 1.5 ml/min to optimize the rate of elution. As shown is Figure 2, the order of elution is as follows: the neutral lipids and PtdH (unresolved), PtdEth, PtdIns, PtdSer, PtdCho and CerPCho. The species are eluted within 30 min. and the column is completely reequilibrated after 20-30 min at 1.5 ml/min with eluent A. The main advantages of this system are that, under these conditions, no plasmalogen hydrolysis takes place and the baseline remains flat, since the two eluents display the same absorptivity at 211 nm. This system is also adequate to separate mixtures of neutral lipids, when only solvent A is used. The order of elution is: 1,2 diglycerides, cholesterol and triglycerides (partially resolved) followed by 1,3-diglycerides. We have been using this system and the same column for over 500 injections without loss in reproducibility, providing that the silica support is periodically regenerated with solvents of graded polarity, as described (68).

TABLE 1. UV-HPLC SYSTEMS FOR THE SEPARATION OF PHOSPHOLIPID CLASSES

HPLC Column	Mobile Phase	Elution	Detection (nm)	Separated Classes	Not Resolved or Contaminated	References
LiChrosorb (10 um)	Hex/ProOH/H_2O	Isocratic	206	PE,PI,PS,LPC,LPE,PC	PC,SPM	Geurts Van Kessel et al (46)
Whatman PXS (10 um)	ACN/MeOH/H_2O	Isocratic	203	PC,PE,PS,LPC,LPE,SPM	PE,PS	Gross & Sobel (49)
*Ultrasil-NH_2 (10 um)	Hex/ProOH/MeOH/H_2O	Gradient	206	PC,PE,LPC	PS,PI	Hanson et al (52)
MicroPak-Si (5 um)	ACN/MeOH/H_2SO_4	Gradient	205	PI,PS,PE,PC,LPE	PC,SPM	Yandrasitz et al (137)
MicroPak-Si (10 um)	ACN/MeOH/H_3PO_4	Isocratic	203	PI,PS,PE,LPE,PC	PA,PG,PC	Chen & Kou (21)
LiChrospher Si-100 (10 um)	Hex/ProOH/EtOH/Phosphate buffer/acetic acid	Isocratic	205	PE,PA,PI,PS,PC, CL,DPG,LPC,SPM	-	Patton et al (96)
MicroPak-Si (5 um)	ACN/MeOH/H_2O/NH_4OH	Gradient	205	PI,PE,PS,PC,SPM,LPC	-	Jungalwala et al (73)
Ultrasphere Si (5 um)	ACN/MeOH/H_2SO_4	Isocratic	202	PI,PE,PS,PC,LPC,SPM	-	Kaduce et al (74)
*LiChrosorb-Diol LiChrosorb-Si (5 um, 50°C)	ACN/Phosphate buffer	Gradient	205	PG,PI,PS,PE,PC,LPC,SPM	-	Kuhnz et al (84)
Zorbax Sil (5-6 um)	Hex/ProOH/H_2O	Gradient	205	ALL	PA,LPE,PS	Dugan et al (35)
**LiChrosorb-Si-60 (7 um)	Hex/ProOH/H_2O	Gradient	206	PE,PI,PS,PC,SPM	-	Ellingson & Zimmerman (37)

*Bonded phase column
**Semi-preparative column (250 x 25 mm) run at 34°C
Hex, Hexane; ProOH, 2-propanol; ACN, acetonitrile; MeOH, Methanol; EtOH, Ethanol

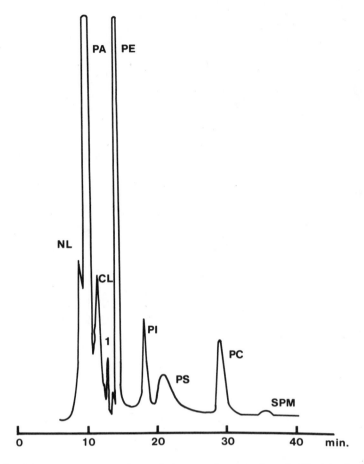

FIGURE 2. One-pump separation of intact brain phospholipid classes at
211 nm. The hippocampal lipid extract was resuspended in
hexane-isopropanol (3:2;v/v) and injected in a volume of
15 ul. Eluent B, consisting of hexane-isopropanol-water-
ethanol-glacial acetic acid (367-490-80-50-0.68;v/v/v/v/v)
was pumped at 0.3 ml/min for the first 10 min, and then rate
was increased to 1.5 ml/min, at room temperature. The
column was re-equilibrated by pumping eluent A which
consisted of hexane-isopropanol-ethanol-water-glacial acetic
acid (367-490-80-50-0.68;v/v/v/v/v) for 20-30 min at 1.5
ml/min. The order of elution is: neutral lipids and PtdH
(unresolved), PtdEth, PtdIns, PtdSer, PtdCho and CerPCho.

2.4.2 PHOSPHOLIPID SUBCLASSES

Nakagawa and Horrocks (92) have described a convenient method for
separating the alkenylacyl (plasmalogens), alkylacyl and diacyl
subclasses of rat brain glycerophospholipids. The fractionated
ethanolamine and choline phospholipids are incubated with
phospholipase C and further acetylated to prevent migration of the 2-
acyl group to the 3-position. From 1-50 umols of the acetyl
derivatives can be separated by UV-HPLC on a silica column by

isocratic elution with a mobile phase consisting of cyclopentane -
hexane - methyl-tert-butyl ether - acetic acid (730:240:30:0.3;
v/v/v/v). According to such studies, the ethanolamine-containing PL
consisted of 41% alkenylacyl, 4% alkylacyl and 55% diacyl types as
compared to over 96% diacyl subclasses in choline-containing PL.
Quantitation was obtained by GLC analysis of the fatty acyl
substituents of the acetylated diglycerides.

To allow for direct quantitation, derivatizing agents with strong
chromogenicity have been used to provide identical absorbance for each
phospholipid subclass. Several routes have been employed including
derivatization to benzoates (15), dinitrobenzoates (79), or to
fluorescent naphthoyl derivatives for detection in the picomole range
(65).

2.4.3 MOLECULAR SPECIES

Each subclass of glycerophospholipids is further composed of a
mixture of different fatty acyl moieties. These differences can be
metabolically very important (64). The molecular species have been
separated and several methods are available, as summarized in Table 2.
Reverse-phase HPLC is the method of choice for separating the
molecular species of derivatized phospholipid subclasses. An isocratic
reverse-phase HPLC system for separation of dimethyl esters of
lysophosphatidic acid has been designed (78) and 6 different species
can be separated in 40 min by using two 100 x 4.6 mm ODS-2 columns in
tandem.

However, in several types of experiments, derivatization may be
prevented and alternate methods must be used. A method to separate
lysophospholipid isomers (sn-1 or sn-2 position), the position (d^6 or
d^9), or the geometric configuration (cis or trans) of the olefin group
in monounsaturated species, has been reported with UV detection at 203
nm without derivatization (29).

In general, separation of the molecular species is more arduous
than the separation of the phospholipid classes or subclasses.

2.5 ANALYSIS OF SPECIFIC PHOSPHOLIPID CLASSES

2.5.1 POLYPHOSPHOINOSITIDES

35 years ago, attention was first directed at PtdIns metabolism
by Hokin and Hokin (62), who showed that acetylcholine stimulated ^{32}P
labeling of PtdH and PtdIns in the exocrine pancreas.

The hydrolysis of membrane phosphatidylinositol 4,5-biphosphate
(PtdInsP$_2$) generates inositol triphosphate (InsP$_{3,4,5}$) and
diacylglycerol (DAG) in response to agonist-stimulation of several
receptors. InsP$_{3,4,5}$ increases cellular Ca^{++}, which promotes activation
of a protein kinase (13), and DAG activates protein kinase C (94).
Generation of DAG by hormone-stimulated phosphatidylinositol
hydrolysis is followed by synthesis of PtdH, from DAG and ATP, and
resynthesis of PtdIns from CDP-DAG and inositol.

Radioisotopic methods are most frequently used to label phosphatidylinositol due to its low contribution to the total cellular
phospholipids (<10%). Dual-labelling of the lipids is possible by
using ortho-[^{33}P] phosphate and ortho-[^{32}P] phosphate. Acidic

TABLE 2. HPLC SYSTEMS FOR THE SEPARATION OF MOLECULAR SPECIES OF PHOSPHOLIPIDS

HPLC Column	Mobile Phase	Elution	Detection	Phospholipid	No. Species Resolved	References
u-Bondapak-C_{18} Nucleosil-5-C_{18} (5 um)	MeOH/phosphatic buffer	isocratic	204 nm	Natural Sphingomyelins	10-12	Jungalwala et al (72)
u-Bondapak C_{18} (10 um) Fatty acid analysis (10 um)	MeOH/H_2O/Chlor	isocratic	Refractive Index	Synthetic lecithins	6-7	Porter et al (99)
Nucleosil-5-C_{18} (5 um)	MeOH/phosphate buffer	isocratic	205 nm	Natural lecithins	11-13	Smith & Jungalwala (81)
Ultrasphere ODS (5 um)	1. MeOH/H_2O/ACN /choline chloride 2. MeOH/ACN/acetic acid/phosphate buffer	isocratic	205 nm	1. Liver PC,PE,PI 2. Liver PS	11(PS,PI) 19-28(PE,PC)	Patton et al (86)
HS-3 ODS-C_{18} (3 um) (50°C)	MeOH/H_2O/ACN /choline chloride	flow gradient	205 nm	Egg PC Bile PC	16 (egg) 10 (bile)	Cantafora et al (18)
Zorbax ODS (5-6 um) (33°C)	ACN/ProOH/M-t-BE /H_2O	isocratic	205 nm	Acetylated diradyl-glycerides from brain PE	22-33	Nakagawa & Horrocks (82)
Ultrasphere-ODS (5 um)	ACN/ProOH	isocratic	230 nm	Diradyl-glycerobenzoates from brain PE	up to 29	Blank et al (15)
Ultrasphere-ODS (5 um)	ACN/ProOH	isocratic	254 nm	Diradyl-glycerodinitro-benzoates	21	Kito et al (79)
PLRP-S (polystyrene, 5 um)	ACN/MeOH/H_2O	isocratic	light scattering detector	Liver lecithin	6	Christie & Hunter (25)

HPLC Column	Mobile Phase	Elution	Detection	Phospholipid	No. Species Resolved	References
Microsorb C$_{18}$ (3 um)	ACN/MeOH /phosphate buffer /choline chloride	ternary gradient	206 nm, mass spectrometry	Natural sphingomyelins	up to 25 (brain)	Teng & Smith (124)
Zorbax SIL (5-6 um) Sepralyte Diol (5 um)	Isooctane/ProOH	isocratic	differential refractometer	Synthetic 1,2 and 1,3 diradyl glycerols		Foglia et al (42)
*Sumipax OA-4100 (5 um)	Hex/ethylene dichloride EtOH		254 nm	1,2 diradyl-glycero 3,5-dinitrophenyl urethane derivatives		Takagi & Itabashi (123)

*The chiral column was comprised of N-(R)-1-(α-naphthyl) ethylaminocarbonyl-(S)-valine bonded to α-aminopropyl silanized silica

Hex, hexane; ProOH, 2-propanol; ACN, acetonitrile; MeOH, methanol; EtOH, ethanol M-t-BE, methyl-tert-butyl ether; chlor, chloroform

conditions of extraction remove the polyphosphoinositides from the cell and thus prevent loss into the aqueous phase. Separation of these phospholipid species has been achieved using HPTLC (9,114), one-dimensional TLC on oxalate-EDTA-impregnated plates (47,66), or using an affinity column of immobilized neomycin (114).

Inositol phospholipids have also been extracted from perchloric acid-precipitated material, deacylated into PtdIns, PtdInsP and PtdInsP$_2$ (28,36), and analyzed by anion-exchange. Hydrolysis of the membrane (PIP$_2$) by a phospholipase C was found to generate several isomers of IP$_3$ (1,4,5 and 1,3,4), IP$_2$ (1,4 and 3,4) and IP (1 and 4) (131). A typical separation of these compounds on Dowex anion exchange column has been described (57).

Alternatively, HPLC anion exchange systems have been used to separate the water soluble isomers (8,31). Separation has been achieved on an analytical Whatman Partisil SAX 10 column using 2 linear gradients with ammonium formate (pH 3.8) from 0 to a final 2M. Inositol 1,2-cyclic 4,5 triphosphate (IcP$_3$) was shown to be formed in thrombin-stimulated platelets (67) and in carbamylcholine-stimulated pancreatic minilobules (11). Dixon and Hokin (34), have separated standards and extracted IcP$_3$ from I(1,3,4)P$_3$ and I(1,4,5)P$_3$ on a Waters Partisil SAX anion exchange radial pack column. Separation of the three species was obtained within 45 min by alternating gradient and isocratic elution, going from water to 1.0 M ammonium formate.

2.5.2 PLATELET-ACTIVATING FACTOR (PAF's)

The separation of PAF's from the major phospholipid classes was developed by Blank and Snyder (14). Separation of 1-O-hexadecyl-2-[^{3}H]acetyl-sn-glycero-3-phosphocholine and acylacetyl-sn-glycero-3-phosphocholine from the phospholipids has been obtained by gradient-driven silica HPLC separation, using three solvent systems comprising different proportions of isopropanol-hexane-water-acetic acid or ammonium hydroxide. The PAF eluted between PtdCho and CerPCho, but before lysoPtdCho.

Jackson et al (68) also used silica HPLC to separate synthetic PAF's (1-O-alkyl-2-acetyl-sn-glycero-3-phosphocholine, AGEPC) and beef heart-derived PAF's from the phospholipid classes, using an isocratic solvent system containing acetonitrile-methanol-phosphoric acid as described in (21). A single peak was observed for beef heart PAF and for mixtures of AGEPC's with alkyl chain lengths ranging from C$_{12}$ to C$_{18}$. Tritiated PAF's were well separated from lyso-PAF's and eluted between PtdCho and lysoPtdCho. C$_{18}$ reverse phase HPLC was subsequently used to fractionate the molecular species; mixtures of PAF's were separated on a 10 um Radial-Pak C18 cartridge (Waters-Millipore) and eluted with methanol-water-acetonitrile (85:10:5;v/v/v) containing choline chloride (20 mM). A very similar method was used for separation of PAF's in human saliva (132).

2.5.3 LIPID PEROXIDATION PRODUCTS

The concept of oxygen-derived free-radicals or reactive oxygen molecules (ROM) in biological systems has emerged as a provocative hypothesis in the etiology of several physiopathological disorders (32), including atherosclerosis (87) and aging (54).

Free-radicals *per se* are extremely short-lived due to their high reactivity with surrounding molecules of which membrane polyunsaturated fatty acids are a preferential target (77,113). Lipid peroxidation products activate phospholipases (90) and may thus contribute to a cascade of irreversible cell injury (45). Quantitation of lipid hydroperoxides is thus required to assess the activity of free-radical reactions and the validity of possible interventions.

Traditionally, total lipid peroxides are assayed in vitro (17, 60). However, several HPLC approaches have been proposed to separate, identify and quantitate hydroperoxide-containing phospholipids. Routinely, absorbance at 234 nm is used to detect the presence of conjuguated dienes, taken as a rearrangement of oxidized fatty acid moieties.

For instance, soy phosphatidylcholine has been separated into its major molecular species on reverse phase (RP)-HPLC with a gradient of aqueous methanol and a uBondapak C_{18} column (27). Oxidized species were differentiated from non-oxidized species by simultaneous detection at 234 and 206 nm. Gas chromatography-mass spectroscopy of the FA moieties revealed the presence of 9 and 13 hydroxy dienes among other oxidized FA that could not be identified (27).

Porter et al (100) have also used RP-HPLC and several isocratic solvent systems to separate the primary lecithin oxidation products (lecithin hydroperoxides). A normal phase system with isocratic elution was used by Reers et al (104) to separate intact PtdEth from its oxidation and hydrolysis products. The column was 4.6 x 50 or 150 mm packed with silicic acid (3 um) and the eluent was hexane-ethanol-water (90:15:0.3;v/v/v). Detection was followed optically at 214 nm or with a flame ionization detector.

The characterization of hydroperoxy and hydroxy derivatives of stearoyllinoleoylphosphatidylcholine (SLPC) produced by Fe^{3+}-ascorbate peroxidation was performed by Ursini et al (128). Oxidized SLPC, consisting mainly of 9 and 13-hydroperoxylinoleic acid derivatives, were separated on an Ultrasphere ODS with a linear gradient of methanol-acetonitrile-aqueous choline chloride at 233 nm (Figure 3). Oxidized FA released by phospholipase A_2 treatment were further characterized on an Ultrasil-NH_2 column followed by gradient elution with hexane-10% orthophosphoric acid-isopropanol-water (85.1:1.2:13.4:0.3;v/v/v/v).

Selective determination of arachidonic acid hydroperoxides such as hydroperoxyeicosatetraenoic acid (HPETE) and its reduced derivative hydroxyeicosatetraenoic acid (HETE) was performed by combining electrochemical detection and UV absorption on RP-HPLC (125).

The reactivity of peroxidized lipids is generally accrued and can undergo further reaction into lipid peroxidation products. The quantitative determination of these products is a very useful tool for assessing lipid peroxidation in biological systems.

4-hydroxy-2,3-transnonenal (HNE), a cytotoxic and highly reactive autoxidation product of n-6 fatty acids such as linoleic, gamma-linolenic and arachidonic acid in biological systems, was analyzed quantitatively (86). The system consisted of a Spherisorb S5 ODS-2 column (Phase Separations, Queensferry, U.K.) eluted with acetonitrile-water (40:60;v/v), with detection at 220 nm. A linear calibration curve was obtained in the range of 0.1 to 500 uM of HNE.

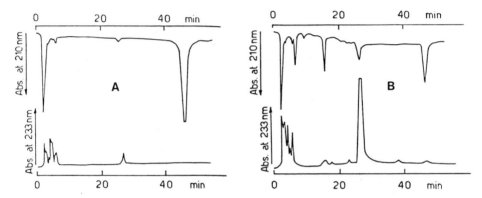

FIGURE 3. Reverse-phase HPLC analysis of stearoyllinolenoyl-phosphatidylcholine (SLPC) and of its peroxidation products. A, HPLC analysis of native SLPC. B, Analysis of the lipid extract after in vitro peroxidation of SLPC in presence of Fe^{3+}-ascorbate. (Reprinted, with permission, from ref. 128)

Benedetti et al (10) characterized all the possible carbonyls produced by lipid peroxidation, other than malonaldehyde, by transforming tissue homogenate species into 2,4 dinitrophenylhydrazone derivatives. Separation was achieved on an RP Zorbax ODS column maintained at 36°C with elution using acetonitrile/water (7:3;v/v) and detection at 350 nm.

Plasma (135) and urinary (80) lipoperoxides can be quantitated by RP-HPLC as malonaldehyde-thiobarbituric acid adducts. In this approach, lipoperoxides are hydrolyzed and reacted with thiobarbituric acid to form the 532 nm-absorbing adducts. The protein-free extract is fractionated on a uBondapak C_{18} column (10 um) with methanol-phosphate buffer and quantitated at 532 nm against a calibration curve of malonaldehyde.

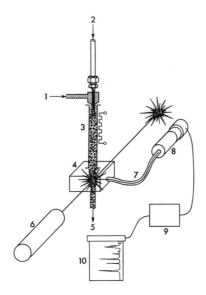

FIGURE 4. Schematic diagram of the evaporative light scattering
detector. 1, Carrier gas stream (constant temperature and
flow rate). 2, LC column effluent. 3, Drift tube. 4, Light
scattering cell. 5, To water ejector. 6, He-Ne Laser
(632.8 nm). 7, Optical fiber. 8, Photomultiplier.
9, Electrometer. 10, Recorder. (reproduced, with permission
from ref. 122)

3.0 METHODS OF THE FUTURE

3.1 HPLC-MASS DETECTION OF PHOSPHOLIPIDS

The lack of a universal HPLC detector for direct measurement of
molar quantities of separated phospholipids is quite restrictive since
only selected solvents invisible to UV can be used. As shown in Tables
1 and 2, these are almost exclusively acetonitrile, hexane and
isopropanol. The coupling of HPLC to field desorption-mass
spectrometry overcomes this difficulty and provides important
stuctural information on small amounts of solutes. The solutes are
directly introduced to the ion source of the magnetic sector or the
quadrupole mass spectrometer by a moving belt interface, which also
removes HPLC solvents. The separation of each class of phospholipids
on silica columns with ammonia-containing solvents and quantitation by
chemical ionization mass spectrometry, was recently reviewed (71).

A laser (light)-scattering mass detector has previously been
described (20,120,122). The principle of the detector is depicted in
Figure 4 and consists of the nebulization of the column effluent in a
gas stream (30-35 °C) where the solvent is totally vaporized in a warm
drift tube (40-45 °C) leaving a cloud of particles to be analyzed.
These particles are carried by the gas stream across a laser beam and
the light diffracted by the particles is collected and transformed by
a photo-multiplier tube into a current which is used as the detector
signal. For quantitative purposes, the detector is not linear but
proportional to some power of the concentration of the compound
analyzed. The response factor depends mostly on the nebulizer design
(121) and very little on the nature of the compound (20,121). Precise
mass determination can be obtained after careful calibration of the
amount of light scattered (121).

The first efforts to integrate this detector for HPLC separation have been reported for triglycerides (121), molecular classes of egg PtdEth and liver PtdCho (119). The detector was more recently used for quantitation of HPLC-separated phospholipid classes of soybean lecithin (129) and from whole tissue lipids (23,24,26,122).

The superiority of this detector resides : (1) in the absence of baseline drift encountered with gradients; (2) in the capability of using any volatile solvent, since solvents are eliminated before the eluate reaches the light source; and (3) in the detection of virtually all solutes and the ability to integrate peak masses (concentration) directly from peak areas.

3.2 PHOSPHORUS NMR ANALYSIS OF PHOSPHOLIPIDS

Phosphorus nuclear magnetic resonance (NMR) spectroscopy can be used for studying biomembranes and model membrane systems due to the property that different phospholipid headgroup ^{31}P nuclei express different chemical shifts.

In early studies (59), resonances of line-width 20-40 Hz obtained from membrane lipids in solvent, in vesicles or in detergents, were too broad to be precisely quantitated and compared with chromatographic data. Later, conditions were improved to provide a significant demonstration that signal widths of less than 1 Hz could be attained with pure phospholipid in a field of 1.8 Tesla (88).

Meneses and Glonek (91) obtained 11 well resolved signals varying between 1.8 to 3.2 Hz at half height in a magnetic field of 11.75 Tesla by analysing ^{31}P NMR profiles of a soybean phospholipid extract preparation. The phospholipid distribution of the soybean extract included all the phospholipid classes and their lyso and plasmalogen derivatives (Figure 5).

Sappey-Marinier et al (108) experimented even further with the possibilities of the NMR in phospholipid analysis. Dried phospholipids from human brain white matter were assayed in D_2O at pH 11.5, with Triton X-100 in a molar ratio detergent/PL's of 5. Under these conditions, linewidths of the chemical shifts of less than 10 Hz allowed quantitative analysis and results were in good agreement with control TLC separation. In parallel, extra white matter tissue was analyzed, but the lipid extraction was omitted and tissue (100 mg) was directly solubilized in buffered detergent and sonicated for 5 min. Phospholipid distribution was within 5% as compared to lipid extract.

^{31}P NMR analysis harbors immense potential as a profiling tool for the characterization and quantitation of diseased tissues (6), and holds many promises for the lipid field.

4.0 CONCLUSION

In this chapter, major areas of actual development in the field of lipid methodology, as well as areas which still require further attention in order to facilitate data acquisition by increasing speed and sensitivity of the analyses, have been underscored. There is no doubt that, in the near future, most of the painstaking labour involved in phospholipid work will be greatly facilitated, as newer modes of analysis become available to the chromatographer.

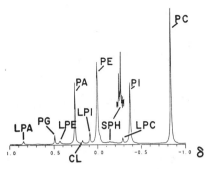

FIGURE 5. ^{31}P NMR spectrum of a crude soybean phospholipid preparation. No filter time-constant was applied and the spectrum was accumulated over 24 hr to demonstrate signal stability. The inset shows the spectrum in the region of the SPH resonance on an expanded (46 x) vertical scale but with the same horizontal scale. The inset is displaced to the right in the illustration (reproduced, with permission, from ref. 91)

ACKNOWLEDGEMENTS

The author is supported by a grant from UCB s.a. Pharmaceutical Sector, Belgium. Thanks to Janice Corey for expert technical assistance, and to Israel Hanin for helpful comments.

REFERENCES

1. Aitzetmuller, K., 1982, Recent progress in the high performance liquid chromatography of lipids, Prog. Lipid Res., 21:171-194.

2. Alvarez, J.G., Levin, S.S., Kleinbart, S., Storey, B.T., and Touchstone, J.C., 1987, Characterization of phosphoglycerides by chemical and enzymatic hydrolysis on thin layer plates in situ, J. Liq. Chromatogr., 10:1687-1705.

3. Alvedano, M.I. and Bazan, N.G., 1975, Differential lipid deacylation during brain ischemia in a homeotherm and a poikilotherm. Content and composition of free fatty acids and triacylglycerols, Brain Res., 100:99-110.

4. Ames, B., 1966, Assay of inorganic phosphate, total phosphate and phosphatases, in: "Methods in Enzymology", S.P. Colowick and N.O. Kaplan, eds, Vol 8, Academic Press, New York, pp. 115-117.

5. Andreesen, P., Modolell, M., Oepke, G.H.F., Common, H., Lohr, G.W. and Munder, P.G., 1982, Studies on various parameters influencing leukemic cell destruction by alkyl-lysophospholipids, Anticancer Res., 2:95-103.

6. Barany, M. and Glonek, T., 1984, Identification of diseased states by P-31 NMR in: "Phosphorus-31 NMR: Principles and

Applications," D. Gorenstein, ed., Academic Press, New-York, pp. 511-545.

7.   Bartlett, G.R., 1969, Phosphorus assay in column chromatography, J. Biol. Chem., 234:466-472.

8.   Batty, I.R., Nahorski, S.R. and Irvine, R.F., 1985, Rapid formation of inositol 1,3,4,5, tetrakisphosphate folowing muscarinic receptor stimulation of rat cerebral cortical slices, Biochem. J., 232:211-215.

9.   Bell, M.E, Peterson, R.G. and Eichberg, J., 1982, Metabolism of phospholipids in peripheral nerve from rats with chronic streptozotocin-induced diabetes: increased turnover of phosphatidylinositol-4,5-biphosphate, J. Neurochem., 39:192-200.

10. Benedetti, A., Pompella, A., Fulceri, R., Romani, A. and Comporti, M., 1986, Detection of 4-hydroxynonenal and other lipid peroxidation products in the liver of bromobenzene-poisoned mice, Biochim. Biophys. Acta, 876:658-666.

11. Benveniste, J., Cochrane, C.G. and Henson, P.M., 1972, Leukocyte-dependent histamine release from rabbit platelets: the role of IgE, basophils and a platelet-activating factor, J. Exp. Med., 136:1356-1359.

12. Berdel, W.E., Griner, E., Fink, U., Stavrou, D., Reichert, A., Rastetter, J., Hoffman, D.R. and Snyder, F., 1981, Cytotoxicity of alkyl-lysophospholipid derivatives and lyso-alkyl-cleavage enzyme activities in rat brain tumor cells, Cancer Res., 43:541-553.

13. Berridge, M.J. and Irvine, R.F., 1984, Inositol triphosphate, a novel second messenger in cellular signal transduction, Nature, 312:315-321.

14. Blank, M.L. and Snyder, F., 1983, Improved high-performance liquid chromatographic method for isolation of platelet-activating factor from other phospholipids, J. Chromatogr., 273:415-420.

15. Blank, M.L., Robinson, M., Fitzgerald, V. and Snyder, F., 1984, Novel quantitative method for determination of molecular species of phospholipids and diglycerides, J. Chromatogr., 298:473-482.

16. Bligh, E.G. and Dyer, W.J., 1959, A rapid method of total lipid extraction and purification, Can. J. Biochem Physiol., 37:911-917.

17. Buege, J.A. and Aust, S.D., 1978, Microsomal lipid peroxidation, in: "Methods in Enzymology", Vol 52, S. Fleischer and L. Packer, eds., Academic Press, New York, pp.302-310.

18. Cantafora, A., Di Biase, A., Alvaro, D., Angelico, M., Marin, M. and Attili, A.F., 1983, High performance liquid chromatographic analysis of molecular species of phosphatidylcholine-development of quantitative assay and its application to human bile, Clin. Chim. Acta, 134:281-295.

19. Cenedella, R.J., Galli, C. and Paoletti, R., 1975, Brain free fatty acids levels in rats sacrificed by decapitation versus focused microwave-irradiation, Lipids, 10:290-293.

20. Charlesworth, J.M., 1978, Evaporative analyzer as a mass detector for liquid chromatography, Anal. Chem., 50:1414-1420.

21. Chen, S.S.H. and Kou, A.Y., 1982, Improved procedure for the separation of phospholipids by high-performance liquid chromatography, J. Chromatogr., 227:25-31.

22. Chignard, M., LeCouedic, J.P., Tence, M., Vargaftig, B.B. and Benveniste, J., 1979, The role of platelet-activating factor in platelet aggregation, Nature, 279:799-800.

23. Christie, W.W., 1985, Rapid separation and quantification of lipid classes by high performance liquid chromatography and mass (light scattering) detection, J. Lipid Res. 26:507-512.

24. Christie, W.W., 1986, Separation of lipid classes by high-performance liquid chromatography with the "mass detector", J. Chromatogr., 361:396-399.

25. Christie, W.W. and Hunter, M.L., 1985, Separation of molecular species of phosphatidylcholine by high-performance liquid chromatography on a PLRP-S column, J. Chromatogr., 325:473-476.

26. Christie, W.W. and Morrison, W.R., 1988, Separation of complex lipids of cereals by high-performance liquid chromatography with mass detection, J. Chromatogr., 383:511-513.

27. Crawford, C.G., Plattner, R.D., Sessa, D.J. and Rackis, J.J., 1980, Separation of oxidized and unoxidized molecular species of phosphatidylcholine by high pressure liquid chromatography, Lipids, 15:91-94.

28. Creba, J.A., Downes, C.P., Hawkins, P.T., Brewster, G., Michell, R.H. and Kirk, C.J., 1983, Rapid breakdown of phosphatidylinositol 4-phosphate and phosphatidylinositol 4,5-biphosphate in rat hepatocytes stimulated by vasoproessin and other Ca^{++}-mobilizing hormones, Biochem. J. 212:733-747.

29. Creer, M.H. and Gross, R.W., 1985, Separation of isomeric lysophospholipids by reverse phase HPLC, Lipids, 20:922-928.

30. Dawson, R.M.C., 1967, Analysis of phosphatides and glycolipids by chromatography of their partial hydrolysis or alcoholysis products, in: "Lipid Chromatographic Analysis", Vol 1, Marinetti, G.V., ed., Marcel Dekker, New York, pp.163-189.

31. Dean, N.M., and Moyer, J.D., 1987, Separation of multiple isomers of inositol phosphates formed in GH_3 cells, Biochem. J., 242:361-366.

32. Del Maestro, R. F., 1980, An approach to free radicals in medicine and biology, Acta Physiol. Scand. suppl., 492:153-168.

33. Dembitski, V.M., 1988, Quantification of plasmalogen, alkylacyl and diacylglycerophospholipids by micro-thin-layer chromatography, J. Chromatogr., 436:467-473.

34. Dixon, J.F. and Hokin, L.E., 1987, Inositol 1, 2-cyclic 4,5-triphosphate concentration relative to inositol 1,4,5-trisphosphate in pancreatic minilobules on stimulation with carbamylcholine in the absence of lithium, J. Biol. Chem., 262:13892-13895.

35. Dugan, L.L., Demediuk, P., Pendley, II, C.E. and Horrocks, L.A., 1986, Separation of phospholipids by high-performance liquid chromatography: all major classes, including ethanolamine and choline plasmalogens, and most minor classes, including lysophosphatidylethanolamine, J. Chromatogr., 378:317-327.

36. Downes, C.P. and Wusteman, M.M., 1983, Breakdown of polyphosphoinositides and not phosphatidylinositol accounts for muscarinic agonist-stimulated inositol phospholipid metabolism in rat parotid glands, Biochem. J., 216:633-640.

37. Ellingson, J.S. and Zimmerman, R.L., 1987, Rapid separation of gram quantities of phospholipids from biological membranes by preparative high performance liquid chromatography, J. Lipid Res., 28:1016-1018.

38. El Tamer, A., Record, M., Fouvel, J., Chap, H. and Douste-Blazy, L., 1984, Studies on ether phospholipids. 1. A new method of determination using phospholipase-Al from guinea-pig pancreas. Application to Krebs-II Ascites-cells, Biochim. Biophys. Acta, 793:213-222.

39. Fewster, M.E., Burns, B.J. and Mead, J.F., 1969, Quantitative densitometric thin-layer chromatography of lipids using copper acetate reagent, J. Chromatogr., 43:120-128.

40. Fine, J.B. and Sprecher, H., 1982, Unidimensional thin-layer chromatography of phospholipids on boric acid-impregnated plates, J. Lipid. Res., 23:660-663.

41. Fisher, S.K. and Agranoff, B.W., 1987, Receptor activation and inositol lipid hydrolysis in neural tissues, J. Neurochem., 48:999-1017.

42. Foglia, T.A., Vail, P.D. and Iwama, T., 1987, High performance liquid chromatographic analysis of 1-alkyl-2-acyl- and 1-alkyl-3-acyl-sn-glycerols, Lipids, 22:362-365.

43. Folch-Pi, J., Lees, M., and Sloane-Stanley, G.H., 1957, A simple method for the isolation and purification of total lipids fromn animal tissues, J. Biol. Chem., 226:497-509.

44. Fowler, S.D., Brown, W.J., Warfeld, J. and Greenspan, P., 1987, Use of nile red for the rapid in situ quantitation of lipids on thin-layer chromatograms, J. Lipid Res., 28:1225-1232.

45. Gamache, D.A., Fawzy, A.A. and Franson, R.C., 1988, Preferential hydrolysis of peroxidized phospholipid by lysosomal phospholipase C, Biochim. Biophys. Acta, 958:116-124.

46. Geurts van Kessel, W.S.M., Hax, W.M.A., Demel, R.A. and De Gier, J., 1977, High performance liquid chromatographic separation and direct ultraviolet detection of phospholipids, Biochim. Biophys. Acta, 486:524-530.

47. Gonzalez-Sastre, P. and Folch-Pi, J., 1968, Thin-layer chromatography of the phosphoinositides, J. Lipid Res., 9:532-535.

48. Goppelt, M. and Resch, K., 1984, Densitometric quantitation of individual phospholipids from natural sources separated by one-dimensional thin-layer chromatography, Anal. Biochem., 140:152-156.

49. Gross, R.W. and Sobel, B.E., 1980, Isocratic high-performance liquid chromatography separation of phosphoglycerides and lysophosphoglycerides, J. Chromatogr., 197:79-85.

50. Gustavsson, L., 1986, Densitometric quantification of individual phospholipids. Improvement and evaluation of a method using molybdenum blue reagent for detection, J. Chromatogr., 375:255-266.

51. Hanahan, D.J., Demopoulos, C.A., Liehr, J. and Pinckard, R.N., 1980, Identification of platelet-activating factor from rabbit basophils as acetyl glycerylether phosphorylcholine, J. Biol. Chem., 255:5514-5521.

52. Hanson, V.L., Park, J.Y., Osborn, T.W. and Kiral, R.M., 1981, High-performance liquid chromatographic analysis of egg yolk phospholipids, J. Chromatogr., 205:393-400.

53. Hara, A., and Radin, N., 1978, Lipid extraction of tissues with a low-toxicity solvent. Anal. Biochem., 90:420-426.

54. Harman, D., 1982, Free radical theory of aging, in: "Free Radicals in Biology," Vol. 5, W.A. Pryor, ed., Academic Press, New York, pp. 258-276.

55. Harrington, C.A., Fenimore D.C. and Eichberg, J., 1980, Fluorometric analysis of polyunsaturated phosphatidylinositol and other phospholipids in the picomole range using high-performance thin-layer chromatography. Anal. Biochem. 106:307-311.

56. Hauser, G. and Eichberg, J., 1973, Improved conditions for the preservation and extraction of polyphosphoinositides, Biochim. Biophys. Acta, 326:201-204.

57. Hawkins, P.T, Stephens, L. and Downes, C.P., 1986, Rapid formation of inositol 1,3,4,5-tetrakisphosphate and inositol 1,3,4-tris phosphate in rat parotid glands may both result indirectly from receptor-stimulated release of inositol 1,4,5-trisphosphate from phosphatidylinositol 4,5-biphosphate, Biochem J. 238:507-517.

58. Heape, A.M., Juguelin, H., Boiron, F. and Cassagne, C., 1985, Improved one-dimensional thin-layer chromatographic technique for polar lipids, J. Chromatogr., 322:391-396.

59. Henderson, T.O., Glonek, T. and Myers, T.C., 1974, Phosphorus-31 nuclear magnetic resonance spectroscopy of phospholipids, <u>Biochemistry</u>, 13:623-628.

60. Hicks, M. and Gebicki, J.M., 1979, A spectrophotometric method for the determination of lipid hydroperoxides, <u>Anal. Biochem.</u>, 99:249-253.

61. Hokin, L.E., 1985, Receptors and phosphoinositide-generated second messengers, <u>Ann. Rev. Biochem.</u>, 54:205-35.

62. Hokin, M.R. and Hokin, L.E., 1953, Enzyme secretion and the incorporation of ^{32}P into phospholipides of pancreas slices, <u>J. Biol. Chem.</u> 203:967-974.

63. Horrocks, L.A., 1968, The alk-1-enyl group content of mammalian myelin phosphoglycerides by quantitative two-dimensional thin-layer chromatography, <u>J. Lipid Res</u>. 9:469-475.

64. Horrocks, L.A., Yeo, Y.Y., Harder, H.W., Mozzi, R. and Goracci, G., 1986, Choline plasmalogens, glycerophospholipid methylation, and receptor-mediated activation of adenylate cyclase, <u>Adv. Cyclic Nucleotide Protein Phosphorylation Res.</u>, 20:263-270.

65. Horrocks, L.A., Dugan, L.L., Flynn, C.J., Goracci, G., Porcellati, S. and Young, Y., 1987, Modern techniques for the fractionation and purification of phospholipids from biological materials <u>in</u>: "Lecithin: Technological, Biological and Therapeutic Aspects", Hanin, I. and Ansell G.B., eds., Plenum Press, New York, pp. 3-16.

66. Irvine, R.F., Letcher, A.J., Meade, C.J. and Dawson, R.M., 1984, One-dimensional thin-layer chromatographic separation of the lipids involved in arachidonic acid metabolism, <u>J. Pharmacol. Methods</u>, 12:171-175.

67. Ishii, H., Connolly, T.M., Bross, T.E. and Majerus, P.W., 1986, Inositol cyclic triphosphate [inositol 1,2-(cyclic)-4,5-triphosphate] is formed upon thrombin stimulation of human platelets, <u>Proc. Natl. Acad. Sci. U.S.A</u>, 83:6397-6401.

68. Jackson, E.M., Mott, G.E., Hoppens, C., McManus, L.M., Weintraub, S.T., Ludwig, J. and Pinckard, R.N., 1984, High performance liquid chromatography of platelet-activating factors, <u>J. Lipid Res.</u>, 25:753-757.

69. Jones, M., Keenan, R.W. and Horowitz, P., 1982, Use of 6-p-toluidino-2-naphthalenesulfonic acid to quantitate lipid after thin-layer chromatography, <u>J. Chromatogr.</u>, 237:522-524.

70. Juaneda, P. and Rocquelin, G., 1985, Rapid and convenient separation of phospholipids and non phosphorus lipids from rat heart using silica cartridges, <u>Lipids</u>, 20:40-41.

71. Jungalwala, F.B., 1985, Recent developments in techniques for phospholipid analysis, <u>in:</u> "Phospholipids in Nervous Tissues," Eichberg, J., ed., John Wiley and Sons, New York, pp. 1-44.

72. Jungalwala, F.B., Hayssen, V., Pasquini, J.M. and McCluer, R.H., 1979, Separation of molecular species of sphingomyelin by reversed-phase high-performance liquid chromatography, J. Lipid Res., 20:579-587.

73. Jungalwala, F. B., Sanyal, S., and LeBaron, F., 1982, Use of HPLC to determine the turnover of molecular species of phospholipids, in: "Phospholipids in the Nervous System," Vol 1, Metabolism, L. Horrocks, G.B. Ansell, G. Porcellati, eds., Raven Press, New York, pp. 91-103.

74. Kaduce, T.L., Norton, K.C. and Spector, A.A., 1983, A rapid, isocratic method for phospholipid separation by high-performance liquid chromatography, J. Lipid Res., 24:1398-1403.

75. Kaitaranta, J.K. and Bessman, S.P., 1981, Determination of phospholipid by a combined liquid chromatography-automated phosphorus analyser, Anal. Chem. 53:1232-1240.

76. Kaluzny, K.A.Q., Duncan, L.A., Merritt, M.V. and Epps, D.E., 1985, Rapid separation of lipid classes in high yield and purity using bonded phase columns, J. Lipid Res., 26:135-140.

77. Kaschnitz, R.M. and Hatefi, Y., 1975, Lipid oxidation in biological membranes, Arch. Biochem. Biophys., 171:292-304.

78. Kennerly, D.A., 1987, Molecular species analysis of lysophospholipids using high-performance liquid chromatography and argentation thin-layer chromatography, J. Chromatogr., 409:291-297.

79. Kito, M., Takamura, H., Narita, H. and Urade, R., 1985, A sensitive method for quantitative analysis of phospholipid molecular species by HPLC. J. Biochem. 98:327-331.

80. Knight, J.A., Smith, S.E., Kinder, V.E. and Pleper, R.K., 1988, Urinary lipoperoxides quantified by liquid chromatography and determination of reference values for adults, Clin. Chem., 34:1107-1110.

81. Kolarovic, L., and Traitler, H., 1985, The application of gradient saturation in unidimensional planar chromatography of polar lipids. Part 1: Presentation of the HPLC system, J. High Resolut. Chromatogr. Chromatogr. Comm., 8:341-346.

82. Kolarovic, L., and Fournier, N.C., 1986, A comparison of extraction methods for the isolation of phospholipids from biological sources, Anal. Biochem., 156:244-250.

83. Korte, K. and Casey, M.L., 1982, Phospholipid and neutral lipid separation by one-dimensional thin-layer chromatography, J. Chromatogr. 232:47-53.

84. Kuhnz, W., Zimmermann, B. and Nau, H., 1985, Improved separation of phospholipids by high-performance liquid chromatography, J. Chromatogr. 344:309-312.

85. Kuksis, A., ed., 1987, "Chromatography of Lipids in Biomedical Research and Clinical Diagnosis", J. Chromatogr. Libr. Vol 37, Elsevier, Amsterdam.

86. Lang J., Celotto, C. and Esterbauer, H., 1985, Quantitative determination of the lipid peroxidation product 4-hydroxynonenal by high-performance liquid chromatography, Anal. Biochem. 150:369-378.

87. Ledwozyw, A., Michalak, J., Stepien, A. and Kadziolka, A., 1986, The relationship between plasma triglycerides, cholesterol, total lipids and lipid peroxidation products during human atherosclerosis, Clin. Chim. Acta 155:275-284.

88. London, E. and Feigenson, G.W., 1979, Phosphorus NMR analysis of phospholipids in detergents, J. Lipid Res., 20:408-412.

89. Macala, L.J., Yu, R.K. and Ando, S., 1983, Analysis of brain lipids by high performance thin-layer chromatography and densitometry, J. Lipid Res., 24:1243-1250.

90. Meerson, F.Z., Kagan, V.E., Kozlov, Y.P., Belkina, L.M. and Arkhipenko, Y.V., 1982, The role of lipid peroxidation in pathogenesis of ischemic damage and the antioxidant protection of the heart, Basic Res. Cardiol. 77:465-485.

91. Meneses, P. and Glonek, T., 1988, High resolution ^{31}P NMR of extracted phospholipids, J. Lipid Res. 29:679-689.

92. Nakagawa, Y. and Horrocks, L.A., 1983, Separation of alkenylacyl, alkylacyl, and diacyl analogues and their molecular species by high performance liquid chromatography, J. Lipid Res., 24:1268-1281.

93. Nelson, G. J., 1975, "Analysis of Lipids and Lipoproteins", E.G. Perkins, ed, Amer. Oil Chem. Soc., Champaign, IL.

94. Nishizuka, Y., 1986, Studies and perspectives of protein kinase C, Science, 233:305-312.

95. Oishi, K., Raynor, R.L., Charp, P.A. and Kuo, J.F., 1988, Regulation of protein kinase C by lysophospholipids, J. Biol. Chem., 263:6865-6871.

96. Patton, G.M., Fasulo, J.M. and Robins, S.J., 1982, Separation of phospholipids and individual molecular species of phospholipids by high-performance liquid chromatography, J. Lipid Res., 23:190-196.

97. Petitou, M., Tuy, F. and Rosenfeld, C., 1978, A simplified procedure for organic phosphorus determination from phospholipids, Anal. Biochem., 91:350-353.

98. Poorthuis, B.J.H.M., Yazaki, P.J. and Hostetler, K.Y., 1976, An improved two-dimensional thin-layer chromatography system for the separation of phosphatidyl glycerol and its derivatives, J. Lipid Res., 17:433-437.

99. Porter, N.A., Wolf, R.A. and Nixon, J.R., 1979, Separation and purification of lechithins by high pressure liquid chromatography, Lipids, 14:20-24.

100. Porter, N.A., Wolf, R.A. and Weenen, H., 1980, The free radical oxidation of polyunsaturated lecithins, Lipids, 15:163-167.

101. Priebe, S.R. and Howell, J.A., 1985, Post-column reaction detection system for the determination of organophosphorus compounds by liquid chromatographpy, J. Chromatogr., 324:53-63.

102. Pucsok, J., Kovacs, L., Zalka, A. and Dobo, R., 1988, Separation of lipids by new thin-layer chromatography and overpressured thin-layer chromatography methods, Clin. Biochem., 21:81-85.

103. Quinn, M.T., Parthasarathy, S. and Steinberg, D., 1988, Lysophosphatidylcholine: a chemotactic factor for human monocytes and its potential role in atherogenesis, Proc. Natl. Acad. Sci. USA, 85:2805-2809.

104. Reers, M., Schmidt, P.C., Erdahl, W.L. and Pfeiffer, D.R., 1986, Separation of phosphatidylethanolamine from its oxidation and hydrolysis products by high-performance liquid chromatography, Chem. Phys. Lipids, 42:315-321.

105. Renkonen, O., 1968, Chromatographic separation of plasmalogenic, alkyl-acyl, and diacyl forms of ethanolamine glycerophosphatides, J. Lipid Res., 9:34-40.

106. Rouser, G., Kritchevsky, G. and Yamamoto, A., 1967, Column chromatographic and associated procedures for separation and determination of phosphatides and glycolipids, in: "Lipid Chromatographic Analysis," Vol. 1, Marinetti, G.V., ed., Marcel Dekker, New York. pp. 147-162.

107. Rouser, G., Kritchevsky, G. and Yamamoto, A., 1976, Column chromatographic and associated procedures for separation and determination of phosphatides and glycolipids. in: "Lipid Chromatographic Analysis," 2nd Edition, Vol 3, Marinetti, G.V., ed., Marcel Dekker, New York, pp. 211-261.

108. Sappey Marinier, D., Letoublon, R. and Delmau, J., 1988, Phosphorus NMR analysis of human white matter in mixed non-ionic detergent micelles, J. Lipid Res., 29:1237-1243.

109. Saunders, R.D. and Horrocks, L.A., 1984, Simultaneous extraction and preparation for high performance liquid chromatography of prostaglandins and phospholipids, Anal. Biochem., 143:71-79.

110. Sax, S., Moore, J., Oley, A., Amenta, J. and Silverman, J., 1982, Liquid-chromatographic estimation of saturated phospholipid palmitate in amniotic fluid compared with a thin-layer chromatographic method for acetone-precipitated lecithin, Clin. Chem., 28:2264-2267.

111. Sekar, M.C., Dixon, J.F. and Hokin, L.E., 1987, The formation of inositol 1,2-cyclic 4,5-triphosphate and inositol 1,2-cyclic 4-biphosphate on stimulation of mouse pancreatic minilobules with carbamylcholine, J. Biol. Chem., 262:340-345.

112. Serrano de la Cruz, D., Santillana, E., Mingo, A., Fuenmayor, G., Pantoja, A. and Fernandez, E., 1988, Improved thin-layer chromatographic determination of phospholipids in gastric

aspirate from newborns for assessment of lung maturity, <u>Clin. Chem.</u>, 34:736-738.

113. Sevanian, A. and Hochstein, P., 1985, Mechanisms and consequences of lipid peroxidation in biological systems, <u>Ann. Rev. Nutr.</u>, 5:365-390.

114. Schacht, J., 1978, Purification of polyphosphoinositides by chromatography on immobilized neomycin, <u>J. Lipid Res.</u>, 19:1063-1067.

115. Shaw, J.O., Pinckard, R.N., Ferrigni, K.S., McManus, L.M. and Hanahan, D.J., 1981, Activation of human neutrophils with 1-O-hexadecyl/octadecyl-2-acetyl-sn-glyceryl-3-phosphorylcholine (platelet activating factor), <u>J. Immunol.</u>, 127:1250-1258.

116. Shukla, V.H.S., 1988, Recent advances in the high performance liquid chromatography of lipids, <u>Prog. Lipid Res.</u>, 27:5-38.

117. Smith, M. and Jungalwala, F.B., 1981, Reversed-phase high performance liquid chromatography of phosphatidylcholine: a simple method for determining relative hydrophobic interaction of various molecular species, <u>J. Lipid Res.</u>, 22:697-704.

118. Snyder, F., Snyder, C., 1975, Glycerolipids and cancer, <u>Prog. Biochem. Pharmacol</u>. 10:1-18.

119. Sotirhos, N., Thorngren, C. and Herslof, B., 1985, Reversed-phase high-performance liquid chromatographic separation and mass detection of individual phospholipid classes, <u>J. Chromatogr</u>. 331:313-320.

120. Stolyhwo, A., Colin, H.M. and Guiochon, G., 1983, Use of light scattering as a detector principle in liquid chromatography, <u>J. Chromatogr.</u>, 265:1-16.

121. Stolyhwo, A., Colin, H. and Guiochon, G., 1985, Analysis of triglycerides in oils and fats by liquid chromatography with the laser light scattering detector, <u>Anal. Chem.</u>, 57:1342-1354.

122. Stolyhwo, A., Martin, M. and Guiochon, G., 1987, Analysis of lipid classes by HPLC with the evaporative light scattering detector, <u>J. Liquid Chromatogr.</u>, 10:1237-1253.

123. Takagi, T. and Itabashi, Y., 1987, Rapid separations of diacyl- and dialkylglycerol enantiomers by high performance liquid chromatography on a chiral stationary phase, <u>Lipids</u>, 22:596-600.

124. Teng, J.I. and Smith, L.L., 1985, Improved high-performance liquid chromatography of sphingomyelin, <u>J. Chromatogr.</u>, 322:240-245.

125. Terao, J., Setsu Shibata, S. and Matsushita, S., 1988, Selective quantification of arachidonic acid hydroperoxides and their hydroxy derivatives in reverse-phase high performance liquid chromatography, <u>Anal. Biochem.</u>, 169:415-423.

126. Touchstone, J.C., Chen, J.C. and Beaver, K.M., 1980, Improved separation of phospholipids in thin layer chromatography, <u>Lipids</u>, 15:61-62.

127. Tyihak, E., Mincsovics, E. and Kalasz, H., 1979, New planar liquid chromatographic technique:overpressured thin-layer chromatography, J. Chromatogr., 174: 75-82.

128. Ursini, F., Bonaldo, L., Maiorino, M. and Gregolin, C., 1983, High-performance liquid chromatography of hydroperoxy derivatives of stearoyllinoleoylphosphatidylcholine and of their enzymatic reduction products, J. Chromatogr., 270:301-308.

129. Van der Meeren, P., Vanderdeelen, J., Huys, M. and Baert, L., 1990, Quantification of soybean phospholipid solublility using an evaporative light scattering mass detector, THIS BOOK.

130. Vidaver, G.A., Ting, A. and Lee, J.W., 1985, Evidence that lysolecithin is an important causal agent of atherosclerosis, J. Theor. Biol., 115:27-41.

131. Volpi, M., Yassin, R., Naccache, P.H. and Sha'afi, R.I., 1983, Chemotactic factor causes rapid decrease in phosphatydylinositol 4,5-biphosphate and phosphatidylinositol 4-monophosphate in rabbit neutrophils, Biochem. Biophys. Res. Comm., 112:957-964.

132. Wardlow, M.L., 1985, Rapid isocratic procedure for the separation of platelet-activating factor from phospholipids in human saliva by high-performance liquid chromatography, J. Chromatogr., 342:380-384.

133. Witkiewicz, Z. and Bladek, J., 1986, Overpressured thin-layer chromatography, J. Chromatogr. (Chromatogr. Rev.), 373:111-140.

134. Witter, B., Gunawan, J. and Debuch, H., 1983, On the phospholipid metabolism of glial cell primary cultures. II. Metabolism of 1-alkyl-glycerophosphoethanolamine during time course, J. Neurochem. 40:64-69.

135. Wong, S.H.Y., Knight, J.A. and Hopfer, S.M., 1987, Lipoperoxides in plasma as measured by liquid-chromatographic separation of malondialdehyde-thiobarbituric acid adduct, Clin. Chem., 33:214-220.

136. Wood, R., ed., 1973, "Tumour Lipids: Biochemistry and Metabolism," American Oil Chemist Society Press, Champaign, Il.

137. Yandrasitz, J.R., Berry, G. and Segal, S., 1981, High-performance liquid chromatography of phospholipids with UV detection: optimization of separations on silica, J. Chromatogr., 225:319-328.

138. Yao, J.K. and Rastetter, G.M., 1985, Microanalysis of complex tissue lipids by high-performance thin-layer chromatography, Anal. Biochem. 150:111-116.

139. Zlatkis, A. and Kaiser, R.E., eds., 1977, "HPTLC-High Performance Thin-Layer Chromatography," Elsevier, Amsterdam.

PHARMACOLOGICAL ACTIONS OF PHOSPHOLIPIDS

G. Pepeu, M.G. Vannucchi and P.L. Di Patre

Department of Pharmacology, University of Florence
Viale Morgagni 65, 50134 Florence, Italy

INTRODUCTION

The biological importance of phospholipids was already clear to Thudicum who in 1884 wrote "Phospholipids are the centre, life, and chemical soul of all bioplasm whatsoever, that of plants as well of animals"[16]. However, for a long time the importance of phospholipids has been considered more structural than functional and pharmacologists have overlooked the possibility that, once released in the extracellular space or administered, they may exert powerful pharmacological actions. These actions may be involved in pathological processes, have toxicological rele-vance or be exploited for therapeutic purposes. The goal of this review is to describe the biological actions of some phospholipids which may be relevant from a pharmacological viewpoint. Details on the mechanisms through which phospholipid interaction with cell membranes brings about some of their actions will be discussed elsewhere (Bruni et al., this symposium).

PHOSPHATIDYLCHOLINE

Phosphatidylcholine is the main component of biological membranes[32]. After the discovery that the loss of cholinergic neurons in the forebrain is an important pathogenetic feature of Alzheimer's disease[9,33],the administration of large doses of choline was proposed in order to stimulate acetylcholine (ACh) synthesis and compensate for the cholinergic deficits. However, gastro-intestinal side effects and an unpleasant fishy odor due to its transformation into trimethylamine by intestinal bacteria make it difficult to administer large amounts of choline[17]. Some of these problems may be overcome by the administration of phosphatidylcholine as a source of choline. Many clinical trials have been carried out in order to evaluate the usefulness of phosphatidylcholine in senile dementia of Alzheimer's type and the results have generally been disappointing[3]. However, the addition of phosphatidylcholine to therapies with cholinesterase inhibitors is still advocated, as shown for instance by the

Phospholipids
Edited by I. Hanin and G. Pepeu
Plenum Press, New York, 1990

clinical trial carried out by Summers et al.[27] on the efficacy of tetrahydroaminoacridine in senile dementia of Alzheimer type. The rationale for associating phosphatidylcholine with anticholinesterase lies in the assumption that stimulated cholinergic neurons, with no adequate choline supply, obtain choline for ACh synthesis by hydrolysis of membrane phosphatidylcholine[34]. Clinical demonstrations of the advantage of associating phosphatidylcholine with anticholinesterase therapy in Alzheimer's disease are needed but are difficult to carry out. However, since phosphatidylcholine may be considered a nutrient more than a drug, a wide margin of discretion can be left to the doctors.

The utilization of phosphatidylcholine as a source of choline for other conditions of choline deficiency is discussed elsewhere in this symposium by Zeisel.

LYSOPHOSPHATIDYLCHOLINE

In the tissues phosphatidylcholine undergoes a constant process of de- and reacylation through which the fatty acid composition of the the phospholipid molecules is "tailored" to the needs of a particular tissue[13].
During this process small amounts of lysophosphatidylcholine are released; this lysophospholipid is therefore present in a very low concentration in almost every biological membrane [2]. It is formed through the actions of phospholipases A on phosphatidylcholine; its concentration is also regulated by acyl-CoA acyltransferases which convert the lysophosphatidyl derivatives into phospholipids.

At variance with phosphatidylcholine, its lyso-derivative exerts many cytotoxic actions due to its detergent-like properties[30]. The biological actions of lysophosphatidylcholine are summarized in Table 1.

The mechanisms through which lysophosphatidylcholine interacts with

Table 1. Actions of Lysophosphatidylcholine

---

1) Intermediate of phosphatidylcholine turnover

2) Surface recognition processes:
- modification of cell shape
- hemolysis
- change of charge distribution
- cell agglutination

3) Effects on membrane-bound enzymes:
- stimulation of glycosyl-and sialyltransferases
(responsable for cell-cell recognition processes)
- inhibition of guanylate and adenylate cyclases
- inhibition of prostaglandin synthesis

---

cell membrane, probably through incorporation in its lipid matrix, have been reviewed by Weltzien[30]. Accumulation of lysophosphatidylcholine may occur in hypoxic and damaged tissues through the mechanisms listed in Table 2. For instance, accumulation of lysophosphatidylcholine in hypoxic cardiac tissue has been demonstrated repeatedly[25]. A linear relationship between severity of ventricular arrhythmias and the increase in lysophosphatidylcholine concentration during the first 3 min of ischemia in cats has been demonstrated[8]. Studies of the electrophysiological modifications induced by lysophospholipids in Purkinje fibers suggest that lysophosphoglycerides could be responsible for the malignant ventricular arrhythmias which account for sudden death in patients with coronary artery disease. In addition, it has been demonstrated that lysophospholipids induce delayed after-depolarization and triggered activity in a dose-dependent manner[22]. Both events are facilitated and enhanced by calcium and catecholamines, which also accumulate in the ischemic myocardial regions.

Table 2. Causes of Accumulation of Lysophosphatidylcholine in Ischemic Myocardial Tissue.

---

1) Enhancement of phospholipase A_2 activity due to the impairment of ATP synthesis.

2) Decrease of lysophospholipase and acyltransferase activity due to the accumulation of acylcarnitine.

3) Decreased washout in ischemic tissue

The accumulated lyso-derivatives are predominantly 2-deacyl isomers containing saturated long-chain fatty acids. The arrhythmias induced by lysophosphatidylcholine can be prevented by albumin which binds the lysophospholipid, and by lidocaine which presumably forms with it a non-lytic complex on the membrane[21].

PHOSPHATIDYLETHANOLAMINE

Phosphatidylethanolamine is the second most abundant phospholipid in cell membranes[10]. It may be used as a precursor of phosphatidylcholine through the base exchange reaction and, to a smaller extent, a triple N-methylation[19]. Exogenous administration of phosphatidylethanolamine has been shown to increase ACh output from rat cerebral cortex "in vivo"[17], and its activity is approximately one third as large as that of phosphatidylserine. An equimolecular mixture of phosphatidylethanolamine and phosphatidylcholine is a powerful activator of factor X of the coagulation cascade. In the physiological process the phospholipid micelles originate from blood platelet membrane and also are involved in the activation of prothrombin to thrombin. Infusion of factor X and the two phospholipids brings about a marked hypercoagulability[31].

PHOSPHATIDYLSERINE

Phosphatidylserine reresents approximately 9% of membrane phospholipids in the brain[10]. It exerts many pharmacological actions which will
be discussed more extensively by Nunzi and Toffano (this Symposium).
Phosphatidylserine enhances histamine release induced by antigens, dextran,
and concanavalin [12,18,26], and when administered in large doses, increases
spontaneous ACh release from rat cerebral cortex[7].

Parenteral administration of small doses of phosphatidylserine for 7
days in aging rats restores the age-dependent decrease in cortical ACh
synthesis and release. At these doses phosphatidylserine is inactive in
young rats. In aging rats the effect begins after at least 3 days of
treatment and lasts for 5 days after interruption[29]. This time course
suggests a gradual incorporation into cell membrane and the activation of
an enzymatic system. In this regard it has been shown that phosphatidylserine is the most effective phospholipid for protein kinase C activation[15]. The possibility that phosphatidylserine may indirectly increase
the pool of choline available for ACh synthesis is supported by
preliminary experiments (Vannucchi and Pepeu, unpublished results). The
recovery in ACh release is associated with an improvement in the age-
dependent impairment of the rat's cognitive behavior.

PLATELET ACTIVATING FACTOR (PAF)

One of the pharmacologically most active phospholipids is 1-O-alkyl-2-
acetyl-sn-glyceryl-3-phosphorylcholine which is released from macrophages,
neutrophils, eosinophils, vascular endothelium and platelets when stimulated under various conditions. It was first indentified by Benveniste et
al. [4] who were investigating the reason for which the antigen-stimulated
histamine release taking place "in vitro" from rabbit platelets required
the presence of leucocytes from the immunized animal. The factor released
from the leucocytes was defined as platelet activating factor (PAF) and
its structure was identified by mass spectrometric analysis[11]. The large
number of investigations defining the actions of PAF and its pathogenetic
role in anaphylaxis, inflammation and asthma have been the object of
several reviews[5,14,24,28]. The biological effects of PAF are summarized in
Table 3.

PAF is present in cells as a precursor, 1-O-alkyl-2-acyl-glycero-3-
pho-sphatidylcholine (alkyl-acyl-GPC), and is released in two steps[1]. The
first step involves the deacylation of alkyl-acyl-GPC to lyso-PAF and is
catalyzed by a phospholipase A_2, of the same type or the same enzyme which
releases arachidonate. In the second step an acetyl transferase acetylates
lyso-PAF into PAF. PAF is effectively removed by de-acylation by the same
phospholipase A responsible for its formation, by an acetylhydrolase.
Lyso-PAF can be recycled to PAF or reincorporated into membrane
phospholipids.

A full understanding of the role of PAF in pathology will be possible
with the discovery of specific PAF antagonists. An extensive search for
these agents is presently in progress in many laboratories [23]. In this
context, it is interesting that alkyl ether derivatives of phosphatidyl-

Table 3. "In Vitro" and "In Vivo" Effects of PAF

1. Platelet shape change, aggregation, and release of granules containing histamine, serotonin and Factor 4.

2. Microvascular leakage and protein extravasation.

3. Bronchoconstriction and long-lasting bronchial hyperresponsiveness.

4. Human eosinophil chemotaxis in vitro and in vivo.

5. Decreased mucociliary clearance and increased protein exudation into the airway lumen.

6. Hypotensive action.

choline with hypotensive actions but no platelet aggregating properties have also been isolated from renal medulla [20].

CONCLUSIONS

This review is by no mean exhaustive and experts in each of the specific fields mentioned will find omissions. Phosphatidylinositol and its derivatives have been omitted because they will be discussed elsewhere (Hawthorne, this Symposium) and they have not yet been proposed either as drugs or as targets of antagonists. Similarly, lysophosphatidylserine has been left out. However, the participants in this symposium have different scientific backgrounds and the aim of the review was to give a bird's eye view of the many biological actions of phospholipids that make them promising as drugs or target for drugs. As Bruni[6] has recently pointed out, the plasma membrane utilizes in many instances its main constituents, the phospholipids, to translate extracellular stimuli in adequate responses at the cellular or a more general level. The pharmacologist may attempt to mimic and exploit the translation or prevent it.

REFERENCES

1. Albert, D.H. and Snyder, F.,1983, Biosynthesis of 1-alkyl-2-acetyl-sn-glycero-3-phosphocholine (platelet-activating factor) from 1-alkyl-2-acyl-sn-glycero-3-phosphocholine by rat alveolar macrophages. Phospholipase A2 and acetyltransferase activities during phagocytosis and ionophore stimulation, J. Biol. Chem., 25: 97-102.

2. Ansell, G.B., 1973, Phospholipids and the nervous system. In: "Form and Function of Phospholipids", G.B. Ansell, J.N. Hawthorne, R.M.C. Dawson, eds, pp 377-422, Elsevier, Amsterdam.

3.Bartus, R.T., Dean, R.L., Beer, B. and Lippa, A.S., 1982, The cholinergic hypothesis of geriatric memory dysfunction, _Science_, 217: 408-417.

4.Benveniste, J.P., Henson, P.M. and Cochrane, C.G., 1972, Leucocyte-dependent histamine release from rabbit platelets. The role of IgE, basophils and a platelet activating factor, _J. Exp. Med._, 136: 1356-1376.

5.Braquet, P. and Rola-Pleszczynski, M., 1987, The role of PAF in immunological responses: a review, _Prostaglandins_, 34: 143-147.

6.Bruni, A., 1988, Autacoids from membrane phospholipids, _Pharmacol. Res. Comm._, 20: 529-543.

7.Casamenti, F., Mantovani, P., Amaducci, L. and Pepeu, G., 1979, Effect of phosphatidylserine on acetylcholine output from the cerebral cortex of the rat, _J. Neurochem._, 32: 529-533.

8.Corr, P.B.,Yamada, K.A., Creer, M.H., Sharma, A.D. and Sobel, B.E.,1987, Lysophosphaglycerides and ventricular fibrillation early after onset of ischemia, _J. Mol. Cell Cardiol._, 19, Suppl. V: 45-53.

9.Davies, P. and Maloney, A.J.F., 1976, Selective loss of central cholinergic neurons in Alzheimer's disease, _Lancet_, 2: 1403.

10.Debuch, H., Witter, B., Illig, K., and Gunawan, J., 1982, On the metabolism of etherphospholipids in glial cell cultures, _in_: "Phospholipids in the Central Nervous System, Vol 1 Metabolism", L.Horrocks, G.B. Ansell and G. Porcellati eds., pp. 199-210, Raven Press, New York.

11.Demopoulos, C.A., Pinckard, R.N, and Hanahan, D.J., 1979, Platelet activating factor. Evidence for 1-0-alkyl-sn-glyceryl-3-phosphorylcholine as the active component (a new class of lipid chemical mediators), _J. Biol. Chem._, 254: 9355-9358.

12.Goth, A., Adams, H.R. and Knoohuizen, M., 1971, Phosphatidylserine: selective enhancer of histamine release, _Science_, 173: 1034-1035.

13.Gurr, M.I. and James, A.T.,1971, "Lipid Biochemistry: An Introduction", Chapman and Hall, London.

14.Handley, D.A. and Saunders, R.N., 1986, Platelets activating factor and inflammation in atherosclerosis: targets for drug development, _Drug. Dev.Res._ 7: 361-375.

15.Hirasawa, K. and Nishizuka, Y.,1981, Phosphatidylinositol turnover in receptor mechanism and signal transduction. _Ann.Rev.Pharmacol.Toxicol._, 25: 147-170.

16. Horrocks, L.A., Ansell, G.B. and Porcellati, G, 1982, Preface, in: "Phospholipids in the Central Nervous System, Vol 1 Metabolism", L. Horrocks, G.B. Ansell, and G. Porcellati eds., Raven Press, New York.

17. Martindale, 1982, "The Extra Pharmacopoeia" 28th Ed, J.E.F.Reynolds ed, pp 1651-1652, The Pharmaceutical Press, London.

18. Mongar, J.L. and Svec, P., 1972, The effect of phospholipids on anaphylactic histamine release, Br. J. Pharmacol., 46: 741-752.

19. Mozzi, R., Goracci, G., Siepi, D., Francescangeli, E., Andreoli, V., Horrocks, L.A. and Porcellati, G, 1982, Phospholipid synthesis by interconversion reactions in brain tissue, in: "Phospholipids in the Central Nervous System. Vol 1 Metabolism", L. Horrocks, G.B. Ansell and G.Porcellati eds., pp 1-12 Raven Press, New York.

20. Muirhead, E.E., Byers, L.W., Desiderio, D., Smith, K.A., Prewitt, R.L. and Brooks, B., 1981, Alkyl ether analogs of phosphatidylcholine are orally active in hypertensive rabbits, Hypertension,3,Suppl.1:107-111.

21. Neufeld, K.J., Lederman, C.L., Choy, P.C. and Man, R.Y.K., 1985, The effect of lidocaine on lysophosphatidylcholine-induced cardiac arrhythmias and cellular disturbances, Can. J. Physiol. Pharmacol.,63: 804-808.

22. Pogwizd, S.M., Onufer, J.R., Kramer, J.B., Sobel, B.E. and Corr, P.B., 1986, Induction of delayed after-depolarizations and triggered activity in canine Purkinje fibers by lysophosphoglycerides, Circ.Res., 59: 416-426.

23. Saunders, R.N. and Handley, D.A., 1987, Platelet-activating factor, Ann. Rev.Pharmacol. Toxicol., 27: 237-255.

24. Snyder, F, 1985, Chemical and biochemical aspects of platelet-activating factor: a novel class of acetylated ether-linked choline-phospholipids, Med.Res.Rev., 5: 107-140.

25. Sobel, B.E., Corr, P.B., Robison, A.K., Goldstein, R.A., Witkowski, F.X. and Klein, M.S., 1978, Accumulation of lysophosphoglycerides with arrhythmogenic properties in ischemic myocardium,J. Clin.Invest., 62:546-553.

26. Sugiyama, K., Sasaki, J. and Yamasaki, H.,1975, Potentiation by phosphatidylserine of calcium-dependent histamine release from rat mast cells induced by concanavalin, Japan. J. Pharmacol. 25: 485-487.

27. Summer, W.K., Majovski, L.V., Marsh, G.M., Tachiki, K. and Kling, A., 1986, Oral tetrahydroaminoacridine in long-term treatment of senile dementia, Alzheimer type, New England J. Med., 315: 1241-1245.

28. Vargaftig, B.B., Lefort, J., Chignard, M. and Benveniste. J, 1980, Platelet-activating factor induces a platelet-dependent bronchoconstriction unrelated to the formation of prostaglandin derivatives, Eur. J. Pharmacol., 65: 185-192.

29. Vannucchi, M.G. and Pepeu, G., 1987, Effect of phosphatidylserine on acetylcholine release and content in cortical slices from aging rats, Neurobiol. Aging, 8: 403-407.

30. Weltzien, H.U., 1979, Cytolytic and membrane-perturbing properties of lysophosphatidylcholine, Biochim.Biophys. Acta 559: 259-287.

31. Wessler, S. and Yin, E.T., 1968, Experimental hypercoagulable state induced by factor X: comparison of non-activated and activated forms, J.Lab.Clin.Med., 72: 256-262.

32. White, D.A., 1973, The phospholipid composition of mammalian tissues, in: "Form and Function of Phospholipids", G.B. Ansell, J.N. Hawthorne, R.M.C. Dawson eds., pp. 441-482, Elsevier, Amsterdam.

33. Whitehouse, P.J., Price, D.L., Struble, R.G., Clark, A.W., Coyle, J.T. and De Long, M.R., 1982, Alzheimer's disease and senile dementia: loss of neurons in the basal forebrain, Science, 215: 1237-1239.

34. Wurtman, R.J., Blusztajn, J.K., Ulus, I.H., Coviella, I.L.G.,Buyukuysal, L., Growdon, J.H. and Slack, B., 1989, Choline metabolism in cholinergic neurons: implications for the pathogenesis of neurodegenerative diseases, in: Advances in Neurology, Vol. 51, "Alzheimer's Disease", Wurtman, R.J., Corkin, S.H., Growdon, J.H. and Ritter-Walker, E. eds., pp. 117-125, Raven, New York.

METABOLISM OF ADRENIC AND ARACHIDONIC ACIDS

IN NERVOUS SYSTEM PHOSPHOLIPIDS

Lloyd A. Horrocks

The Ohio State University
Department of Physiological Chemistry
1645 Neil Ave., Room 214
Columbus, OH 43210, U.S.A.

CONCENTRATIONS, FUNCTIONS AND SOURCES OF POLYUNSATURATED FATTY ACIDS IN BRAIN

Concentrations

High concentrations of polyunsaturated fatty acids of the (n-6) and (n-3) series are found in the glycerophospholipids of the nervous system. The most plentiful are docosahexaenoic acid, 22:6 n-3, adrenic acid, 22:4 n-6, and arachidonic acid, 20:4 n-6. These fatty acids are distributed very differently in different glycerophospholipids (Tables 1 and 2). Generally, the ethanolamine lipids have higher proportions of the polyunsaturated fatty acids and lower proportion of the saturated fatty acid, palmitic acid. The comparison of human myelin with mouse brain compositions shows lower proportions of saturated and 22:6 fatty acids in myelin and higher proportions of monounsaturated fatty acids, mostly in oleic acid, 18:1. In human myelin the principal phospholipid is ethanolamine plasmalogen. This phospholipid contains about twice as much adrenic acid as arachidonic acid. The contents of these two fatty acids are about equal in the phosphatidylethanolamine. The mouse brain has lower proportions of adrenic acid than does the human brain. The proportion of adrenic acid in mouse brain ethanolamine glycerophospho- lipids is particularly important in the ether-linked types, ethanolamine plasmalogen and the alkylacyl type of ethanolamine glycerophospholipid (Fig. 1). In contrast, in the choline glycerophospholipids of mouse brain adrenic acid is almost completely missing whereas some arachidonic acid is present.

Functions

The function of adrenic acid in brain tissue is unknown, even though it is the principal polyunsaturated fatty acid in the myelin. One possibility is that adrenic acid could be acted on by cyclooxygenase and lipoxygenase. In order to investigate this possibility, [^{14}C]adrenic acid was synthesized. It was incubated with human platelets (12). The homologues of thromboxane B$_2$ and of HHT, dihomothromboxane B$_2$ and 14-hydroxy-7,10,12-nonadecatrienoic, acid were formed at a rate about one fifth of that found for the conversion of arachidonic acid. The homologue of 12-HETE, 14-hydroxy-7,10,12,16-docosatetraenoic acid, was

also found. The formation of thromboxane and HETE homologues by platelets suggests that these compounds, together with prostaglandin and
leukotriene homologues, may also be made in white matter of the brain.
If so, these eicosanoid homologues made from adrenic acid could have
unique functions due to their 2-carbon greater length. Several dihomo-
prostaglandins were detected in the metabolites of rabbit renal medulla
microsomes by Sprecher and colleagues (11). Otherwise, the possible
function of adrenic acid by forming metabolites through cyclooxygenase
and lipoxygenase pathways has been completely overlooked.

TABLE 1

ACYL GROUP COMPOSITIONS OF ETHANOLAMINE GLYCEROPHOSPHOLIPIDS IN BRAIN.

| | Mouse brain | | | Human myelin | |
	PtdEtn	PlsEtn	PakEtn	PtdEtn	PlsEtn
16:0	8	6	7	5	3
18:0	32	0	2	20	3
18:1	12	21	12	44	45
20:1	1	8	7	5	8
20:4 n-6	14	15	12	8	10
22:4 n-6	4	14	19	9	20
22:6 n-3	27	32	37	4	2

Principal fatty acids are given as their percentage. Human myelin
compositions, age range 36-50 years, were reported earlier (4).
Abbreviations, Ptd, phosphatidyl; Pls, plasmenyl; Pak, phosphalkanyl; and
Etn, ethanolamine. The compositions of the PlsEtn and PakEtn are for the
acyl groups from the 2-position only, whereas the compositions of the
PtdEtn are from both the 1 and 2-positions.

TABLE 2

ACYL GROUP COMPOSITIONS OF CHOLINE GLYCEROPHOSPHOLIPIDS.

| | Mouse brain | | Human myelin | |
	PtdCho	PlsCho	PakCho	PtdCho
16:0	46	29	50	29
18:0	13	6	4	14
18:1	28	30	27	48
20:1	1	9	5	1
20:4 n-6	6	11	6	3
22:4 n-6	1	2	1	1
22:6	5	7	4	0

Principal fatty acids are given as their percentage. Human myelin
compositions, age range 36-50 years, were reported earlier (4).
Abbreviations, Ptd, phosphatidyl; Pls, plasmenyl; Pak, phosphalkanyl; and
Cho, choline. The compositions of the PlsCho and PakCho are for the acyl
groups from the 2-position only, whereas the compositions of the PtdCho
are from both the 1 and 2-positions.

Uptake From Blood

Adrenic, arachidonic, and docosahexaenoic acids may be obtained from
the diet and transported to the brain, or they may be synthesized from
linoleic and linoleic acids in the liver or brain. Unesterified fatty
acids in the blood are mostly complexed with albumin (2, 8). A small part
of these fatty acids is always dissociated from albumin in the blood and
are taken up by the brain at a relatively rapid rate because they are
soluble in the membranes of the blood brain barrier and are esterified
into glycerophospholipids within seconds after entering a brain cell
(5). The rate of uptake of palmitic acid is between 0.5 and 3 nmol/g·min
into gray matter (6, 10). The rate of uptake into white matter is some-

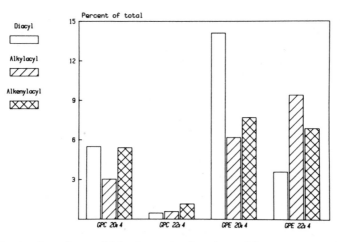

Fig. 1. Contents of arachidonic and adrenic acids in diradyl classes in choline and ethanolamine glycerophospholipids (GPC and GPE respectively) from mouse brain.

what lower. Quantitative radioautography of rat brain slices was used to evaluate the uptake of [1-^{14}C]arachidonic acid from the blood after intravenous injection (1). The highest incorporation in the brain was at 15 min after injection with 89% of the label in lipids. The incorporation into brain structures was not homogenous (5).

Recovery From Dietary Deficiency

Another approach to the investigation of the turnover of polyunsaturated fatty acids in brain is to induce a dietary deficiency, and then to replace the fatty acids with the normal fatty acids. A deficiency of n-3 fatty acids can be induced by including sunflower oil as the only fatty acid source in the diet (13). When rats were made deficient in n-3 fatty acids, then changed to soya oil at 15 days of age, recovery of the normal composition in brain subcellular fractions was found after two months of this diet (13). When the same deficiency was carried through to 60 days of age, the recovery of the normal composition in the capillaries and the choroid plexus required two months for the decrease of 22:5 n-6 and 2.5 months for the increase of 22:6 n-3 (3). The capillary cells play an important role in the uptake of fatty acids into the brain from the blood. References to earlier studies on deficiency of (n-6) fatty acids are given in the papers cited in this paragraph.

SYNTHESIS AND TURNOVER OF MOLECULAR SPECIES OF GLYCEROPHOSPHOLIPIDS IN MOUSE BRAIN

Methods

[^{3}H]Arachidonic acid and [1-^{14}C]adrenic acid were injected into the lateral ventricles of two month old mice. The injection was made through a guide cannula that had been put in place six days previously. The injection of 80 pmol arachidonate and 30 nmol adrenate in five μl was delivered over 2.5 min without anesthesia. The mice were frozen in liquid nitrogen before the brain was removed and lipids were extracted.

Preparative thin layer chromatography was used to separate diacylglycerol, triacylglycerol, and ethanolamine, choline and inositol glycerophospholipid fractions. The glycerophospholipids were converted into
diradylglyceroacetates and types of glycerophospholipids and the molecular species were separated (7). The specific radioactivity of each
separated molecular specie was determined by assay of the radioactivity
and of the content of arachidonate or adrenate from gas liquid chromatography with an internal standard. In this way, the specific radioactivity
values were based on the actual content of that specific fatty acid in
the molecular specie. Usually, studies with labeled arachidonate have
used total counts without separation of arachidonate and adrenate. This
separation is important because there is considerable rapid elongation of
arachidonate to adrenate immediately after injection. The rapid diffusion of arachidonate throughout the brain after intracerebral injection
was previously demonstrated (9).

Incorporation and Turnover of Arachidonic acid

Among the diacyl type of glycerophospholipids, specific activities of
[^{3}H]20:4 were highest for 16:0-20:4 and 18:1-20:4 species of phosphatidylcholine with peak values at 60 min after injection (Fig. 2). The 24 hour
values were more than 70% lower indicating half-lives of 12 hours or
less. The 18:0-20:4 specie had less than half of the specific radioactivity of the other two species at 60 min. The specific radioactivity
values of these species of phosphatidylinositol were similar but slightly
lower than the corresponding values for phosphatidylcholine. Also, the
turnover rate of the former was slightly slower. As judged by specific
radioactivity values, the metabolism of arachidonic acid in phosphatidylethanolamine was much more sluggish, particularly in the 18:0-20:4
specie, the major component of this glycerophospholipid. The diacyl
types were pulse-labeled, but the ether-linked glycerophospholipids had a
lag in uptake (Fig. 3). The latter types are labeled by energy and
CoA-independent transfer of arachidonic acid from phosphatidylcholine.
The specific radioactivity values of the 16:0-20:4 and 18:1-20:4 species
of the alk-1-enylacylglycerophosphocholine were greater than those for
the diacyl type at 30 and 60 min after injection. These values increased

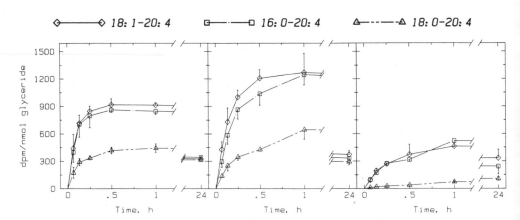

Fig. 2. Specific radioactivity of ^{3}H in molecular species containing
arachidonic acid in diacyl-glycerophosphoinositol (left), diacyl-glycer-
ophosphocholine (center), and diacyl-glycerophosphoethanolamine (right).
Reprinted with permission of N.Y. Acad. Sci. (5).

between 1 and 24 hours. These results may be due to the release of arachidonic acid from the choline plasmalogens followed by replenishment from phosphatidylcholine.

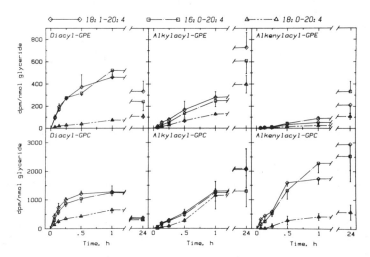

Fig. 3. Specific radioactivity of ^{3}H in arachidonic acid isolated from molecular species of ethanolamine glycerophospholipids and choline glycerophospholipids. Abbreviations, GPE, glycerophosphoethanolamine; GPC, glycerophosphocholine. Reprinted with permission of N.Y. Acad. Sci. (5).

Incorporation and Turnover of Adrenic Acid

Part of the injected [^{3}H]arachidonic acid was immediately elongated to [^{3}H]adrenic acid. The uptake of adrenic acid was much faster into phosphatidylinositol than into choline or ethanolamine phospholipids at early times after injection (Fig. 4, Fig. 5, Fig. 6, Table 3). There was a considerable turnover of the [^{14}C]adrenate in the phosphatidylinositol between 1 and 24 hours. Part of this appeared to be transferred to the phosphalkanylethanolamine. By 24 hours after injection the phosphalkanyl-ethanolamine showed a preferential uptake of the adrenic acid. All three adrenate molecular species showed a very rapid turnover in the phosphatidylinositol. At 30 minutes the 18:0-22:4 species had less than half the specific radioactivity of the other two molecular species. The turnover of this species was also less than that of the others. The turnover of the 18:1-22:4 specie of phosphatidylinositol was particularly rapid with an indicated half-life of less than 8 hours. For the phosphatidylcholine and the phosphatidylethanolamine both the 18:1-22:4 and 16:0-22:4 showed a good uptake of [^{14}C]adrenic acid and considerable turnover between 1 and 24 hours with half-lives of less than 24 hours. In contrast the 18:0-22:4 species of phosphatidylcholine and phosphatidylethanolamine gained considerably in specific radioactivity between 60 min and 24 hours. For the phosphatidylcholine, this species had the highest specific radioactivity at 24 hours. The specific radioactivity of [^{14}C]22:4 was much higher in the phosphalkanylethanolamine than in the ethanolamine plasmalogen. Between 1 and 24 hours, all species of the ethanolamine plasmalogen more than doubled their specific radioactivity whereas only small changes were seen in the phosphalkanylethanolamine. At 60 min after injection all of the phosphalkanylethanolamine molecular species were at least 7-fold greater in specific radioactivity when compared with the ethanolamine plasmalogen molecular species.

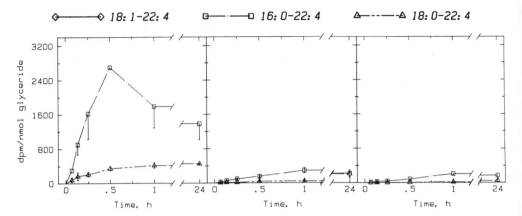

Fig. 4. Specific radioactivity of ³H in molecular species containing adrenic acid in diacyl-GPI (left), diacyl-GPC (center), and diacyl-GPE (right).

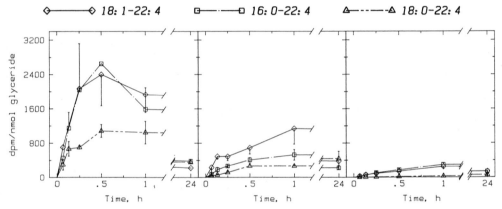

Fig. 5. Specific radioactivity of ¹⁴C in molecular species containing adrenic acid in diacyl-GPI (left), diacyl-GPC (center), and diacyl-GPE (right).

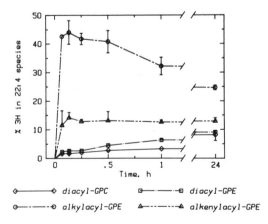

Fig. 6. The proportion of ³H recovered in adrenic acid in different glycerophospholipid types after injection of [³H]arachidonic acid.

TABLE 3
THE INCORPORATION OF [^{14}C]ADRENIC ACID
INTO GLYCEROPHOSPHOLIPIDS OF MOUSE BRAIN.

Time	PtdIns	PtdCho	PlsCho	PakCho	PtdEtn	PlsEtn	PakEtn
4 min	12.0±1.0	1.5±0.5	N.D.	0.4±0.5	0.8±0.4	0.1±0.03	1.5±1.2
8 min	22.2±0.7	3.5±0.7	0.5±0.5	0.9±0.1	1.8±0.5	0.3±0.1	3.8±0.6
15 min	26.9±1.3	5.8±0.5	1.4±0.3	1.9±0.3	2.8±0.1	0.5±0.02	7.2±0.5
30 min	33.9±1.1	7.9±1.2	2.5±0.4	3.9±0.3	4.4±0.7	0.8±0.2	9.7±3.6
60 min	28.9±2.1	11.9±0.4	6.9±7.0	5.1±1.5	6.8±0.4	1.1±0.3	15.1±1.9
24 hr	11.8±0.6	5.2±0.9	5.1±1.5	9.4±1.1	5.4±1.6	3.1±0.6	13.4±4.0

Values are dpm/nmol lipid P ± S.D., n=4. N.D. is non-detectable.
Abbreviations, Ptd, phosphatidyl; Pls, plasmenyl; Pak, phosphalkanyl;
Ins, inositol; Cho, choline; and Etn, ethanolamine.

CONCLUSIONS

     The polyunsaturated fatty acids in brain are found in different
molecular species. The metabolism is different for each molecular specie
and depends on the type of linkage at the 1-position, the side chains at
the 1- and 2-positions, and the compound attached to phosphoric acid at
the 3 position. It is necessary to understand the metabolism of all of
these molecular species in order to understand the metabolism of
polyunsaturated fatty acids in brain.

REFERENCES

1. DeGeorge, J.J., Noronha, J.G., Rapoport, S.I. and Lapetina, E.G.
    (1988) Incorporation of intravascular [1-^{14}C]arachidonate into rat
    brain. Trans. Am. Soc. Neurochem. 19: 108.

2. Dhopeshwarkar, G.A. and Mead, J.F. (1973) Uptake and transport of
    fatty acids into the brain and the role of the blood-brain barrier
    system. Adv. Lipid Res. 11: 109-142.

3. Homayoun, P., Durand, G., Pascal, G. and Bourre, J.M. (1988)
    Alteration in fatty acid composition of adult rat brain capillaries
    and choroid plexus induced by a diet deficient in n-3 fatty acids:
    Slow recovery after substitution with a nondeficient diet. J.
    Neurochem. 51: 45-48.

4. Horrocks, L.A., VanRollins, M. and Yates, A.J. 1981. Lipid changes
    in the ageing brain. In, "The Molecular Basis of Neuropathology",
    ed. by A.N. Davison and R.H.S. Thompson. Edward Arnold (Publishers)
    Ltd., London, pp. 601-630.

5. Horrocks, L.A. (1989) Sources of brain arachidonic acid uptake and
    turnover in glycerophospholipids. Ann. N.Y. Acad. Sci. 559: 17-24.

6. Kimes, A.S., Sweeney, D., London, E.D. and Rapoport, S.I. (1983)
    Palmitate incorporation into different brain regions in the awake
    rat. Brain Res. 274: 291-301.

7. Nakagawa, Y. and Horrocks, L.A. 1983. Separation of alkenylacyl,
    alkylacyl, and diacyl analogues and their molecular species by high
    performance liquid chromatography. J. Lipid Res. 24: 1268-1275.

8. Pardridge, W.M. and Mietus, L.M. (1980) Palmitate and cholesterol transport through the blood-brain barrier. J. Neurochem. 34: 463-466.

9. Pediconi, M.F., Rodriguez de Turco, E.B. and Bazan, N.G. (1982) Diffusion of intracerebrally injected [1-^{14}C]arachidonic acid and [2-^{3}H]glycerol in the mouse brain. Neurochem. Res. 7: 1453-1463.

10. Reddy, P.V. and Drewes, L.R. (1988) Transport of palmitate from blood to brain and its incorporation into lipids. FASEB J. 2: A1790.

11. Sprecher, H., VanRollins, M., Sun, F., Wyche, A. and Needleman, P. (1982) Dihomo-prostaglandins and -thromboxane. J. Biol. Chem. 257: 3912-3918.

12. VanRollins, M., Horrocks, L., and Sprecher, H. (1985) Metabolism of 7,10,13,16-docosatetraenoic acid to dihomo-thromboxane, 14-hydroxy-7,10,12-nonadecatrienoic acid and hydroxy fatty acids by human platelets. Biochim. Biophys. Acta 833: 272-280.

13. Youyou, A., Durand, G., Pascal, G., Piciotti, M., Dumont, O. and Bourre, J.M. (1986) Recovery of altered fatty acid composition induced by a diet devoid of n-3 fatty acids in myelin, synaptosomes, mitochondria, and microsomes of developing rat brain. J. Neurochem. 46: 224-228.

PHOSPHOLIPID ABSORPTION AND DIFFUSION THROUGH MEMBRANES

Alessandro Bruni,[*] Fabrizio Bellini, Lucia Mietto, Elena
Boarato and Giampietro Viola

[*]Department of Pharmacology, University of Padova, Padova,
Italy and Fidia Research Laboratories, Abano Terme, Italy

INTRODUCTION

Phospholipids are constituents of cell membranes. A main field of
investigation is the study of mechanisms directing their synthesis and
their selected distribution in different membranes of the cell. However, it is now widely appreciated that, together with a structural
function, phospholipids have regulatory activity inside the cell,
within the plasma membrane, as well as outside the cell (9). In their
natural position, phospholipids modulate the activity of membrane-bound
enzymes. Furthermore, the operation of selected synthetic pathways
yields phospholipid derivatives behaving as second messengers. Thus, it
is found that phosphatidic acid (PA), a phospholipid produced during
the activation of signalling mechanisms, from the phosphorylation of
diacyl glycerol (DAG) or by the action ₒf phospholipase D, may activate
the superoxide production in neutrophils (2). Also, a lysophosphatidylserine analogue bearing an ether group at the sn-1 position of
glycerol, acts as a negative modulator of the interaction between
steroids and their intracellular receptors (5). Within the plasma
membrane phospholipids serve as a substrate for the receptor-dependent
phospholipases involved in the transduction of external stimuli into
intracellular signals. Second messengers such as DAG are generated by
these reactions. Outside the cells phospholipids behave as auto-
pharmacological agents (autacoids, 6). For example, phosphatidylserine
exposed to the extracellular environment activates the monocyte-macro-
phage system (30), exogenous PA is translocated inside the cell in the
form of DAG (25), lysophosphatidylserine activates mouse mast cells (4)
and phosphatidylserine (PS) vesicles interact with lymphocytes (20).
The purpose of this review is to outline unifying aspects of these
different actions of phospholipids.

Phospholipid uptake in parenteral administrations

Initial attempts to administer hydrosoluble compounds entrapped in
liposomes by parenteral routes met with failure due to the impact of
phospholipid vesicles with plasma lipoproteins and their subsequent
fragmentation. Adequate adjustment of liposome composition (i.e. the
inclusion of sphingomyelin (SP) or cholesterol (28), yields stable lipid
particles which, however, are promptly cleared from the circulation by
the activity of the reticuloendothelial system through a non-specific

process of phagocytosis. The recognition of lipid vesicles by phagocytes can be prevented by including a monosialoganglioside (GM_1) into
the lipid bilayer (1). This effect of GM_1 can however be overcome by
the addition of PS which restores the liposome uptake by the phagocytes. The action of PS cannot be reproduced by other negatively
charged phospholipids. Thus, PS is able to change the non specific
ingestion by phagocytes into a specific process driven by the phosphorylserine head-group of this phospholipid. These findings confirm
previous observations showing that macrophages recognize and internalize red cells coated with PS or red cells in which the asymmetric
distribution of PS is lost (26). As predicted by the properties of PS
we find that vesicles of this phospholipid injected by the i.v. route
are promptly removed from plasma (Fig. 1A). However, evidence from in
vitro experiments suggests that plasma clearance of PS is not due
exclusively to macrophages. At variance from lipid mixtures containing
PS as a minor component, vesicles made only of this phospholipid
promote PS internalization also by lymphocytes (20).

Similarly to PS the lyso derivative (lysoPS) has a short plasma
half-life (Fig. 1B). In this case the plasma decay reflects the interaction of phospholipid with cells other than those of the reticuloendothelial system. Lysophospholipids do not form organized structures,
which are stable in plasma. Rather, they circulate as a complex with
plasma albumin and distribute in several organs, mainly the liver. Due
to its affinity for connective tissue mast cells, lysoPS elicits
pharmacological effects during its short life span preceding its

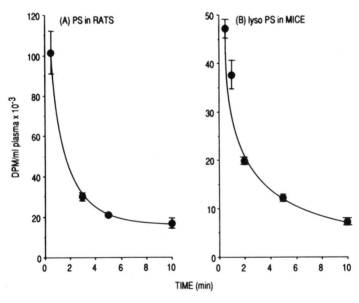

Fig. 1. Plasma-decay curve of serine phospholipids in rodents. A. Phos-
phatidyl-[^{14}C]serine vesicles (2 μg/g; 8400 DPM/g) were injected i.v.
into rats weighing 290 g. B. Lysophosphatidyl-[^{14}C]serine (2.5 μg/g;
6900 DPM/g) was injected i.v. into mice weighing 37 g. The data (means
± S.E.) refer to the total radioactivity of plasma in 3 rats or 8-10
groups of 3 mice. As shown by appropriate tests, the metabolism of
injected phospholipids did not alter the curves during this short time
of observation.

removal from plasma (7). If the ingestion of liposomes by phagocytes is a true form of elimination yielding metabolites to be utilized only for _de novo_ synthesis, the uptake of lysophospholipids by tissues is a first step toward their immediate utilization (12). As shown in Fig. 2, the incorporation of lysoPS into rat leukocytes is coincident with its conversion to PS (19). The uptake of lysoPS and the reacylation reaction were both enhanced when cell activation was induced by the phorbol ester, tetradecanoyl phorbol acetate. Lysophospholipid incorporation into a cell proceeds by the incorporation into the outer layer of plasma membrane, its translocation to the inner layer, and its acylation into the corresponding diacyl derivative (21). For lysoPS, the translocation step may be accelerated by the translocase for the aminophospholipids (11).

A particular interaction between circulating phospholipids and cells is the transfer of these compounds to parasites infecting the erythrocytes. As shown by recent studies, Plasmodium knowlesi which is present in red cells, may use extracellular phosphatidylcholine (PC), PS and phosphatidylethanolamine (PE) to assemble their membranes (22). The incubation of infected cells with PS and PE is followed by conversion of the phospholipids into PE and PC, respectively. Since PS decarboxylase and PE methylating enzymes are absent in erythrocytes,

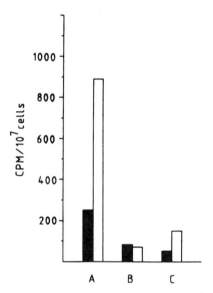

Fig. 2. Metabolism of lysoPS in rat leukocytes (37% polymorphonuclear and 33% mononuclear cells). 10^7 cells were incubated for 15 min at 37°C with 1 nmol lysophosphatidyl-[U-^{14}C]serine (2700 CPM/nmol) in 1 ml of buffered saline solution supplemented with 10 mM glucose and 1 mM Ca^{2+}. After centrifugation and washing the phospholipids were extracted with chloroform-methanol and separated by two-dimensional thin layer chromatography. Black bars, no addition; white bars, 8 nM tetradecanoyl phorbol acetate (6). The data are from a representative experiment and show that the main fraction of labeled phospholipids was formed by PS (A) and not lysoPS (B). The hydrosoluble fraction is shown in (C). From ref. 6.

these observations indicate that intact phospholipid molecules migrate from the extracellular environment to the parasite, crossing the plasma membrane of host cells.

Interaction with cells during the intestinal absorption of phospholipids

The uptake of phospholipids by the enterocyte and their conversion into the components of circulating lipoproteins is an example of controlled phospholipid metabolism. Phospholipids escaping complete degradation in the intestinal lumen are absorbed in the form of lyso derivatives. Their transfer into the enterocyte likely proceeds through the same steps outlined above (21): the incorporation into the outer layer of plasma membrane, followed by the energy-independent translocation into the inner layer. The rate of translocation is not dependent on the concentration of external lysophospholipid. Rather, it is a function of the amount incorporated in the outer layer of plasma membrane. Although occurring by passive diffusion, the translocation requires a temperature higher than 20°C. When the lysophospholipid reaches the inner layer of plasma membrane it is promptly reacylated. The acylation of lysoPC is one of the two pathways by which the mucosal epithelial cell provides chylomicrons with PC, the main phospholipid component of these lipoproteins. An alternative pathway is the de novo synthesis that is the main source of chylomicron PC in the absence of phospholipids in the intestinal lumen. At 1.5-2.5 mM PC in the lumen, the two pathways are approximately equivalent; at 10 mM PC only the acylation of lysoPC is operative (17,18). The possibility of regulating the relative contribution of these two pathways has the advantage of saving DAG at high phospholipid input. Furthermore, it provides a means of saving energy when food supply is available. Indeed, much less high-energy phosphate bonds are required when PC is generated from lysoPC, when compared with its de novo synthesis (18). Recently, the existence of intracellular pools of phospholipids with different availability for lipoproteins synthesis has been demonstrated in the liver (31). For example, PE generated via PS decarboxylase is released in preference to PE generated via the CDP-ethanolamine pathway.

At variance from other phospholipids, the fate of PS in the intestinal tract, has been less studied. Our tests (Table 1) show that after oral administration PS appears in the enterocytes within 1 hour, most likely after conversion into lysoPS and subsequent reacylation. In the successive 3 hours PS is slowly decarboxylated into PE. Of interest is the observation that PS gains access not only into the mucosal epithelial cells but also into Peyer's patches, a lymphoid tissue distributed in this organ.

First messenger effect of phospholipids

It is well established that compounds originating from the hydrolysis of membrane phospholipids, interact with plasma membrane receptors of target cells, activating the sequence of signal transducing systems. These compounds are named lipid autacoids (6). Eicosanoids and platelet activating factor are examples of lipid autacoids. Our studies show that lysoPS can be added to this list, based on extensive investigations on the typical target cell for this phospholipid, the rodent mast cell. In rat peritoneal mast cells, lysoPS elicits histamine release if added together with specific mast cell ligands (e.g. nerve growth factor; 8). The two compounds are ineffective when added separately. A survey of phospholipid metabolism during the stimulation induced by lysoPS plus nerve growth factor shows that the stimulus is delivered inside the cell by the phospholipase C-phosphoinositide system (3). Similar results are obtained with mouse peritoneal mast cells where lysoPS elicits histamine release in the absence of synergistic stimuli (4). As shown in Table 2 the addition of lysoPS to mouse mast cells preincubated with $^{32}PO_4$ causes an increase of PA labeling, a result suggesting phosphorylation of newly formed DAG. In agreement with the activation of a phospholipase C acting on PC and PI the labeling of these phospholipids is also increased under the action of lysoPS. Structure-activity relationship in the effect of lysoPS indicates that the mast cell activation is due to the structure of the phosphorylserine head group. The amino and the carboxyl groups linked to the serine in an L-configuration are required for full lysoPS activity (4,10).

TABLE 1

PHOSPHATIDYLSERINE ABSORPTION BY INTESTINAL CELLS

Phospholipid	Time (h)	DPM/100 mg wet weight	
		Intestine	Peyer's patches
PS	1	3179 $\pm$ 182	2964 $\pm$ 153
PE	1	1841 $\pm$ 65	1655 $\pm$ 69
PS/PE ratio	1	1.72 $\pm$ 0.06	1.79 $\pm$ 0.04
PS	4	7757 $\pm$ 869	7626 $\pm$ 1061
PE	4	12235 $\pm$ 1170	12076 $\pm$ 1290
PS/PE ratio	4	0.64 $\pm$ 0.04	0.63 $\pm$ 0.03

Radiolabeled PS (50 mg/kg; 2.9×10^6 DPM/rat) dispersed in a buffered saline solution (pH 7.2) was given by oral route to male albino rats weighing 230 g. At the indicated time 5 fragments of the intestinal tissue and 5 Peyer's patches were taken in the tract between the pylorus and the middle part of small intestine. The samples were extracted with chloroform-methanol (2:1 v/v) and the phospholipids resolved by thin layer chromatography. PS and PE were the only phospholipids showing significant radioactivity. The data are means $\pm$ S.E.

TABLE 2

LYSOPS-INDUCED LABELING OF MAST CELL PHOSPHOLIPIDS

Phospholipid	$^{32}PO_4$ incorporation (DPM/10^5 cells)		
	None	1 µM lysoPS	ratio
Phosphatidic acid	870	1900	2.2
Phosphatidylcholine	170	350	2.0
Phosphatidylinositol	120	290	2.4

10^5 purified mouse mast cells were preincubated for 5 min at 37°C with 16 µCi of $^{32}PO_4$ in a final volume of 100 µl of phosphate-free saline solution, buffered to pH 7.2 with Hepes (3). Ten min after the addition of lysoPS the phospholipids were extracted and counted.

Phospholipid internalization by cells

Internalization of diacylphospholipids by host cells follows three main routes: endocytosis, phospholipid hydrolysis with internalization of a diffusible specie, translocation catalysed by specific proteins. A major contribution to these studies has been achieved by the use of fluorescent lipid analogues with which it has been show that various mechanisms may operate in the same cell (25). Thus, PC analogues are internalized by endocytosis but at the same time are partially hydrolysed to lysoPC after their binding to plasma membrane. PE analogues together with endocytosis may utilize the energy-dependent translocation pathway catalysed by the lipid-transfer protein discovered by Seigneuret and Devaux (27). Also, PA is degraded to DAG that diffuses across the plasma membrane to be converted into triacylglycerol (TG), PA and other phospholipids inside the cell. A limitation in these studies is that the physical-chemical properties of fluorescent phospholipid analogues are similar to lysophospholipids rather than natural diacylphospholipids. Convection through the aqueous phase and incorporation into the membrane are therefore facilitated. Furthermore, the monomer concentration produced by PS and PE analogues is sufficient to activate the lipid transporting system (the estimated Km is equal to 5.4 μM; 33). By contrast, long chain natural diacylphospholipids show little inter-bilayer exchange and yield a low monomer concentration (the critical micellar concentration of PS is approx. 10^{-8} M). The study of the translocation of PS with acyl chain of variable length supports these considerations (11).

The possibility of phospholipid metabolism before the translocation across the plasma membrane is a point requiring further investigations, since phospholipases are believed to be located inside the cell. This view may however require reconsideration as it has been demonstrated that the Fc receptor in lymphocytes is endowed with phospholipase A_2 activity (29). The interaction of PS with the inflammatory cells accumulating in the peritoneal cavity of rats after the local injection of casein, conforms to these general rules (20). As shown in Fig. 3, PS preferentially interacts with the macrophages and lymphocytes when these are compared to polymorphonuclear cells. Furthermore, separate experiments have shown that PS has greater activity than PC and phosphatidylinositol (PI) in the binding to these mononuclear cells. As shown by the combined addition of PS and radiolabeled arachidonate, part of PS uptake is due to hydrolysis of the phospholipid with generation of lysoPS, a diffusible intermediate penetrating the plasma membrane of recipient cells. Partly, it is the result of endocytosis. This has been demonstrated by the inhibition induced by metabolic inhibitors and by cytochalasin B. Unexpectedly, endocytosis during PS uptake is more active in lymphocytes than macrophages. Endocytosis may lead to the internalization of PS vesicles; alternatively, the endocytic vesicles internalize the excess of PS coming from the acylation of the incorporated lysoPS.

CONCLUDING REMARKS

The studies reviewed in this paper deal with the absorption of exogenous phospholipids by cells, occurring when these compounds are administered by parenteral routes or when ingested phospholipids reach the mucosal epithelial cells of the intestinal tract. In both cases the interaction of phospholipid with the cell can not simply be considered as the passive transfer of a lipid molecule across the plasma membrane. Nor it can be regarded as a means to induce a stable enrichment of the cell membrane with exogenous phospholipids. The impact with the cell invariably results in the activation of systems involved in the regulation of phospholipid metabolism. For selected phospholipids such as lysoPS, which bears a polar head-group suitable to activate receptor sites, the first consequence of the impact is the activation of metabolic sequences yielding second messengers inside the cell (3). Furthermore, based on the incorporation of external PC and lysoPC in a mutant of chinese hamster ovary cell defective in CDP-choline synthetase, it has been suggested that a "phospholipid sensor" is present in animal cells, which regulate the absorption of exogenous phospholipids in relation to the efficiency of the endogenous synthesis (13). Conversely, when external phospholipids are incorporated, the endogenous synthesis is decreased. This is best demonstrated by the analysis of the intestinal absorption of phospholipids showing that lysoPC incorporation inhibits the endogenous synthesis of PC in the enterocytes (18). Much in the same way, the absorption of PS by the chinese ovary cell prevents the endogenous synthesis of this phospholipid (24).

The models used to study the interaction of phospholipids with the cells stress the role of lysophospholipid in this process. Since fusion between vesicles of diacylphospholipid and cells is a limited event and considering that the low monomer concentration formed by these compounds is not sufficient to fully activate the lipid transporting

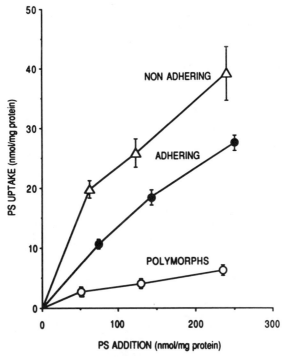

Fig. 3. PS uptake by different populations of casein-elicited rat peritoneal cells. Polymorphonuclear cells (mainly neutrophils) containing 400 μg of proteins, adhering mononuclears (macrophages) containing 300 μg of proteins and non-adhering mononuclears (mainly lymphocytes) 300 μg of proteins were incubated for 1 h at 37°C with the indicated amount of phosphatidyl-[^{14}C]serine vesicles. After washing, the radioactivity was measured in the cell sediment. Thin layer chromatography of extracted phospholipids showed that 80% of the bound radioactivity was in the form of PS (from ref. 20).

proteins (34), endocytosis becomes the major process of internalization of diacylphospholipids. The degradation of endocytic vesicle content only supplies materials for de novo synthesis of phospholipids. By contrast, the acylation of the incorporated lysophospholipids causes the increase of cell phospholipid content with subsequent stimulation of mechanisms regulating the phospholipid composition of membranes. Furthermore, lysophospholipids may be reacylated by transacylation with the membrane phospholipids, a process causing further perturbation of membrane composition (16).

The decrease of endogenous phospholipid metabolism induced by the newly incorporated phospholipids is not without consequence for the cell activity. Although this aspect has not been fully explored, the first indications suggest that both activation and inhibition is produced. Thus, the internalization of PS and lysoPS by macrophages reduces microbicidal activity (14,15) and the ingestion of phospholipid vesicles containing PI inhibits phagocytosis and PI turnover in these cells (32). By contrast, cell growth is observed under the influence of PA (23) and after long term exposure to PS (33). Finally, when the phospholipid-cell interaction involves the binding to specialized receptors, activation of a specific cell function (histamine secretion) may result as a consequence (3).

REFERENCES

1. Allen, T. M., Williamson, P., and Schlegel, R. A., 1988, Phosphatidylserine as a determinant of reticuloendothelial recognition of liposome models of the erythrocyte surface, Proc. Natl. Acad. Sci. USA 85:8067-8071
2. Bellavite, P., Corso, F., Dusi, S., Grzeskowiak, M., Della Bianca, V., and Rossi, F., 1988, Activation of NADPH-dependent superoxide production in plasma membrane extracts of pig neutrophils by phosphatidic acid, J. Biol. Chem. 263:8210-8214.
3. Bellini, F., Toffano, G., Bruni, A., 1988, Activation of phosphoinositide hydrolysis by nerve growth factor and lysophosphatidylserine in rat peritoneal mast cells, Biochim. Biophys. Acta 970:187-193
4. Boarato, E., Mietto, L., Toffano, G., Bigon, E., and Bruni, A., 1984, Different responses of rodent mast cells to phosphatidylserine, Agents & Actions 14:613-618
5. Bodine, P. V., and Litwack, G., 1988, Purification and structural analysis of the modulator of the glucocorticoid-receptor complex. Evidence that the modulator is a novel phosphoglyceride, J. Biol. Chem. 263:3501-3512
6. Bruni, A., 1988, Autacoids from membrane phospholipids, Pharmacol. Res. Commun. 20:529-544
7. Bruni, A., Bigon, E., Battistella, A., Boarato, E., Mietto, L., and Toffano, G., 1984, Lysophosphatidylserine as histamine releaser in mice and rats, Agents & Actions 14:619-625
8. Bruni, A., Bigon, E., Boarato, E., Mietto, L., Leon, A., and Toffano, G., 1982, Interaction between nerve growth factor and lysophosphatidylserine on rat peritoneal mast cells, FEBS Letters 138:190-192
9. Bruni, A., Mietto, L., Battistella, A., Boarato, E., Palatini P., and Toffano, G., 1986, Serine phospholipids in cell communication, in: "Phospholipids Research and the Nervous System," L. Horrocks, L. Freysz and G. Toffano, eds., Fidia Research Series, Vol. 14, Liviana Press, Padova, pp. 217-223
10. Chang, H. W., Inoue, K., Bruni, A., Boarato, E., Toffano, G., 1988, Stereoselective effects of lysophosphatidylserine in rodents, Br. J. Pharmac. 93:647-653

11. Daleke, D. L., and Huestis, W. H., 1985, Incorporation and translocation of aminophospholipids in human erythrocytes, _Biochemistry_
    24:5406-5416
12. Esko, J. D., and Matsuoka, K. Y., 1983, Biosynthesis of phosphatidylcholine from serum phospholipids in chinese hamsters ovary
    cells deprived of choline, _J. Biol. Chem._ 258:3051-3057
13. Esko, J. D., Nishijima, M., and Raetz, C. R. H., 1982, Animal cells
    dependent on exogenous phosphatidylcholine for membrane biogenesis,
    Proc. Natl. Acad. Sci. USA 79:1698-1702
14. Gilbreath, M. J., Hoover, D. L., Alving, C. R., Swartz, G. M.,
    Meltzer, M. S., 1986, Inhibition ,of lymphokine-induced macrophage
    microbicidal activity against Leishmania major by liposomes:
    characterization of the physicochemical requirements for liposome
    inhibition, _J. Immunol._ 137:1681-1687.
15. Gilbreath, M. J., Nacy, C.A., Hoover, D. L., Alving, C. R., Swartz,
    G. M., and Meltzer, M. S., 1985, Macrophage activation for
    microbicidal activity against Leishmania major: inhibition of
    lymphokine activation by phosphatidylcholine-phosphatidylserine
    liposomes, _J. Immunol._ 134:3420-3425
16. Irvine, R. F., and Dawson, R. M. C., 1979, Transfer of arachidonic
    acid between phospholipids in rat liver microsomes, _Biochem._
    Biophys. Res. Commun. 91:1399-1405
17. Mansbach, C. M. II, 1977, The origin of chylomicron phosphatidylcholine in the rat, _J. Clin. Invest._ 60:411-420
18. Mansbach, C. M. II, and Parthasarathy, S., 1979, Regulation of de
    novo phosphatidylcholine synthesis in rat intestine, _J. Biol. Chem._
    254:9688-9694
19. Mietto, L., Boarato, E., Toffano, G., and Bruni, A., 1987, Lysophosphatidylserine-dependent interaction between rat leukocytes and
    mast cells, _Biochim. Biophys. Acta_ 930:145-153
20. Mietto, L., Boarato, E., Toffano, G., and Bruni, A., 1989, Internalization of phosphatidylserine by adherent and non-adherent rat
    mononuclear cells. _Biochim. Biophys. Acta_ 1013:1-6
21. Mohandas, N., Wyatt, J., Mel, S. F., Rossi, M. E., and Shohet, S.
    B., 1982, Lipid translocation across the human erythrocyte
    membrane. Regulatory factors, _J. Biol. Chem._ 257:6537-6543
22. Moll, G. N., Vial, H. J., Ancelin, M. L., Op den Kamp, J. A. F.,
    Roelofsen, B., and van Deenen, L. L. M., 1988, Phospholipid uptake
    by Plasmodium knowlesi infected erythrocytes, _FEBS Lett._ 232:341-346
23. Moolenaar, W. H., Kruijer, W., Tilly, B. C., Verlaan, I., Bierman,
    A. J., and de Laat, S.W., 1986, Growth factor-like action of
    phosphatidic acid, _Nature_ 323:171-173
24. Nishijima, M., Kuge, O., and Akamatsu, Y., 1986, Phosphatidylserine
    biosynthesis in cultured chinese hamster ovary cells I. Inhibition
    of de novo phosphatidylserine biosynthesys by exogenous phosphatidylserine and its efficient incorporation, _J. Biol. Chem._
    261:5784-5789
25. Pagano, R. E., and Sleight, R. G., 1985, Defining lipid transport
    pathways in animal cells, _Science_ 229:1051-1057
26. Schroit, A. J., Tanaka, Y., Madsen, J., and Fidler, I. J., 1984,
    The recognition of red blood cells by macrophages: role of phosphatidylserine and possible implications of membrane phospholipid
    asymmetry, _Biol. Cell_ 51:227-238
27. Seigneuret, M., and Devaux, P.F., 1984, ATP-dependent asymmetric
    distribution of spin-labeled phospholipids in the erythrocyte
    membrane: relation to shape changes, _Proc. Natl. Acad. Sci. USA_
    81:3751-3755
28. Senior, J., and Gregoriadis, G., 1982, Stability of small unilamellar liposomes in serum and clearance from the circulation: the
    effect of the phospholipid and cholesterol components, _Life Sci._
    30:2123-2136

29. Suzuki, T., Saito-Taki, T., Sadasivan, R., and Nitta, T., 1982, Biochemical signal transmitted by Fcγ receptors: phospholipase A$_2$ activity of Fcγ2b receptor of murine macrophage cell line P388D$_1$, Proc. Natl. Acad. Sci. USA 79:591-595

30. Tanaka, Y., and Schroit, A. J., 1983, Insertion of fluorescent phosphatidylserine into the plasma membrane of red blood cells. Recognition by autologous macrophages, J. Biol. Chem. 258:11335-11343

31. Vance, J. E., 1988, Compartmentalization of phospholipids for lipoprotein assembly on the basis of molecular species and biosynthetic origin, Biochim. Biophys. Acta 963:70-81

32. Wassef, N. M., Roerdink, F., Richardson, E. C., and Alving, C. R., 1984, Suppression of phagocytic function and phospholipid metabolism in macrophages by phosphatidylinositol liposomes, Proc. Natl. Acad. Sci. USA 81:2655-2659

33. Yui, S., and Yamazaki, M., 1987, Relationship of ability of phospholipids to stimulate growth and bind to macrophages, J. Leukocyte Biol. 41:392-399

34. Zachowski, A., Favre, E., Cribier, S., Hervé, P., and Devaux, P. F., 1986, Outside-inside translocation of aminophospholipids in the human erythrocyte membrane is mediated by a specific enzyme, Biochemistry 25:2585-2590

PHOSPHOLIPID STABILISED EMULSIONS FOR PARENTERAL NUTRITION

AND DRUG DELIVERY

Stanley S. Davis

Department of Pharmaceutical Sciences
University of Nottingham, University Park
Nottingham, NG7 2RD, U.K.

INTRODUCTION

Emulsions have been used in medical practice from the very earliest times. They comprise the dispersion of one immiscible liquid in another, and consequently can take different physical forms.[7] For example, the dispersion of an oil in water provides the oil-in-water (o/w) emulsion system whereas the converse is the water-in-oil system (w/o). The physical nature of the system often dictates its route of administration into the body. Water-in-oil emulsions are normally given topically to the skin or the eye, or used as controlled release injectables (intramuscular or subcutaneous). Oil-in-water emulsions are those that can be given orally or by injection into the bloodstream. More complicated types of emulsion also exist, where a prototype system is reemulsified to give a so-called multiple emulsion system. Some of these, in the form of water-oil-water (w/o/w) systems, have been used as vaccine carriers.[10] While these multiple emulsions may have advantage in terms of controlled release, they are difficult to stabilise and therefore have found limited use in medical and pharmaceutical practice. One further class of emulsions is known as microemulsions. These are thermodynamically stable solubilized systems and are rarely employed in pharmaceutical practice.[14] As far as the author is aware, they have not been used as injectable systems. Unfortunately, within the pharmaceutical world, there is presently confusion over the term "microemulsion". Emulsions can be produced that have a small particle size, for example less than 500 nm average diameter. If needed, such systems can be referred to as "fine" emulsions but they should not be termed "microemulsions". This confusion seems to have arisen from the promotional activities of certain manufacturers of homogenisation equipment.

The present article will focus on the use of parenteral oil-in-water emulsions, especially those used for parenteral nutrition and drug delivery. Understandably, attention will be given to those systems stabilized by phospholipid emulsifiers. In reality there are a few emulsifying agents that can be used in parenteral products and phospholipids (either of egg or soya origin) are normally employed.

Phospholipids
Edited by I. Hanin and G. Pepeu
Plenum Press, New York, 1990

Attempts to administer nutrients via the intravenous route have covered a variety of different options.[19] Concentrated amino acid and glucose systems together with added electrolytes can be employed. However, the high tonicity of such systems necessitates their being given by a central vein. It was appreciated many years ago that an emulsion system would provide a large amount of energy per gramme of administered material. Moreover, the reduced tonicity of such a system would allow it to be given by a peripheral line. An early product was based upon purified cotton-seed oil emulsified by a block copolymer (poloxamer F68). This product was in use for a number of years, but unfortunately some patients suffered from adverse reactions. These reactions were associated with components in the oil and possibly also in the emulsifying agent. As a consequence the product was withdrawn.

An alternative type of product was developed in Sweden by Wretlind and others based upon purified soyabean oil, with purified egg lecithin as the emulsifier. This product, known as Intralipid, has now become the standard emulsion product in parenteral nutrition. Today, egg lecithin is normally preferred as the emulsifier for various parenteral fat emulsions, largely because of putative adverse effects found with soy-phosphatides. These effects included changes in blood pressure and have been associated with phosphatidylinositol.[34] However, one would expect that a suitably purified soya product would be satisfactory.

The composition of the phospholipid emulsifier can have a very important bearing upon the stability of a resultant fat emulsion. For example it is well known that the use of highly purified phosphatidylcholine/phosphatidylethanolamine mixtures (PC/PE) makes it difficult to produce stable fat emulsions.[29] The presence of minor components, that can ionize at the oil/water interface, in the form of phosphatidylserine, phosphatidic acid, together with the lysophosphatidyl derivatives of PC and PE play a critical role.

Fat emulsion stabilized by suitable phospholipids achieve their stability by two mechanisms; the phospholipid layer provides a mechanical barrier to coalescence, but more importantly, it also provides a charged repulsion barrier.[6] At a pH of 7.4, a suitable egg lecithin for the preparation of fat emulsions (i.e. one that contains ionizable minor components) will provide a charge of approximately -50 mV on the droplets of a fat emulsion prepared using a vegetable oil. This stabilizing charge is more or less independent of pH between 4 and 8. In contrast, if a fat emulsion was to be prepared using pure PC/PE it would carry almost no stabilizing charge in this pH range. The magnitude of the charge on a fat emulsion can have an important influence on its use in mixed systems intended for total parenteral nutrition (TPN). In many countries it is a common practice to administer an intravenous fat emulsion, in a combined system of perhaps 3 litres total volume containing amino acids, electrolyte and carbohydrate.[6] A typical formulation is given in Table 1. This admixture of the fat emulsion with other components can lead to instability through changes in pH and ionic strength. Provided the pH is kept above 5, the charge on the droplets will be maintained above a critical value for stability. In our work this critical value has been found to lie between 15 and 20 mV. If the surface charge drops below this value the emulsion can flocculate (aggregate) and in this condition coalescence of the oil droplets can be greatly enhanced.

Table 1.   A mixture intended for total parenteral nutrition

Component	Volume
Fat emulsion (e.g. 10% Intralipid)	500 ml
Glucose 20%	500 ml
Amino acid solution	500 ml
(e.g. Synthamin, Vamin, Freamine, Aminoplex)	
Water soluble vitamins	q.s.
Fat soluble vitamins	q.s.

Added electrolytes can have a very important effect upon emulsion stability. The extent of the effect is dependent upon the concentration and valency of the added counter ion. The reducing effect on surface charge is related to the valency of the counter ion raised to the power six. Furthermore, divalent ions, particularly calcium, can undergo specific interactions with phospholipid emulsifiers so that at high ionic strength instead of the charge being totally suppressed and close to zero, a charge reversal phenomenon can occur.[6] Provided that this charge reversal leads to a high enough positive charge (for example greater than +15 mV) such emulsion systems recover their stability characteristics.[32] A diagram showing the surface charge and related flocculation behaviour of an intravenous fat emulsion, in the presence of calcium is shown in Figure 1. The various stages of no flocculation, flocculation and subsequent deflocculation are demonstrated.

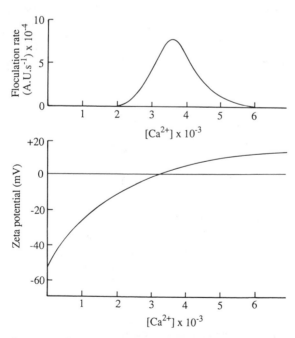

Fig. 1.   Flocculation process for Intralipid – Calcium salt System. Flocculation rate (upper graph) and zeta potential (surface charge) (lower graph) (after Washington and Davis, ref. 32).

Through an appropriate knowledge of colloid science, it is now possible to produce mixed TPN systems that have shelf-life extending to many days. As expected, those systems with high electrolyte content tend to demonstrate poor stability. It has also been found that the presence of amino acids can have a stabilizing and/or de-stabilizing effect in TPN systems depending upon the amino acid composition present and the time period over which the mixed emulsion systems are stored.[1,6] Certain amino acids seem to be able to interact with the interfacial film; the acidic and basic amino acids are important in this respect.

A suitable emulsifier system for a typical fat emulsion will consist of major and minor components as shown in Table 2. Various commercial manufacturers are now able to provide emulsifiers that are acceptable to regulatory authorities. The fat emulsions can be produced by homogenization equipment such as that provided by the company Manton-Gaulin (APV). Small scale equipment is also available. We have also found the Microfluidizer apparatus available from Microfluidics Corporation in the United States to be a particularly useful method for making small batches of emulsion for test purposes.[33]

Preparation of small batches using ultrasonics is another possibility, however, with this method the particle size distribution produced is often different to that obtained using conventional homogenization. As a consequence, scale-up may not be a simple matter. Whatever the technique used for emulsion preparation the final particle size of a fat emulsion will be dictated largely by the physical chemistry of the system, (i.e. the nature and concentration of the emulsifier and the chemical nature of the oil used). Thus, it is not altogether suprising that commercially available fat emulsions prepared from egg lecithin and soybean oil, all have very similar particle sizes (about 300 nm average diameter for 10% oil and about 400 nm for those with 20% oil). For some commercial products the stability of the system is enhanced by the deliberate addition of a small amount of oleic acid. An improvement in the stability of emulsions intended for parenteral nutrition can also occur during production, autoclaving and storage, by the generation of free fatty acids and lysophospholipids.

Table 2. Phosphatide emulsifiers: Major and minor components

Major

Phosphatidylcholine
Phosphatidylethanolamine

Minor

Phosphatidic acid
Phosphatidylserine
Phosphatidylinositol
Sphingomyelin
Lysophosphatides

The major difference between the available commercial fat emulsion products is now largely one of price, although it is known that subtle changes in phospholipid composition can affect the subsequent fate of the emulsion once it is injected into the body.[20] Fat emulsions are intended to be copies of the natural fat particles, the chylomicrons. It is therefore essential that upon administration into the bloodstream, the fat emulsion takes onto its surface the required apoproteins CII and CIII. These then have a role with the enzyme lipoprotein lipase in the metabolism of the fat emulsion droplets at peripheral sites in the capillary network.[3] If the uptake of apoproteins is not encouraged by the nature of the phospholipid surface (for example due to differences in chemical nature or phase transition properties) then the emulsion may have undesirable clearance characteristics.[35] In some cases the emulsion may be recognised as being foreign and then will be sequestered by the macrophages in the vascular system, especially the Kupffer cells in the liver.[9] Not withstanding the success of emulsions based upon soybean oil and egg lecithin, newer types of fat emulsion are under active investigation. These include the medium chain triglyceride systems, used either on their own or in conjunction with long chain triglycerides.[27]

PERFLUOCHEMICAL EMULSIONS USED AS BLOOD SUBSTANCES

A similar type of product to that available for parenteral nutrition, is the perfluorochemical emulsion intended as a blood substitute.[15] Many of the techniques used in the development of intravenous fat emulsions have been used with some degree of success for this interesting class of material. It is well known that perfluochemicals have favourable gas (oxygen and carbon dioxide) exchange properties and consequently, many attempts have been made to produce emulsions of these materials to act as blood substitutes. The perfluochemical called perfluodecalin has been a popular choice, and mixtures of egg lecithin and block copolymer surfactants have been employed as emulsifying agents.[23] These emulsions are still being assessed in the clinic and besides being used in blood replacement they also are being examined for the treatment of strokes, in organ preservation (transplants) and as potential drug carriers. For toxicological reasons it would be an advantage to have such systems stabilised solely by phospholipid emulsifiers. Unfortunately as yet, such products tend to be unstable unless large quantities of phospholipids are employed. Furthermore, an alternative pathway for emulsion destabilization has been identified (e.g. Ostwald Ripening) and methods have been proposed for its elimination.[2]

FAT EMULSIONS FOR DRUG DELIVERY AND DRUG TARGETING

The type of emulsion used in parenteral nutrition can also find utility as a drug carrier system. Indeed, provided that the drug in question has reasonable lipid solubility, then an emulsion system can be an extremely useful delivery option.[7] Many workers have now shown that the emulsion is far preferable to system based upon drug solubilization, for example using mixtures of non-ionic surfactants or bile salt derivatives.[11] In using fat emulsions for drug delivery, one is essentially doing what the body does, since it is known that lipid soluble drugs are associated with the apoproteins in the bloodstream. The advantages and disadvantages of using emulsion systems as drug carriers are given in Table 3. In Europe, two successful commercial products are based on the Intralipid system; these being emulsion systems for diazepam[5] and the anaesthetic agent propofol (Diprivan).[4] This latter product has been especially successful in the market. Originally attempts were made to solubilize the active principle (diisopropylphenol) using surfactant systems (Cremophor). While good solubilization was obtained, the product was associated with adverse reactions, including the release of histamine which may be a contributing factor to the well known anaphylactic reaction associated with Cremophor surfactants[16] (Figure 2).

Various groups have made extensive use of fat emulsion systems as drug carriers. The pioneering work of Jeppssen and others,[21] provided the guidelines for the introduction of the two commercial products mentioned above. Jeppssen and colleagues have also described many more examples of emulsion systems that can be used for the successful intravenous administration of pharmacological agents.[21] In the field of cancer chemotherapy, emulsions have been used on many occasions as a means of administering highly water insoluble compounds.[28,31] Drugs given in this way included valinomycin, hexamethylmelamine and lomustine. Not only does an emulsion system greatly facilitate the administration of water insoluble drugs, it can also have a dramatic effect upon their stability. Repta[28] has shown that the use of an emulsion system can enhance the stability of a drug undergoing hydrolysis. The enhanced stability effect can be calculated from a knowledge of the oil/water partition coefficient of the compound and the phase volume (quantity of dispersed oil phase) of the emulsion system. For instance, if the drug had an oil/water partition coefficient of 100 and if the emulsion had a phase volume of 22% then the half-life of the drug would be extended by a factor of 23 fold over that for a simple aqueous system.

Table 3. Emulsions as drug delivery systems for parenteral administration

Advantages

1. Easy to manufacture - using proven technology.
2. Fairly stable (e.g. 1-2 years at 4°C).
3. Oil phase acts as solubilizer of lipophilic compounds.
4. Oil phase acts to stabilize drug against hydrolysis.
5. Well tolerated in body - resemble physiological carriers, chylomicrons.
6. Lower incidence of side effects observed (e.g. precipitation of drug, pain on injection, thrombophlebitis).
7. Altered pharmacokinetics and distribution - possibility for site specific delivery and drug targeting.

Disadvantages

1. Water soluble component cannot be entrapped.
2. Drug may decrease emulsion stability.
3. Drug may alter surface character of emulsion.

Mitzushima and colleagues in Japan, in collaboration with the Green Cross Corporation, have recently published a number of papers describing how emulsion systems can be used to deliver lipid soluble drugs or drug derivatives known as prodrugs.[24,25] Their papers have described how oil-in-water emulsions based upon a triglyceride/phosphatide emulsified combination have been used successfully with prostaglandin El (PGE1), steroid ester, indomethacin prodrugs, and more complex molecules such as coenzyme Q10 (Ubiquinone). The first two systems are now the basis of commercial products in Japan. For the case of emulsions containing

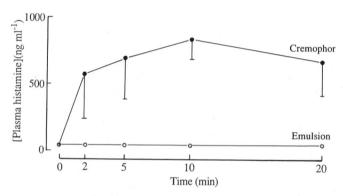

Fig. 2. Toxicological studies on lipid emulsions containing drugs. Measurement of plasma histamine in dogs after the administration of emulsion and Cremophor formulations of propofol (7.5 mg Kg^{-1}). Mean values ± S.D. n = 4). (After Glen and Hunter, ref. 16).

steroids (dexamethasone palmitate) selective delivery of the emulsion to an induced site of inflammation was shown. However, in their publication the authors claimed this to be an example of drug targeting. Such a claim is perhaps over ambitious since less than 0.2% of the total dose arrived at the designated target site! One can question "to where in the body was the remainder of the dose delivered?" Inadvertent delivery to the liver could result in adverse reactions. Such considerations are particularly important when one is to use emulsions for the delivery of anticancer agents. It is laudable to achieve a higher amount of drug at to say a tumour site but if the remainder of the dose has been delivered to the liver or the spleen, then this could result in an enhancement of toxicity and a reduction of the therapeutic index.

Attempts have been made to show that the use of an emulsion can
change the half-life of a drug in the circulation; for example PGEl.
However, the data were obtained with labelled compound and it is well
known that PGEl is rapidly metabolised. Thus, it is necessary not only
to follow the presence of the free drug but also the metabolites.
Certainly with the PGEl system, significant differences in thrombus
occurrence were demonstrated between treatment with emulsion and solution
systems.[30]

The question of drug release from the emulsion system is an
important one. One would expect from physico-chemical considerations
that upon administration of a drug emulsion into the systemic
circulation, the drug will be able to diffuse rapidly from the oil phase
into the blood. Only when the drug has a very high partition coefficient
in favour of the oil phase, or when the oil phase is surrounded by some
form of thick diffusional barrier or when the oil phase itself is solid-
like, would one expect the drug to be released slowly. The studies
conducted on diazepam and propofol emulsion formulations have
demonstrated that there is no significant difference in the
pharmacokinetics of the administered drug whether it is given as an
emulsion formulation or in a solubilizing vehicle.[18] Early suggestions
of sustained release affects with low molecular weight drugs dissolved in
oil-in-water emulsions can be better explained by the effect of the
emulsion on the metabolism of the drug (barbituate).

The release of some drugs from oil-in-water emulsion systems may be
slower than expected because the drug is not within the oil phase but
trapped in the interfacial layer.   The drug amphotericin-B is a good
example of an amphiphilic species that is incorporated into emulsion
droplets in this way.[13] Amphotericin-B is a very useful drug in the
treatment of systemic candidiasis but is normally administered in a
solubilized (bile salt) formulation. Unfortunately, this formulation
gives rise to toxicity, particularly to the kidneys. Various
formulations of amphotericin-B in phospholipid liposomes have been
described.[22] These systems are able to reduce the toxicity of
amphotericin by the intercalation of the drug into the phospholipid
bilayer. The drug, while still retaining its antifungal activity, no
longer has a strong affinity for mammalian cells. We have prepared an
emulsion formulation similar in properties to the liposomal product.
Here the drug is trapped in the emulsion (phospholipid) surface. Tests
have shown that our emulsion formulation of amphotericin-B appears to
have all the benefits of a liposomal formulation (Figure 3) but should
have the advantage of being able to be prepared using conventional
equipment and to be sterilizable by terminal heating.

IN VITRO EVALUATION OF EMULSIONS

The important colloidal characteristics of an emulsion system
include its particle size, surface charge, and more subtle factors such
as surface hydrophobicity. In our work we have employed photon
correlation spectroscopy, (PCS), laser diffraction, Mie scattering
Coulter counter and microscope techniques for particle size analysis.[12]
Typically, in a commercial fat emulsion intended for parenteral nutrition
or drug delivery, particles can range in size from 10 nm to 5 um.
Consequently it is important to decide upon the best size analysis method
by taking into account the intended use of the product. For example,
while photon correlation spectroscopy is an excellent means for studying
the smaller particles in an emulsion system, it will by necessity provide
a biased value. In clinical use, it is likely that the presence of small
numbers of large particles will have a more dramatic effect. Therefore,

any particle size analysis method should be chosen to take into account not only the small but also the large particles. The presence of free oil on the surface of emulsions can be detected using staining techniques with appropriate dyes (Sudan III, Sudan IV).

Surface charge is normally measured by a microelectrophoresis technique. We have employed successfully the Malvern Laser Doppler Anemometer.[26] A knowledge of surface charge is extremely helpful, not only for following changes in emulsion systems due to emulsifier properties, pH or ionic strength (for example in TPN mixtures) but also when formulating drug containing emulsions. It should be remembered in these systems, the drug can have a dramatic effect on surface charge, particularly if the drug is able to ionize. Likewise, the pH of the environment can be an important critical factor.

Surface hydrophobicity is a more difficult parameter to characterize but could have important implications for the fate of the emulsion in a biological milieu. We have employed hydrophobic interaction chromatography (HIC) and two-phase partitioning.[26] Both methods have been useful in characterising the properties of emulsions.

Emulsions are inherently unstable systems thermodynamically (except for microemulsions) and therefore one would expect them to undergo change in their particle size with time. The effects of electrolytes on emulsion flocculation have been discussed above. This is a reversible phenomenom as compared to coalescence where the particles increase in size through interparticular collisions.

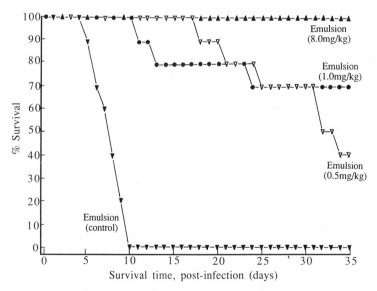

Fig. 3. Emulsion formulations of amphotericin-B. Survival time of mice injected with <u>Candida albicans</u> (after Davis et al. ref. 11).

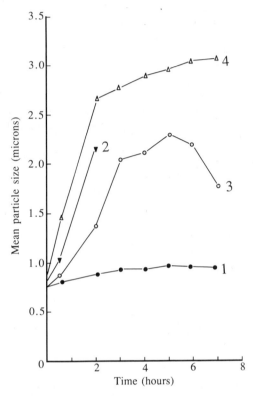

Fig. 4.   Accelerated testing of fat emulsion mixtures.
          Particle growth under shaking test, 40°C,
          100 cycles/min.

          Legend.

          1. 10 ml Intralipid + 5 ml Sodium Chloride
              (75 mMol/ℓ).
          2. 10 ml Intralipid + 5 ml Calcium Chloride
              (1.5 mMol/ℓ). (Emulsion cracked at 3 hours).
          3. 10 ml Travamulsion + 5 ml Calcium Chloride
              (75 mMol/ℓ).
          4. 10 ml Travamulsion + 5 ml Calcium Chloride
              (1.5 mMol/ℓ).

          Particle size analysis by laser diffractometry.
          (Davis, S.S. and Jones, H.D., unpublished results).

It is useful in any stability testing programme to have ways of accelerating the breakdown of a product. For emulsions a variety of different techniques has been investigated including centrifugation, freeze-thaw cycling, autoclaving and shaking tests. For simple emulsion systems, the last three methods can be useful, although in the case of freeze-thaw cycles and autoclaving, the exact relevance of the tests to normal storage conditions can be disputed. The shaking test is a useful way of assessing differences between emulsion systems and also to evaluate long term stability. By placing an emulsion sample in a shaking water bath, one is able to study differences in emulsion stability over a relatively short period of time (for example 6-12 hours) (Figure 4). Even a normally stable emulsion such as Intralipid, which will maintain its particle size over a period of two years or more, can be caused to undergo coalescence and separation free oil after shaking for ten hours or less under the conditions described above.

Drug release from carrier emulsions is clearly a desirable property to follow, however, practically this is extremely difficult unless the drug is released slowly. One has to resort to methods such as rapid centrifugation, filtration, dialysis techniques, etc. Unfortunately, in many cases the release rate is much faster than the processing stage! As a consequence, it is desirable to have analytical methods available where concentration measurements can be taken rapidly within the diluted emulsion system. Polarographic methods as well as UV spectroscopy (with appropriate correction for scatter due to emulsion droplets) can sometimes be employed for this purpose.

IN VIVO EVALUATION OF EMULSIONS

The performance of emulsion systems in vivo can be evaluated by standard tests examining the blood level versus time profiles for the emulsion system itself or for drug containing emulsions, a comparison of blood drug content with for example, a solubilized system. The more general fate and clearance of an emulsion system can be evaluated in animal experiments. In one of our recent studies, the drug and the emulsion vehicle were labelled with a radioisotope and the distribution of both the drug and carrier within the bloodstream and at organ sites followed by gamma scintigraphy and organ level determinations.[11] This method has been used for a variety of emulsion systems, to include those intended for parenteral nutrition and for drug delivery. We have shown that it is possible to modify the uptake of an emulsion in the liver by the judicious choice of the emulsifying agent.[11] Emulsions stabilized by phospholipids demonstrate sequestration by the liver, whereas emulsification using the block copolymer Tetronic 908 (that provides the emulsion with a hydrophilic steric barrier thereby preventing interaction with macrophages) effectively keeps the emulsion system within the vascular compartment.[17] Enhanced uptake in the liver can be achieved using a gelatin co-emulsifier as a recognition factor (Figure 5). Such differential uptake could allow emulsions to be used for the more selective delivery of therapeutic agents.

Further knowledge of the interaction of emulsions with the biological milieu can be obtained using various cultured cells. We have shown previously that cultured macrophages (and indeed amoeba) can be used to study the uptake of emulsions and to develop correlations between surface properties and uptake rate to include factors such as surface charge.[8] More recently we have employed isolated liver macrophages (Kupffer cells) as a way of evaluating intravenous fat emulsions. As expected, it was apparent in these experiments that the nature of the phospholipid emulsifier had an important effect on recognition and subsequent phagocytic uptake.

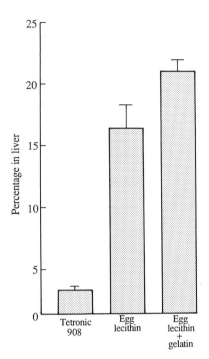

Fig. 5. Liver uptake of intravenous fat emulsions prepared
with different emulsifying agents (mean ± S.E.,
n = 3). (Date corrected for liver blood pool
concentration) (after Davis et al. ref. 11).

REFERENCES

1. Burnham, W.R., Hansrani, P.K., Knott, C.E., Cook, J.A. and
    Davis, S.S., 1983, Stability of a fat emulsion based intravenous
    feeding mixture, Int. J. Pharm. 13:9.
2. Buscall, R., Davis, S.S. and Potts, D.C., 1979, The effect of
    long chain alkanes on the stability of oil-in-water emulsions.
    The significance of Ostwald ripening, Colloid Polym. Sci.
    257:636.
3. Carlson, L.A., 1980, Studies on the fat emulsion intralipid.
    I. Association of Serum Proteins to Intralipid Triglyceride
    Particles (ITP), Scand. J. Clin. Lab. Invest. 40:139.
4. Cummings, G.C. and Spence, A.A., 1985, Comparison of propofol in
    emulsion with althesin for induction of anaesthesia, Br. J.
    Anaeth. 57:234.
5. Dardel, O. von., Mebius, C., Mossberg, T. and Svensson, B., 1983,
    Fat emulsion as a vehicle for diazepam. A study in 9492
    patients, Brit. J. Anaesth. 55:41.
6. Davis, S.S. and Hansrani, P.K., 1985, The influence of emulsifying
    agents on the phagocytosis of lipid emulsions, Int. J. Pharm.
    23:69.

7. Davis, S.S. and Walker, I.M., 1987, Multiple Emulsions as
    Targetable Delivery Systems, In "Drug and Enzyme Targeting", R.
    Green and K.J. Widder, eds. Methods in Enzymology, Vol. 149, p.
    51, Academic Press, New York.
8. Davis, S.S., 1983, The stability of fat emulsions for
    intravenous administration, In "Advances in Clinical Nutrition",
    I.D.A. Johnson, ed. p.213, M.T.P. Press, Lancaster.
9. Davis, S.S., Hadgraft, J. and Palin, K., 1985, Pharmaceutical
    Emulsions, In "Encyclopedia of Emulsion Technology", Vol. 2, P.
    Becher, ed., p.159, Dekker, New York.
10. Davis, S.S., Illum, L., West, P. and Galloway, M., 1987, Studies
    on the fate of fat emulsions following intravenous administration
    to rabbits and the effect of added electrolyte, Clin. Nutr. 6:13.
11. Davis, S.S., Washington, C., Illum, L., Liversidge, G., Sternson, L.
    and Kirsh, R., 1987, Ann. N.Y. Acad. Sci. 507:75.
12. Douglas, S.J. and Davis, S.S., 1986, Particle size analysis of
    colloidal systems by photon correlation spectroscopy, In
    "Targeting of drugs with synthetic systems", G. Gregoriadis, J.
    Senior and G. Poste, eds. p.265, Plenum Press, New York.
13. Forster, D., Washington, C. and Davis, S.S., 1988, Toxicity of
    solubilized and colloidal amphotericin B formulations to human
    erythrocytes, J. Pharm. Pharmac. 40:325.
14. Friberg, S.E. and Bothorel, P. (editors), 1987,
    Microemulsions: Structure and Dynamics, CRC Press, Boca Raton.
15. Fujita, T., Sumaya, T., and Yokohama, K., 1971, Fluorocarbon
    emulsion as a candidate for artificial blood. Correlation
    between particle size of the emulsion and acute toxicity. Europ.
    Surg. Res. 3:436.
16. Glen, J.B. and Hunter, S.C., 1984, Pharmacology of an emulsion
    formulation of ICI 35868, Brit. J. Anaesth. 56:617.
17. Illum, L., Davis, S.S., Muller, R.H., Mak, E. and West, P., 1987,
    The organ distribution and circulation time of intravenously
    injected colloidal carriers sterically stabilized with a block
    copolymer poloxamine 908, Life Sci. 40:367.
18. Jeppsson, R.I., 1976, Plasma levels of diazepam in the dog
    and rabbit after two different injection formulations, emulsion
    and solution, J. Clin. Pharm. 1:181.
19. Johnson, I.D.A. (editor), 1983, "Advances in Clinical
    Nutrition", MTP Press, Lancaster.
20. Lenzo, N.P., Martins, I., Mortimer, B.C. and Redgrave, T.G., 1988,
    Effects of phospholipid composition on the metabolism of
    triacylglycerol, cholesterol ester and phosphatidylcholine from
    lipid emulsions injected intravenously in rats, Biochim. Biophys.
    Acta 960:111.
21. Ljungberg, S. and Jeppsson, R., 1970, Intravenous
    administration of lipid soluble drugs, Acta Pharm. Suec. 7:435.
22. Lopez-Berestein, G., Mehta, R., Hopfer, R.L., Mills, K.,
    Kasi, L., Mehta, K., Fainstein, U., Luna, M., Hersh, E.M. and
    Juliano, R.L., 1983, Treatment and prophylaxis of disseminated
    candida albicans injections in mice with liposome encapsulated
    amphotericin B, J. Infec. Dis. 147:939.
23. Mitsuno, T., Ohyanagi, H. and Naito, R., 1982, Clinical studies of a
    perfluorochemical whole blood substitute (Fluosol DA) Summary of
    186 cases. Ann. Surg. 195:60.
24. Mizushima, Y., 1985, Lipid microspheres as novel drug
    carriers, Drugs Expl. Clin. Res. 9:595.
25. Mizushima, Y., Shoji, Y., Fukushima, M. and Kurozumi, S., 1986, Use
    of lipid microspheres as a drug carrier for anti-tumour drugs,
    J. Pharm. Pharmacol. 38:132.

26. Muller, R.H., Davis, S.S., Illum, L. and Mak, E., 1986, Particle charge and surface hydrophobicity of colloidal drug carriers, In "Targeting of Drugs with Synthetic Systems", G. Gregoriadis, J. Senior and G. Poste, eds., p.239, Plenum Press, New York.

27. Muller, R.H., Nehne, J., Davis, S.S. and Heinemann, S., 1988, Characterisation of fat emulsions for parenteral nutrition, Acta Pharm. Technol., 34:235.

28. Repta, A.J., 1981, Formulation of investigational anticancer drugs, In "Topics in Pharmaceutical Sciences", D.D. Breimer and P. Speiser, eds. p.131, Elsevier, Amsterdam.

29. Rydhag, L., 1979, The importance of the phase behaviour of phospholipids for emulsion stability, Fette Seif. Anstrich. 81:168.

30. Sim, A.K., McCraw, A.P., Cleland, M.E., Aihara, H., Otomo, S. and Hosoda, K., 1986, The effect of prostaglandin El incorporated in lipid microspheres on thrombus formation and disaggregation and its potential to target to the site of vascular lesions, Arzn. Forsch. 36:1206.

31. Tarr, B.D., Sambandan, T.G. and Yalkowsky, S.H., 1987, A new parenteral emulsion for the administration of Taxol, Pharm. Res., 4:162.

32. Washington, C. and Davis, S.S., 1988, Stability evaluation of total parenteral nutrition mixtures, In "Clinical Progress in Nutrition Research", Sitges-Serra, Sitges-Creus, Schwartz-Riera, eds., p.189, Karger, Basel.

33. Washington, C. and Davis, S.S., 1988, The production of parenteral feeding emulsions by microfluidizer, Int. J. Pharm. 44:169.

34. Wretlind, A., 1976, Current Status of Intralipid and other Fat Emulsions, In "Fat Emulsions in Parenteral Nutrition", H.C. Meng and D.W. Wilmore, eds., p.109, American Medical Association, Chicago.

35. Ziak, E., Pristautz, H., Brandt, D., Schaupp, K. and Musil, E., 1984, Vergleichsstudie zwischen drei fetteemulsionen zur parenteralen Ernahrung, Ernahrung 8:617.

PHOSPHATIDYL CHOLINE AND LYSOPHOSPHATIDYL CHOLINE IN MIXED LIPID

MICELLES AS NOVEL DRUG DELIVERY SYSTEMS.

David W. Yesair

BioMolecular Products, Inc.
P.O.Box 347
Byfield, Massachusetts 01922 USA

I. INTRODUCTION

Since the early nineteen seventies, considerable research effort has been focused on the merits of using delivery systems for improving the efficiency and safety of drugs and thus realizing a higher percentage of the commercial market. Some of the projected benefits of drug delivery systems are summarized in Table 1. Of primary importance, drug delivery systems must enhance the quality of practiced medicine. For example, drug delivery systems can increase the therapeutic benefits by minimizing side effects, by increasing the efficacies, by decreasing the amount and frequency of dosing, by enhancing absorption and by affecting parameters such as the palatability and stability of the drug. Further, the drug delivery system can affect the biodegradation of drugs within the milieu of the G.I. tract and during passage through the intestine and liver. Lastly, drug delivery systems can make drugs categorically more useful.

Table 1
PROJECTED BENEFITS OF DRUG DELIVERY SYSTEMS

MEDICINE:
    o Increase therapeutic benefits
    o Protect against biodegradation
    o Make drugs more useful

RESEARCH & DEVELOPMENT:
    o Revitalize one's chemical library
    o Develop new lipophilic drug analogs
    o Deliver genetic-engineered drugs

MARKET:
    o Patent protection
    o Expansion of market shares
    o Unique and novel sales promotion

FINANCES:
    o Greater return on R&D investment
    o Greater and more profitable sales
    o License income

Phospholipids
Edited by I. Hanin and G. Pepeu
Plenum Press, New York, 1990

Most drug companies have historically emphasized the development of new chemical entities and chemical modifications of existing drugs. As a consequence the chemical libraries of most companies number within the hundreds of thousands. Thus, drug delivery systems have the theoretical potential of revitalizing this chemical treasury per se and some unique delivery systems would also be most useful with new lipophilic analogs of drugs. The delivery of genetic-engineered drugs, especially polypeptides, represents the greatest challenge for developing new drug delivery systems.

The successful union of a drug and drug delivery system has many advantages within the competitive market and for the economics of this market. Briefly, a unique delivery system provides, in many instances, patent protection, expansion of market share and development of unique and novel sales promotion. The bottom line, whether it is academic, government or industry, is a greater return on investment; not only in terms of money but in time, in research and development and in licensing the product.

Having had the opportunity in recent years to search the patent literature, the author has categorized drug delivery systems into three general classifications (Table 2). Briefly, delivery systems can control the

Table 2
DRUG DELIVERY SYSTEMS

CONTROLLED RELEASE:
- o First generation
- o Greatest number of patents
- o Oral, topical, injectable
- o Limited future benefits

MODIFY SYSTEMIC DISTRIBUTION:
- o Second generation
- o Modest number of patents
- o Bright future

TARGET RECOGNITION:
- o Third generation
- o Greatest potential
- o Decade before commercial success

release of drugs, can affect a change in the systemic distribution of drugs in situ, and can have the potential of directing the drug to some pre-destined site. Controlled release of the drug has been the most actively pursued system and is used primarily for oral, topical and injectable drug delivery systems. These delivery systems are primarily composed of co-valent matrix materials; e.g. carboxymethyl cellulose, polyvinyl alcohol, hydrogel, gelatin, polylactic, polyglycolic and polyamino acids, and organo-polysiloxane rubber. When greater success is seen for second and third generation systems (those that modify the systemic distribution of drugs and those that recognize specific targets) controlled release may well become obsolete. The remainder of this paper will focus on several systems that modify the systemic distribution of drugs. The success with these systems will then provide an impetus for attaching cell and organ specific recognition factors to the delivery vehicle for the third generation of drug delivery systems.

The physical chemical characteristics of phospholipids that are ubiquitous in living systems promote unique, ordered molecular organizations with other lipid components and the resultant ordered structures have been developed as novel drug delivery systems (Table 3).

Table 3
NATURE'S LIPOPHILIC DRUG DELIVERY SYSTEMS
THAT MODIFY THE SYSTEMIC DISTRIBUTION

LIPOSOMES:
- o Phospholipids and cholesterol
- o IV., dermal and oral
- o Use and production patents
      Liposome Technology, Inc.
      The Liposome Company

CIRCULATING MICRORESERVOIRS (CMRs):
- o Phosphatidyl choline and cholesterol esters
- o IV. and dermal
- o Composition and use patent, 1981
      Arthur D. Little, Inc.
- o Production patent under development
      BioMolecular Products Inc.

LYM-X-SORB (L-X-S):
- o Lysophosphatidyl choline, monoglycerides and fatty acids
- o Oral and dermal
- o Composition and use patent issued and pending in 21
      countries

Phosphatidyl choline (PC) is one of only a few phospholipids that forms a monolayer at an air/water interphase. For example, dipalmitoyl phosphatidyl choline (DPPC) is the principal surfactant lipid of the lung[5] and is functionally required for the transfer of oxygen from inhaled air into the pulmonary blood stream[10,36]. The inclusion of cholesterol and other phospholipids into the DPPC monolayer can affect the surface properties of phospholipids in relationship to those of lung extracts[32,47].

Phosphatidyl choline (PC) also forms bilayers in aqueous systems, with the polar head groups of PC oriented towards water. Using heat and some shearing forces, the PC bilayers can be physically reduced in size. The resultant particles have been termed liposomes. As noted for the lung surfactant, DPPC, the addition of other phospholipids and cholesterol into the bilayers affect modification in the transition temperature, stability, size etc. of the resultant liposomes. Liposomes were first described by Bangham et al[3] in 1965. Since then, Bangham[2], Chapman[8], and Huang and Thompson[21] have further described the physical characteristics of multilamellar and unilamellar liposomes, and Patel and Ryman[35], Gregoriadis[19] and Papahadjopoulos[34] have principally focused on using liposomes in the delivery of drugs by several different routes; namely, intravenously and dermally. The oral route has also been evaluated, but with little or no success. Today, in the United States there are two companies which, as their names imply (see Table 3), are attempting to commercialize the liposome technology.

Nature has also provided another unique lipid system that exploits the physical chemical characteristics of phosphatidyl choline (PC), namely high density lipoproteins (HDL). The lipids in HDL are principally PC and cholesterol esters (CE)[45]. These esters (CE) have very ordered crystalline structures, which transmit polarized light upon melting, are extremely water insoluble, and are incompatible with PC. Thus the CE domains in HDL are surrounded by a PC monolayer which has the acyl group of PC oriented towards CE and its polar head group towards the aqueous phase. The physical characteristics of PC and CE have been exploited in the development of a new drug delivery system termed circulating micro-reservoirs (CMRs). The CMRs are uniquely suited for intravenous and dermal applications.

The physical characteristics of lipids, in general, make them incompatible with aqueous systems. However, the physiological processing of orally consumed triglycerides and lecithin (PC) results in an aqueous compatible lipid organization. Briefly, triglycerides are processed by pancreatic lipase to yield a bile salt micelle of monoglycerides (MG) and fatty acids (FA). Lecithin (PC) is enzymatically processed by a phospholipase to yield lysophosphatidyl choline (LPC). Lysophosphatidyl choline (LPC), unlike PC, is water soluble and produces an ordered molecular organization with defined molar ratios of monoglycerides (MG) and fatty acids (FA). This technology has been exploited as a drug delivery system for the oral absorption of drugs via the lymphatic route and has been termed LYM-X-SORB, i.e. the lymphatic absorption of xenobiotics.

The characterization of these lipophilic drug delivery systems will now be described in more detail.

II. LIPOPHILIC DRUG DELIVERY SYSTEMS

A.   LIPOSOMES AND CIRCULATING MICRORESERVOIRS (CMRs)

The preparation of liposomes and CMRs has many common features (Figure 1). The components of liposomes, namely phospholipids and cholesterol, and of CMRs, lecithin (PC) and cholesterol esters (CE), are dissolved in organic solvents. Drugs, if soluble in the solvent, can be added at this step. After removing the solvents in vacuo, an aqueous buffer and drug is added, heated and sonicated to yield the respective liposomes and CMRs.

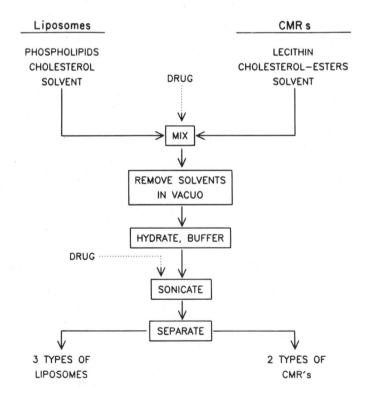

Fig. 1.   Preparation of liposomes and circulating microreservoirs (CMRs).

Three types of liposomes are produced by vigorous mixing or sonication and two types of CMRs can be made by controlling the mole ratios of PC and CE. These are visualized in Figure 2.

ⴼ PHOSPHATIDYL CHOLINE
█ CHOLESTEROL
ⴽ CHOLESTEROL ESTER

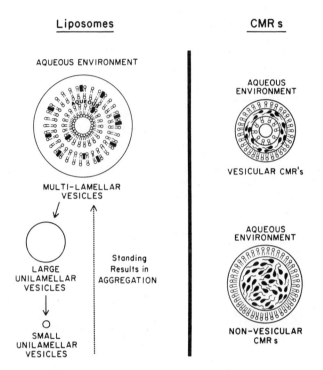

Fig. 2.   Types of liposomes and circulating microreservoirs in an aqueous environment.

Multilamellar vesicles (MLV) have been described, principally by Bangham et al.[3], following vigorous mixing, they range in size of 0.2 to 10 microns (2000-100,000 angstroms). The entrapped water between the PC bilayers is approximately 4 microliters per milligram of liposome. The use of sonification, rather than vigorous mixing, yields large unilamellar vesicles (LUV), ranging in size of 0.2 - 1 micron (2000-10,000 angstroms) and having a captured aqueous volume of approximately 9 ul/mg. Using heat and more extensive sonication, small unilamellar vesicles (SUV) are produced with a size range of 0.02-0.05 micron (200-500 angstroms) but with a very small captured aqueous volume (0.5 ul/mg)[21]. Since kinetic energy is required to produce a reduction in size, then the absence of this energy permits the PC bilayers to merge upon contact and thus there is an aggregation of the SUV and LUV to make the more stable multilamellar vesicles. The most useful liposomes have been the LUV because they are fairly stable and these can capture meaningful quantities of hydrophilic, as well as lipophilic, drugs.

Circulating microreservoirs (CMRs) are also shown in Figure 2. Two types of CMRs are obtained by using different mole ratios of PC and CE. Principally using PC concentrations between 67 and 97 mole %, sonication yields vesicular CMRs (VCMRs) whose diameter ranges from about 0.02 to 0.03 microns (200 to 300 angstroms). The captured aqueous volume in vesicular CMRs is comparable to SUV liposomes and has little utility for capturing hydrophilic drugs. Non-vesicular CMRs (NVCMRs) are produced when about 50 mole % and greater amounts of CE are used. The overall size is dependent on the amount of CE used, but will range from about 0.03 to 0.1 microns (300 to 1000 angstroms). The driving force to maintain these sub-micron organized particles is the unique characteristic of the cholesterol esters, being immiscible with PC. These CMRs are stable in aqueous environments (years at refrigerator temperatures), can be frozen and thawed and can be lyophilized and reconstituted in buffered solutions. These favorable characteristics of the CMRs (maintenance of submicron sizes in an aqueous milieu) have many advantages especially when compared to liposomes.

It is well-known[23] that intravenously administered liposomes are removed rapidly and nearly quantitatively from the systemic circulation of the liver, spleen and lung of mammals (Figure 3). On the other hand CMRs, containing cholesterol-$[C^{14}]$-oleate and administered intravenously to rats, are initially (2 minutes post dosing) found within the systemic circulation. The calculated initial volume of distribution of the $[C^{14}]$-labelled CMRs corresponds to plasma volume and this conclusion is confirmed by whole body radioautographic studies. The initial decay in the plasma kinetics of the $[C^{14}]$-labeled CMRs has a $t_{1/2}$ of about 1 hour, which, based upon pharmaco-kinetics calculations and whole body radioautographic studies, corresponds to the distribution of CMRs from the plasma compartment into the intra-cellular space, i.e. extracellular fluid of all organs. The second slope has a $t_{1/2}$ of about 6 hours and corresponds to the metabolism of the cholesteryl $[^{14}C]$-oleate to $[^{14}C]O_2$. At the end of four (4) hours post dosing approximately half of the administered dose continues to circulate within the plasma compartment, greatly contrasting with the extensive and rapid removal of the liposomes from the systemic circulation. Thus the CMRs, due to their small size and long circulation time within the plasma compartment have the potential of being employed as drug delivery systems and, in the future, as target-recognition delivery systems.

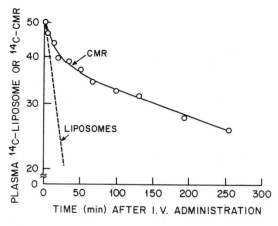

Fig. 3.   Plasma kinetics of $[^{14}C]$-labeled-liposomes or $[^{14}C]$-labeled CMRs in animals.

In Figure 4 CMRs have been evaluated as a delivery vehicle for imidocarb. Imidocarb is a cationic compound with antileukemic, tuberculostatic and antitrypanosomal activities and was developed from the phthalanilides[39]. Imidocarb is extremely active against Anaplasma marginal[37], an organism that is primarily confined to the blood compartment and which is a fatal disease in cattle. Even though a cure for this disease is available for cattle using imidocarb, imidocarb is extensively and rapidly removed from the plasma compartment and deposited into tissues[49]. The $t_{1/2}$ of imidocarb in muscle is in years, thus compromising its use for treating meat and milk producing cattle.

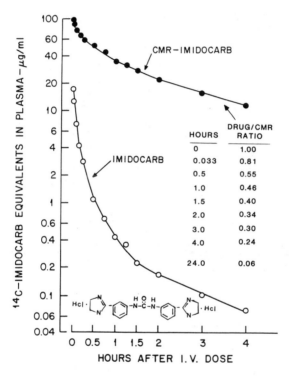

Fig. 4.   Kinetics of Imidocarb in Plasma of Rats.

The rapid removal of intravenously administered imidocarb from the plasma of rats is shown in Figure 4. The dose of imidocarb was chosen to yield a 125 ug/ml concentration in plasma, if the imidocarb were retained by the plasma compartment. Within 2 minutes greater than 80% of intravenously administered imidocarb was removed and deposited into tissues. Within 4 hours more than 99.9% of the imidocarb had been primarily deposited into tissues. Imidocarb administered in CMRs shows a plasma profile that contrasts with intravenously administered imidocarb. The plasma kinetics of the CMRs imidocarb appear similar to the plasma kinetics of CMRs without drug (Figure 3). However, the imidocarb/CMR ratio, as shown as a function of hours post dosing, indicates that imidocarb is effluxing from the CMRs faster than the removal of the CMRs from the plasma compartment.

For example, at 4 hours the concentration of imidocarb in plasma was approximately 10 ug/ml or approximately 8% of the theoretical 125 ug/ml initial concentration. At 4 hours the concentration of CMRs in the plasma compartment represents about 50% of its initial concentration (Figure 3). The drug/CMR ratio at 4 hours was 0.24, indicating that approximately 75% of the drug had effluxed from the CMRs. If the CMRs had retained the imidocarb its concentration in plasma would have been about 40 ug/ml. Thus the selection of the drug or its analog for incorporating into the CMRs needs to be assessed.

A method for evaluating the efflux of drug in a formulated vesicular CMR is shown in Figure 5. The preformed drug-VCMR is stirred with large multilamellar liposomes (MLV) and aliquots are removed at selected time intervals. During the mixing interval the drug has the opportunity to efflux from the VCMRs to the liposomes since the surface composition of both species is primarily lecithin (PC). The large liposomes are centrifuged out of the solution and the solution is monitored for both drug and VCMR.

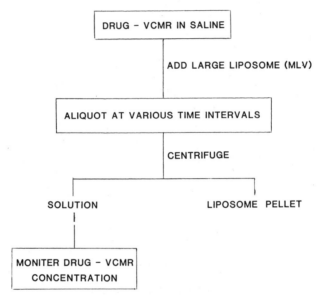

Fig. 5.   Method for evaluating efflux of drug in formulated circulating microreservoirs (CMRs).

Using this technique, the efflux characteristics of several anthra-cyline analogs from CMRs to large multilamellar liposomes were compared. These are shown in Figure 6. These analogs differ by their constituents on the 14th carbon of the aglycone (R_1) and the amine substituent (R_2) on daunosamine. Adriamycin (14-OH, N-H) and daunomycin (14-H, N-H) are the drugs that are clinically used as antineoplastic drugs. They only differ structurally by the substitution of a hydroxyl group on the 14-carbon, yet 80% of the adriamycin effluxes from the CMRs to the liposomes, versus 50% for daunomycin. The esterification of the 14th hydroxyl group of adriamycin decreases its efflux from CMRs; however, the acetylation of daunosamine, as demonstrated with AD-32, enhances its efflux from the CMRs. The enhanced efflux by acetylation of the daunosamine was also apparent for the very lipophilic anthracyline analogs (AD-74 & AD-45), which show little efflux from the CMRs. The AD numbered analogs were provided by M. Israel[28].

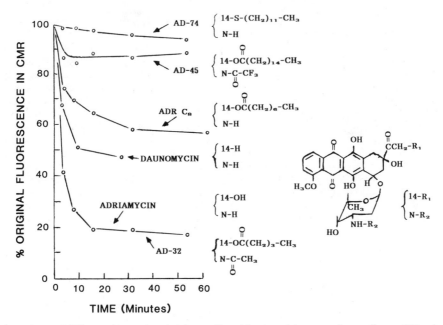

Fig. 6.   Efflux characteristics of anthracycline analogs from CMRs to large liposomes.

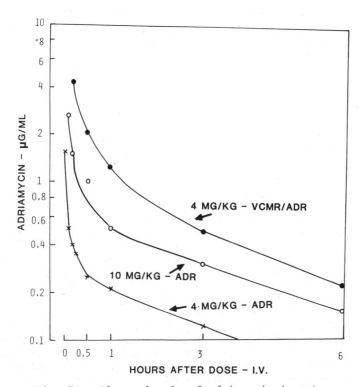

Fig. 7.   Plasma levels of adriamycin in rats.

The plasma levels of adriamycin (ADR) administered as a vesicular CMR (VCMR) dosage form is shown in Figure 7. It is readily apparent that the VCMR-ADR maintains higher plasma concentrations than a comparable dose of 4 mg/kg-ADR and a higher dose of 10 mg/Kg-ADR. However, the magnitude of the plasma concentration after administration of VCMR/ADR is consistent with the anticipated extensive efflux of ADR from the CMRs (Figure 6). Nevertheless, even though adriamycin can efflux from the VCMR, there are some beneficial effects as shown by the decreased acute toxicity of adriamycin when administered in a VCMR formulation (Figure 8). Briefly, VCMR-ADR at 7.1 mg/Kg resulted in 9 survivors out of 10, in contrast to zero survivors for ADR at the same dose. Also, survivors were seen at all doses of the VCMR-ADR indicating that the CMRs, by retaining some ADR and by slowly releasing most of the ADR from the CMRs, can have therapeutic value. In addition, an increased therapeutic index was observed for the antineoplastic drugs, adriamycin, daunomycin and AD-32, when formulated in vesicular circulating microreservoirs (VCMRs).

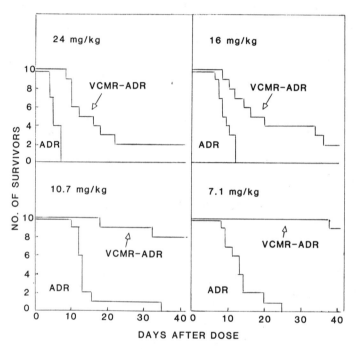

Fig. 8.   Toxicity of free vs. VCMR-adriamycin (I.P. dose to BDF$_1$ mice).

In summary, the circulating microreservoirs were so named because they maintain a small size (0.02 to 0.2 microns) in vitro and in vivo, because they circulate within the systemic circulation of mammals for extended periods of time, and because they retain their formulated drug as a reservoir for subsequent release. The efflux of the drug can be multiphasic, due in part to the regionalization of the PC and CE and to the differential partitioning of the drug between these phases. The CMRs can affect the distribution of a drug after its intravenous administration and, as a result, can enhance therapy of antineoplastic drugs as well as decrease their toxicity. The CMRs, although remaining within the plasma compartment, will release the drug at a rate that is faster than the removal of the CMRs by metabolism. The chemical make-up of the CMRs and their small size also represent desirable characteristics for use as a dermal delivery system.

B. LYM-X-SORB (L-X-S)

As shown in Figure 9, drugs taken orally are principally absorbed from the ileum of the G.I. tract and transported via the portal blood to the liver prior to reaching the systemic circulation. Thus, first pass metabolism of the drug can be extensive within the intestine and liver. Another absorption pathway exists from the G.I. tract, namely absorption into and transport via the thoracic lymph which connects with the venous blood at the juncture of the subclavian vein. The lymphatic absorption route will bypass the first pass metabolism by the liver which would be advantageous for many drugs. The rate of thoracic lymph flow in humans greatly contrasts with blood flow, being about 1/3000 the rate of blood flow[38,51]. Therefore drug absorption via the lymphatic route will be comparatively slow and sustained over a longer period of time, thus representing another advantage. Lastly, if this lymphatic route of absorption of drugs were feasible many therapeutic benefits would ensue.

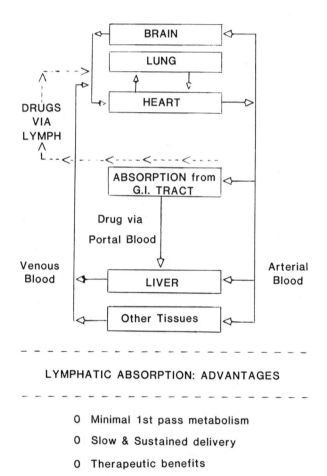

LYMPHATIC ABSORPTION: ADVANTAGES

O Minimal 1st pass metabolism

O Slow & Sustained delivery

O Therapeutic benefits

Fig. 9.   Drug absorption from the gastro-intestinal tract.

Most biological and physiological scientists are well aware that the various food constituents are processed in the stomach and intestines at different rates and are absorbed at different sites[6,20,25]. For example, carbohydrates, and subsequently proteins, are enzymatically processed and the resultant sugars and amino acids are absorbed in the ileum and further absorbed via the portal blood. Fats, primarily triglycerides, are enzymatically processed last and are absorbed in the jejunum region (segment of the intestine prior to the ileum). These absorbed fats are reconstituted as chylomicrons and are transported via the thoracic lymph. Thus, the lymphatic route might be a viable route for the absorption of drugs.

In Table 4 we have included a number of substances which are preferentially absorbed from the intestine via the lymphatic route. Many of the smaller molecules have lipid soluble characteristics; e.g. cholesterol, fatty acids such as palmitic and oleic acids, monoglycerides of comparable fatty acids and esters of vitamin A. Somewhat surprising is the absorption of small quantities of several native proteins via the lymphatics. It is the absorption of the intact native protein that results in serious medical difficulties. Many other compounds are quantitatively absorbed via the lymphatics if administered in triglycerides, for example hexachlorobenzene.

Table 4

LYMPHATICS: A SIGNIFICANT ROUTE OF ABSORPTION FROM INTESTINE

SUBSTANCES(S)	REFERENCES
Cholesterol & its esters	Mueller[30]
Fatty Acids	Bloom et al.[4]
Monoglycerides	Clark et al.[9]
Vitamin A esters	Drummond et al.[16]
Native Proteins	
Egg white albumin	Alexander et al.[1]
Colostrum globulin	Comline et al.[12]
Clos. bot. Type A toxin	May & Whaler.[29]
Hexachlorobenzene	Iatropoulos et al.[22]

A partial list of drugs whose absorption is improved by lipids and/or whose lymphatic absorption was improved by lipids is summarized in Table 5. The drugs fall into several different therapeutic categories and, not too surprisingly, there are several examples of the absorption of polypeptides via the lymphatic route. The chemical characteristics of the drugs vary widely but most are characterized by their low solubility in an aqueous medium. The formulations that have been employed are primarily those which themselves are absorbed per se via the lymphatic route or are metabolically processed to yield components that are absorbed via the lymphatic route. Two notable exceptions to this general conclusion are hydrogenated castor oil and sugar esters which are themselves not absorbed per se from the G.I. tract.

Since most of these formulations are lipid in character, the enzymatic processing and absorption of lipids from the intestines is summarized in Figure 10. Triglycerides, which were the principal lipids for formulating those drugs in Table 5, are enzymatically processed to yield mixed lipid micelles composed of bile salts, monoglycerides and fatty acids[41]. This type of mixed lipid micelle was evaluated for those drugs in Table 5, however without great success. We believe the reason for minimal absorption of drug from these lipid compositions is due to other processes which are critical for fat absorption. It has been shown that lysophosphatidyl choline (LPC or LYSO PC) is an absolute requirement for the absorption of fat[33,46]. Thus, the lecithin and bile salts are extensively secreted in bile and lecithin is enzymatically processed to lyso PC. The resultant organized structure of Lyso PC, monoglycerides and fatty acids was discovered by the author and represents the basis of the LYM-X-SORB drug delivery system. These lipids are absorbed in the region of the jejunum, are reconstituted into chylomycrons and are transported via the thoracic

Table 5
DRUGS, WHOSE MINIMAL LYMPHATIC ABSORPTION IS IMPROVED WITH LIPIDS

THERAPEUTIC CATEGORIES	FORMULATION
Anti-infectives:	
Griseofulvin	High fat meal[13,14]
	Triglycerides[27]
	Corn oil/polysorbate[7,18,24]
Streptomycin & Gentamycin	Monoolein/bile salts[31]
	Oleic acid/bile salts[31]
Tetracycline	Tripalmitin[15]
p-Amino salicylic acid	Tripalmitin[15]
Hormones:	
Ethynyl Estradiol-3-cyclopentyl ether	Sesame oil[17] Sesame oil/monoolein[17]
Testosterone undecanoate	Peanut oil[11]
Progesterone	Cholesterol and its esters[26]
Estradiol	Fatty acids, monoglycerides and bile salts[48]
Polypeptides:	
Human leukocytes interferon	Linoleic and HCO-60[52]
Cyclosporin A	HCO-60 > sugar esters > sesame oil > linoleic acid[44]

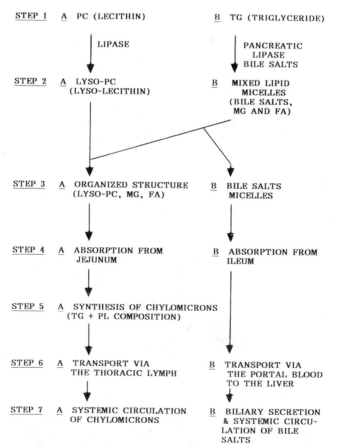

STEP 1   A   PC (LECITHIN)   B   TG (TRIGLYCERIDE)

LIPASE   PANCREATIC LIPASE BILE SALTS

STEP 2   A   LYSO-PC (LYSO-LECITHIN)   B   MIXED LIPID MICELLES (BILE SALTS, MG AND FA)

STEP 3   A   ORGANIZED STRUCTURE (LYSO-PC, MG, FA)   B   BILE SALTS MICELLES

STEP 4   A   ABSORPTION FROM JEJUNUM   B   ABSORPTION FROM ILEUM

STEP 5   A   SYNTHESIS OF CHYLOMICRONS (TG + PL COMPOSITION)

STEP 6   A   TRANSPORT VIA THE THORACIC LYMPH   B   TRANSPORT VIA THE PORTAL BLOOD TO THE LIVER

STEP 7   A   SYSTEMIC CIRCULATION OF CHYLOMICRONS   B   BILIARY SECRETION & SYSTEMIC CIRCULATION OF BILE SALTS

Fig. 10. Enzymatic processing and absorption of lipids from the intestine.

lymph to the systemic circulation. Most of the bile salts are absorbed from the ileum, transported by the portal blood and preferentially secreted by the liver back into bile.

The remainder of this chapter will focus on the molecular organization of the lyso PC, MG and FA constituting the lipid components of the Lym-X-Sorb delivery system (TABLE 6). The molecular organization of these components is such that drugs can reside within an inclusion space. This inclusion space is surrounded by the acyl chains and the polar head group of lyso PC. Thus, the drug and acyl chains can affect a hydrophobic bonding whereas the phosphate and choline of LPC can affect ionic binding with appropriate groupings on the drug. Cationic metal chelates between drug and fatty acids as well as the bound water of hydration that is associated with the polar head group of lyso PC can stabilize the molecular associa-tion of drug with the L-X-S complex. The PCT patent has been published, the U.S. claims have been allowed and the patent has either been issued or is pending in 21 countries.

TABLE 6
LYM-X-SORB DELIVERY SYSTEM

Multi-lipid Components:
        Lysophosphatidyl choline (LPC, Lyso-PC)
        Monoglycerides (MG)
        Fatty acids (FA)

Molecular Association with Drug:
        Inclusion complex
        Hydrophobic bonding
        Ionic binding: cationic & anionic
        Amphoteric polar head group (LPC)
            Bound water of hydration
        Cationic metal chelates

Patent Status:
        PCT WO 86/05694, 9 October 1986
        U.S. Patent 4,874,795; October 17, 1989
        Patent issued & pending in 21 countries

The molecular organization of these three components, namely LPC, MG and FA has been derived from interpretation of data from differential scanning calorimetry (Figure 11) and viscometry (Figure 12) after varying the mole ratios among these three components.

The differential scanning calorimetry (DSC) of L-X-S delivery system and its components is shown in Figure 11. Palmitic acid (16:0 FA) was a single melting species at 66.9°C, whereas, the monopalmitin had two melting components, 42.6°C and 69.6°C. The synthetic LPC contained multiple melting species, which probably represents different degrees of water of hydration. The L-X-S formulation, using defined mole ratios of these three components, yields a single melting eutectic which strongly indicates that there was a specific molecular organization. Interpretation of the viscosity of liquid L-X-S formulations provides insight into the molecular organization of the L-X-S components.

In Figure 12 the viscosity of the individual components and various combinations is summarized. Monoglycerides (MG), composed primarily of monoolein, is liquid at 35°C and has a viscosity of about 130 centipoise (CPS). Oleic acid (FA) is much less viscous, having a viscosity of 20 CPS at 35°C. Mixing MG (triangular heads) and FA (circles) at equal concen-trations results in a viscosity of 70 CPS, which is proportional to the

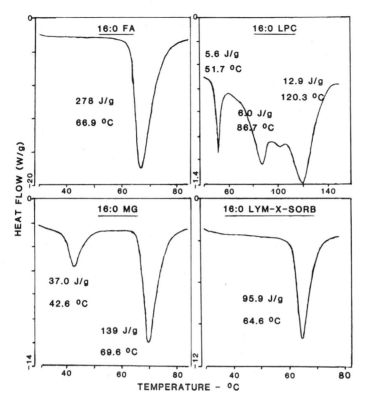

Fig. 11. Differential scanning calorimeter of L-X-S and its components.

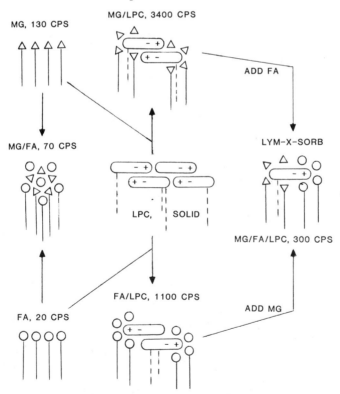

Fig. 12. Interpretation of viscosity data for the LYM-X-SORB components.

mole ratio of the two components. Lyso PC (LPC) is a solid at 35°C and when added to monoglycerides or fatty acids, results in a marked increase in viscosity to 3400 CPS for MG/LPC and 1100 CPS for FA/LPC. Lyso PC always has bound water of hydration, thus the electrostatic interaction between LPC is probably maintained in the MG/LPC arrangement whereas the carbonyl group of the fatty acids interacts with the choline group of LPC to minimize the electrostatic interaction among the lyso PC (LPC). When fatty acids are added to the MG/LPC system the viscosity is decreased by an order of magnitude; i.e. 3400 to 300 CPS, whereas, the addition of MG to the FA/LPC mixture only shows a 3-fold decrease in viscosity. The organization as shown here represents the monomeric LYM-X-SORB drug delivery system.

Molecular models of the individual lipids and the molecular associations in the Lym-X-Sorb formulation are shown in Figure 13. The individual components of the L-X-S drug delivery system, 16:0 MG, 16:0 LPC and 16:0 FA, are shown respectively from left to right in segment A. In the Lym-X-Sorb formulation as shown in segment B, top view, and in segment C, side view, the MG are arranged around the acyl grouping of the LPC and the FA surrounds the choline moiety of LPC. In segment D one MG and one FA have been removed to show the inclusion space that resides beneath the polar head group, phosphoryl choline, of LPC. In segment E a steroid has been shown occupying the inclusion space and it should be noted that the steroid was also present in the top (segment B) and side views (segment C) of the L-X-S formulation. In segment F, a steroid is shown with LPC showing in more detail that drugs can reside beneath the polar head groups of LPC.

The bioavailability of a drug, fenretinamide, is shown in Figure 14. Fenretinamide is poorly soluble in a corn oil/polysorbate formulation and when given as a suspension along with food shows a 15 to 20% absolute bio-availability. Fenretinamide is soluble in a L-X-S formulation and, most interestingly, the solubility of fenretinamide is directly proportional to the mole ratio between fenretinamide and LPC. Also, fenretinamide affects a doubling of the viscosity of the L-X-S formulation. When this mole ratio exceeds 1, fenretinamide remains as a solid suspension, indicating that only one fenretinamide molecule can reside within the inclusion space of the L-X-S formulation and will affect a molecular rearrangement among the L-X-S components.

This bioavailability of fenretinamide in the L-X-S formulation in fed animals shows a 5-fold increased peak concentration and a four-plus-fold greater area under the curve (AUC). In the absence of food there is about a 3-fold increase in the bioavailability. Thus the L-X-S formulation, by maintaining the fenretinamide within the inclusion space, enhances its co-absorption with lipids. At present we do not have direct evidence that fenretinamide is being delivered via the lymphatic route to the systemic solution, but the coincidence of the absorption of both lipids and fenretinamide strongly indicates the lymphatic route.

The bioavailability of a second drug is shown in Figure 15. Briefly, the $t_{1/2}$ of 1.3 hours and 11-12 hours is similar to the intravenously dosed drugs. The bioavailability of this formulation was about 40% with a fairly narrow range in the AUCs. The actual composition of the formulation is shown at the bottom of the figure. Because of the necessity to increase the amount of delivered dose, and because the drug complexes with 2 moles of fatty acids, a L-X-S composition was chosen which solubilized the drug. Only later did we recognize that lyso PC could only accommodate a finite number of fatty acids and monoglycerides. Thus, if one reinterprets the composition as noted here, the drug could be either associated with the L-X-S or with the fatty acids. However, based upon the difficulty of incorporating the drug within the L-X-S formulation and the relative ease

Fig. 13. Corey-Pauling-Koltun Precision Molecular Models of molecular organization of LYM-X-SORB and its components: (A) 16:0 MG, 16:0 LPC & 16:0 FA; (B) Top view of L-X-S monomeric organization; (C) Side view of L-X-S formulation; (D) Side view of L-X-S, less one MG & FA, showing the inclusion space beneath the LPC polar head group; (E) Same view as (D), but containing a steroid within the inclusion space; and (F) 18:2 LPC and a steroid.

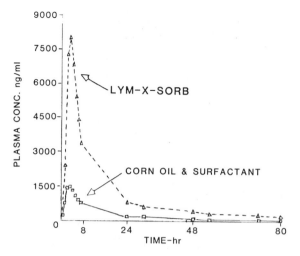

Fig. 14. Bioavailability of L-X-S formulation of fenretinamide in dogs.

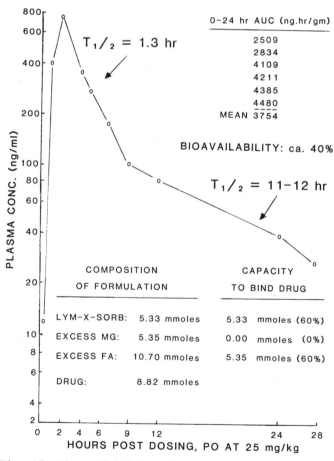

Fig. 15. Bioavailability of LYM-X-SORB & drug in dogs.

of forming the drug-fatty acid complex, we suspect that only 40% of the drug becomes associated with the L-X-S formulation. This would correspond with the observed 40% bioavailability of the drug. This experiment thus indicates that excess fatty acids and monoglycerides may not be beneficial if the drug forms complexes with other components.

In Table 7, another important feature of the L-X-S/drug formulation is shown. This L-X-S formulation was similar to that shown in Figure 11 and was composed of 16:0 MG, FA and LPC. The eutectic melting point was 63.5°C. As one incorporates the polypeptide into the inclusion space of the L-X-S formulation, one observed initially a shoulder on the L-X-S DSC melting profile with a subsequent lowering of the melting temperature. As the molar composition of the polypeptide is increased, the yield in this shoulder increases proportionally and eventually becomes the lower melting specie with the L-X-S representing the shoulder.

Table 7
LYM-X-SORB FORMULATION OF A POLYPEPTIDE

Molar Composition		DSC MPT. °C	Est. % yield by DSC	
L-X-S	Polypeptide		L-X-S	L-X-S(Pep.)
1	0.00	63.5	100	0
1	0.26	59.0	71	29
1	0.38	57.4	59	41
1	0.52	55.4	41	59
1	0.64	51.1	28	72
1	0.80	50.6	18	82

A question that is frequently asked concerns the drug capacity of the L-X-S formulation in terms of weight percentage. In all examples that we have evaluated to date, a 1:1 molar ratio of L-X-S:drug is a practical limit for maximum absorption. Thus, the per cent weight capacity of the L-X-S is reflected in the molecular weight of the drug. For fenretinamide, which has a molecular weight of 391, the weight percentage is about 14%. In the example of the polypeptide, a 1:0.8 molar ratio corresponds to approximately 31 percent by weight. In all drugs that have been evaluated to date drug capacity has not been a limiting factor.

Most drugs can be formulated in the Lym-X-Sorb formulation. Several types of drugs, however, have created some problems. As noted in Figure 15, excess fatty acid which can complex with the drug can compromise absorption. Fatty acid will also compete with anionic drugs, such as cromolyn, and the cromolyn will precipitate from a liquid Lym-X-Sorb formulation. Other drugs, such as aminoglycosides, will complex with the fatty acids and produce formulations that will not disintegrate in the aqueous milieu of the G.I. tract.

This leads to the next series of questions that concern the characteristics of the L-X-S formulation in aqueous solutions, shown in Table 8. Briefly, the L-X-S formulation will swell in distilled water and at a concentration of 5% by weight, a gelatinous matrix is observed. At 1.2% by weight smaller gelatinous crystals are observed under a polarized light microscope but the crystal sizes are too large to pass a 0.2 micron filter. The L-X-S formulation in 0.1N HCl, corresponding to gastric juices, and in 0.1M phosphate buffers at pH 7, corresponding to duodenal conditions, remains as large solid or liquid particles. The formulations do not swell or have any characteristics which would indicate that they would disintegrate in acidic or basic solutions. Most interesting is the finding that the L-X-S formulations swell in bile salt solutions and

Table 8
CHARACTERISTICS OF L-X-S FORMULATIONS IN AQUEOUS SOLUTIONS

Aqueous System	Characteristics
Distilled water	Gelatinous, crystalline (polarized light microscope)
0.1N Hcl (gastric)	Large oil/solid particles
0.1M Phosphate, pH 7 (duodenum)	Large oil/solid particles
Bile salt solutions	Micron sized particles, crystalline (polarized light microscope)

disintegrate into a wide range of crystal sizes, including sizes that pass through a 0.2 micron filter. This finding correlates with the greater absorption of L-X-S formulated drug when administered with food for both dogs and humans. Drugs that are incorporated into the L-X-S formulation are stable in all these aqueous milieus.

III. CONCLUSIONS

The highlights of both the Circulating MicroReservoirs (CMRs) (Figure 16) and LYM-X-SORB (L-X-S) (Figure 17) as drug delivery systems are summarized below.

As shown in Figure 16, the CMRs are composed of a water compatible monolayer, namely lecithin (PC), with an inner stabilizing core composed of cholesterol esters. These CMRs are small, ranging between 0.02 to 0.05 microns (200 to 500 angstroms) and are extremely stable in aqueous solutions. Due to the small size a large surface area can be covered for the controlled release of drug and/or fragrance for dermal uses. The sub-micron size of the CMRs is reflected in a long half life in plasma and is also useful for administering drugs intravenously. The composition of the CMRs is such that the efflux of drug from the CMRs will depend on where they reside in the multiphasic composition and the efflux of drug will generally be multiphasic. The CMRs, by affecting the distribution of drugs, can enhance therapy as well as reduce toxicity. More information on the CMRs can be found in the issued world wide patents[40].

The LYM-X-SORB (L-X-S) formulation[50] (Figure 17) is composed of lysophosphatidyl choline (LPC), monoglycerides (MG) and fatty acids (FA). In the presence of water, the L-X-S monomeric components organize themselves as a bilayer which swells in the presence of water and, in the presence of bile salts, becomes reduced in size to submicron crystals. The organization of the L-X-S formulation as a bilayer, terminated by bile salts (B.S.), is analogous to the lecithin/bile salt bilayer[42,43]. The drug becomes entrapped within the inclusion space beneath the polar head of LPC. In comparison to most oils, the L-X-S formulation has an increased capacity for drugs, as well as providing a very stable micro-environment for drugs. Since the L-X-S formulation remains intact in acidic and basic solutions a protection of the drug to pH and enzymes within the GI milieu would be anticipated. The co-absorption of drug and lipids will preferentially deliver the drug via the lymphatics, which will also provide enhanced therapeutic benefits. The L-X-S formulation is compatible with many types of drugs and peptides.

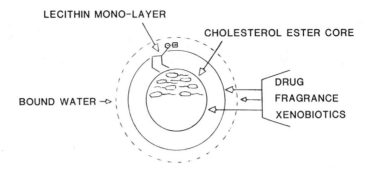

LECITHIN MONO-LAYER

CHOLESTEROL ESTER CORE

BOUND WATER

DRUG
FRAGRANCE
XENOBIOTICS

O SMALL, 200-500 A⁰

O STABLE in Aqueous Solution

O COVERAGE of large surface area (DERMAL)

O LONG plasma half-life (IV)

O Potential MULTI-PHASIC efflux rates

O U.S. PATENT: 4,298,594 NOV. 3,1981

Fig. 16.   CONCLUSION: Circulating MicroReservoirs (CMRs) Delivery System.

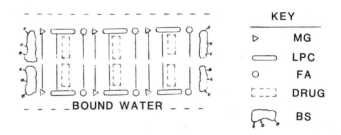

BOUND WATER

KEY

▷   MG

⊂⊃   LPC

○   FA

⌐_⌐   DRUG

       BS

O   DRUG ENTRAPMENT WITHIN INCLUSION SPACE

O   INCREASED DRUG CAPACITY & STABILITY

O   PROTECT DRUGS IN G.I. MILIEU, pH & ENZYMES

O   LYMPHATIC ABSORPTION OF DRUGS

O   ENHANCED THERAPEUTIC BENEFITS

O   COMPATIBLE WITH MANY TYPES OF DRUGS
        & PEPTIDES

O   PCT WO 86/05694 OCT.9,1986

Fig. 17.   CONCLUSION: LYM-X-SORB (L-X-S) DELIVERY SYSTEM.

The opportunities for the Lym-X-Sorb formulation as a drug delivery system are shown in Table 9. Many therapeutic benefits are plausible with drug delivery systems and some examples of these benefits are shown at the top. Examples of therapeutic categories and specific drugs that would benefit using a drug delivery system are shown at the left. A plus (+) indicates that there is a therapeutic need for this drug and that the L-X-S formulation will favorably address this therapeutic need. It is apparent that the L-X-S formulation can be very beneficial to a large number of compounds in all therapeutic categories. Just as importantly, many drugs have the potential of being made into lipophilic analogs in which the L-X-S formulation as well as CMRs and liposomes would be most useful.

Table 9

OPPORTUNITIES FOR LYM-X-SORB DRUG DELIVERY SYSTEM

THERAPEUTIC CATEGORIES	STABLE	PALATABLE	ABSORPTION	METABOLISM	LONG ACTING	EFFICACY	SIDE EFFECT	LIPOPHILIC ANALOG
ANTI-INFECTIVES								
CEPHALOSPORINS	+		+		+	+		+
CHLORAMPHENICOL		+		+	+	+		+
ERYTHROMYCINS	+	+	+		+	+		+
PENICILLINS	+		+		+	+		+
TETRACYCLINES			+		+	+		+
AMINOGLYCOSIDES			+		+	+		+
CARDIOVASCULAR								
PROPRANOLOL	+	+	+	+	+			+
DIGITALIS			+	+	+	+	+	
QUINIDINE			+	+			+	
METHYL DOPA	+		+	+	+	+	+	+
PRAZOSIN			+	+		+		
NIFEDIPINE	+	+		+	+			+
NITROGLYCERIN					+	+		
ISOSORBIDE DINITRATE			+			+	+	
NADOLOL			+			+		+
VERAPAMIL		+			+	+	+	

In conclusion, both the lipophilic CMRs and the L-X-S drug delivery systems represent a unique technology which exploits the unique physical and chemical characteristics of phosphatidyl choline and its lyso analog.

IV . REFERENCES

1. Alexander, H.L., Shirley, K. and Allen, D., 1936, The route of ingested egg white to the systemic circulation, J. Clin. Invest. 15:16.
2. Bangham, A.D., DeGier, J. and Greville, G.D., 1967, Osmotic properties and water permeability of phospholipid liquid crystals, Chem. Phys. Lipid 1:225.
3. Bangham, A.D., Standish, M.M. and Watkins, J.C., 1965, Diffusion of univalent ions acros the lamellae of swollen phospholipids, J. Mol. Biol. 13: 238.
4. Bloom, B., Chaikoff, I.L. and Reinhardt, W.O., 1951, Intestinal lymph as pathway for transport of absorbed fatty acids of different chain length, Am. J. Physiol. 166:451.
5. Brown, E.S., 1964, Isolation and assay of dipalmityl lecithin in lung extracts, Am. J. Physiol. 207:402.
6. Carey, M.C., Small, D.M. and Bliss, C.M., 1983, Lipid digestion and absorption, Ann. Rev. Physiol. 45:651.
7. Carrigan, P.J. and Bates, T.R., 1973, Biopharmaceutics of drugs administered in lipid-containing dosage forms I:GI absorption of griseofulvin from an oil-in-water emulsion in the rat, J. Pharm Sci. 62:1476.
8. Chapman, D., Flunck, D.J., Penkett, S.A. and Shipley, G.G., 1968, Physical studies of phospholipids. X. The effect of sonication on aqueous dispersions of egg yolk lecithin, Biochem. Biophys. Acta 163:255.
9. Clark, B. and Hubscher, G., 1963, Monoglyceride transacylase of rat intestinal muscosa, Biochim. Biophys. Acta 70: 43.
10. Clements, J.A., 1957, Surface tension of lung extracts, Proc. Soc. Exptl. Biol. Med. 95:170.
11. Coert, A., Geelan, J., deVisser, J. and van der Vies, J., 1975, The pharmacology and metabolism of testosterone undecanoate (TU), a new orally active androgen, Acta Endocrinol. 79:789.
12. Comline, R.S., Roberts, H.E. and Titchen, D.A., 1951, Route of absorption of colostrum globulin in the newborn animal, Nature 167:561.
13. Crounse, R.G., 1961, Human pharmacology of griseofulvin: The effect of fat intake on gastrointestinal absorption, J. Invest. Dermatol. 37:529.
14. Crounse, R.G., 1963, Effective Use of Grisiolvin, Arch. Dermatol. 87:17.
15. DeMarco, T.J. and Levine, R.R., 1969, Role of the lymphatics in the intestinal absorption and distribution of drugs, J. Pharmacol. Expt. Therap. 169:142.
16. Drummond, J.C., Bell, M.E. and Palmer, E.T., 1935, Observations of the absorption of carotene and vitamin A, Brit. Med. J. 1: 1208.
17. Grammina, T., Steinetz, B.G. and Meli, A., 1966, Pathway of absorption of orally administered ethynylestradiol-3-cyclopentyl ether in the rat as influenced by vehicle of administration, Proc. Soc. Exptl. Biol. Med. 121:1175.
18. Greco, G.A., Moss, Jr., E.L. and Foley, E.J., 1959-60, Observations on treatment of fungus infections of animals with griseofulvin, Antibiot. Ann. 663.
19. Gregoriadis, G., 1976, The carrier potential of liposomes in biology and medicine, New England J. Med. 295:704.
20. Hinder, R.A. and Kelly, K.A., 1977, Canine gastric emptying of solids and liquids, Am. J. Physiol. 233:E335.
21. Huang, C. and Thompson, T.E., 1974, Preparation of homogeneous, single-walled phosphatidylcholine vesicles, Methods in Enz. 32 (Biomembranes, Part b):485.

22. Iatropoulos, M.J., Milling, A., Muller, W.F., Nohynek, G., Rozman, K., Coulston, F. and Korte, F., 1975, Absorption, transport and organotropism of dichlorobiphenyl (DCB), Dieldrin, and hexachlorobenzene (HCB) in rats, Envrionmental Res. 10:384.

23. Juliano, R.J. and Stamp, D., 1975, The effect of particle size and charge on the clearance rates of liposomes and liposome encapsulated drug, Biophys. Biochem. Res. Communs. 63:651.

24. Kabasakalian, P., Katz, M., Rosenkrantz, B. and Townley, E., 1970, Parameters affecting absorption of griseofulvin in a human subject using urinary metabolite excretion data, J. Pharm. Sci. 59:595.

25. Kelly, K., 1981, Mobility of the stomach and gastroduodenal junction, in: "Physiology of the Gastrointestinal Tract," L. R. Johnson, ed., Raven, New York.

26. Kincl, F.A., Ciaccio, L.A. and Benagiano, G., 1978, Increasing oral bioavailability of progesterone by formulation, J. Steroid Biochem. 9:83.

27. Kraml, M., Dubuc, J. and Beall, D., 1962, Gastrointestinal absorption of griseofulvin. 1. Effect of particle size, addition of surfactants, and corn oil on the level of griseolfulvin in the serum of rats, Canad. J. Biochem. Physiol. 40:1449.

28. Lazarus, H., Yuan, G., Tan, E. and Isreal, M., 1978, Comparative inhibitory effects of adriamycin, AD32, and related compounds on in vitro growth and macromolecular synthesis, Proc. Am. Assoc. Cancer Res. 19:159.

29. May, A.J. and Whaler, B.C., 1958, The absorption of Clostridiums botulinum type A toxin from the alimentary canal, Brit. J. Exp. Pathol. 39:307.

30. Mueller, H., 1915, The assimilation of cholesterol and its esters, J. Biol. Chem. 22:1.

31. Muraniski, S., Muranuski, N. and Sezaki, H., 1979, Improvement of absolute bioavailability of normally poorly absorbed drugs: inducement of the intestinal absorption of streptomycin and gentamycin by lipid-bile salt mixed micelles in rat and rabbit, Internat. J. Pharmaceut. 2: 101.

32. Notter, R.H., Holcomb, S. and Mavis, R.D., 1980, Dynamic surface properties of phosphatidylglycerol-dipalmitoryl phosphatidylcholine mixed films, Chem. Phys. Lipids 27:305.

33. O'Doherty, P.J.A., Kakis, G. and Kuksis, A., 1973, Role of the lumina lecithin in intestinal fat absorption, Lipids 8:249.

34. Papahadjopoulos, D., 1978, Liposomes and their use in biology and medicine, Ann. N.Y. Acad. Sci. 308.

35. Patel, H.M. and Ryman, B.E., 1976, Oral administration of insulin by encapsulation within liposomes. FEBS Letters 62:60.

36. Pattle, R.E., 1955, Properties, function and origin of alveoler lining layer, Nature 175:1125.

37. Roby, T.O. and Mazzola, V., 1972, Elimination of the carrier state of bovine anaplasmosis with imidocarb, Am. J. Vet. Res. 33:1931.

38. Rusznyak, I., Foldi, M. and Szabo, S., 1967, Lymphatics and Lymph Circulation, Physiology and Pathology, Pergamon Press, Oxford.

39. Schmidt, G., 1965, Uber die trypanocide wirksamkeit von terephthalaniliden, Experientia 21:276.

40. Sears, B. and Yesair, D.W., 1981, Xenobiotic Delivery Vehicles, Method of Using Them, U. S. Patent 4,298,594, 3 November.

41. Senior, J.R., 1964, Intestinal absorption of fats, J. Lipid Res. 5:495.

42. Small, D.M., 1968, A classification of biologic lipids based upon their interaction in aqueous systems, J. Am. Oil Chem. Soc. 45:108.

43. Small, D.M., Penkett, S.A. and Chapman, D., 1969, Studies on simple and mixed bile salts micelles by nuclear resonance spectroscopy, Biochim. Biophys. Acta 176:178.

44. Takada, K., Yoshimura, H., Yoshikawa. H., Muraniski, S., Yasumura, T. and Oka, T., 1986, Enhanced selective lymphatic delivery of cyclosporin A by solubilizers and intensified immunosuppressive activity against mice skin allograft, Pharmaceut. Res. 3:48.

45. Tall, A.R. and Small, D.M., 1978, Plasma high-density lipoproteins, New England J. Med. 299:1232.

46. Tso, P., Kendrick, H., Balint, J.A. and Simmonds, W.J., 1981, Role of biliary phosphatidylcholine in the absorption and transport of dietary triolein in the rat, Gastroenterology 80:60.

47. Watkins, J.C., 1968, The surface properties of pure phospholipids in relation to those of lung extracts, Biochem. Biophys. Acta 152:293.

48. Yesair, D.W., Micellular drug delivery system, Patent Cooperation Treaty (PCT), Internat. Pub. No. 83/00294, 3 February, 1983.

49. Yesair, D.W., 1985, Lipid macromolecules as chemotherapeutic targets, in: "Experimental and Clinical Progress in Cancer Chemotherapy," F. M. Muggia, ed., Martrinus Nijhoff Publishers, Boston.

50. Yesair, D.W. Composition for delivery of orally administered substances, Patent Cooperation Treaty (PCT), Internat. Pub. No. 86/05694, 9 October 1986.

51. Yoffey, J.M. and Courtice, F.C., 1970, Lymphatics, Lymph and the Lymphomyeloid Complex, Academic Press, London.

52. Yoshikawa, H., Takada, K., Satok, Y., Naruse, N. and Muraniski, S., 1985, Potential of enteral absorption of human interferon alpha and selective transfer into lymphatics in rats, Pharmaceut Res. 2:249.

PHOSPHOLIPIDS IN COSMETICS:

IN VITRO TECHNIQUE BASED ON PHOSPHOLIPIDS MEMBRANE TO PREDICT THE IN VIVO

EFFECT OF SURFACTANTS

Gérard Redziniak

Parfums Christian Dior
P.O Box 58
45804 St Jean de Braye cedex - France

OVERVIEW

In cosmetology, phospholipids in terms of "lecithin extract" (mixture of phospholipids alone or with triglycerids) are used as coemulsifiers and sometimes as moisturizing substances [3,4,7,10,11,12,14,15,19].

In the last ten years extensive work has been performed in using phospholipids :

- as <u>carriers</u> of active ingredients [13,18]
- as <u>membrane models</u> to predict the in vivo effect of surface active agents [21,9]

In this paper we present our study which has employed one of these approaches.

INTRODUCTION

At the present time the DRAIZE test is the most convenient to evaluate the eye irritation provoked by surfactants or detergents [5,6]. However, application of the DRAIZE test still presents many problems. Specifically, this test requires skillful techniques of objective judgement. Besides, a large number of animals must be used. Not unexpectedly, animal rights activists object to this procedure. Consequently various in vitro methods have been proposed instead of animal tests [2,20]. The principle of these methods lies in the determination of the toxic effects of test materials on cells under specific conditions.

Liposomes have gained wide acceptance in chemotherapy, immunotherapy and topical application as targetable drug carriers [1,8]. Moreover liposomes (cells membrane model) may also be available as tools to investigate the pharmacological activity at the cell membrane level. In the study reported here, liposomes prepared from soya phospholipids and B-sitosterol were used as models of cell membrane. Degradation by various surface active agents was determined by measuring the degree of release of 5-6-carboxyfluorescein (5-6-C-F) from treated liposomes. These results were compared with chemical effects on the eye, as evaluated by means of the DRAIZE eye irritation test [5,6].

Phospholipids
Edited by I. Hanin and G. Pepeu
Plenum Press, New York, 1990

MATERIALS AND METHODS

Materials

- Soya phospholipids were obtained from American Lecithin Co. (Atlanta - Georgia) or Lucas Meyer (EPIKURON 200 - Hamburg - West Germany).
- B-sitosterol was purchased from Sigma Chemical Co. (St Louis, MO) and 5-6-C-F from Eastmann Kodak (Rochester - NY).
- The surface active agents (cationic, anionic, nonionic, amphoteric) were obtained commercially (see Table 1).

TABLE 1

CHEMICAL NAME AND IN VIVO CLASSIFICATION OF THE DIFFERENT SURFACTANTS TESTED

Chemical name	Code	In vivo effect
Alkylamino betaine	A	Non irritant[a]
Polyoxypropylene-poly oxyethylene esters	B	Non irritant
Protein fatty acid condensation product	C	Very slightly irritant
Sodium N.laurylimido-diproprionate	D	Slightly irritant
Sodium lauryl sulfate	E	Irritant
Propylamido betaine	F	Irritant
Na C12-18 alkyl sulfate	G	Irritant
Na Lauryl ether sulfate	H	Irritant
Na Imidazolin-dicarboxyl sulfate	I	Irritant
Alkyl monoethanolamide	J	Very irritant[a]

[a] : indicated by the supplier.

- All other chemicals were commercial products of reagent grade.

In Vivo test

The DRAIZE eye irritation test [5] was carried out in order to evaluate the eye irritancy power of the test materials. The test was generally run at ROC Toxicological Laboratory (Colombes - France). Some data were given by the suppliers (see Table 1).

Preparation of liposomes

Conventional small unilamellar vesicles (SUVs) containing 5-6-C-F (50 mM) in the interior phase were prepared in Tris-HCl buffer (0.1 M ; pH

7.4) according to the same method as previously described [16,17]. These
5-6-C-F liposomes were diluted with empty liposomes of the same
composition (soya PC/B-sitosterol, 8/2-M/M ; without 5-6-C-F) in Tris-HCl
buffer (0.1 M ; pH = 7.4) to obtain a final concentration of phospholipids
of about 400 ug/ml. The phospholipid concentration was determined by an
enzymatic-colorimetric procedure (phospholipids BIOLYON KIT). The size of
the liposomes was evaluated by light scattering (Coulter N4 - Coultronics)
and their structure by Electron Microscopy after freeze fracture.

5-6-C-F-release from liposomes destabilized by surfactants

Fluorescence measurements were performed on a Fluoroscan II multiwells
fluorophotometer (Flow Lab.). In a 96 wells multidish (NUNC) the
surfactants were diluted with Tris-HCl buffer solution containing
different concentrations (0.02 mg to 10 mg/ml) of the surface active
agents. In each well we added 100 ul of the liposome suspension (5-6-C-F
liposomes plus empty liposomes). After 3 hours incubation at room
temperature, we measured the fluorescence intensity at 520 nm on
excitation at 488 nm. The percent of 5-6-C-F-release was calculated by
means of the following equation :

$$\% \text{ 5-6-C-F-release} = \frac{S - B}{T - B} \times 100$$

where : B was the fluorescence intensity of liposomes incubated without
surfactant.

T was the fluorescence intensity of liposomes completely
destroyed by a Triton X-100 solution in Tris-HCl buffer (2% v/v).

S was the fluorescence intensity of liposomes in the presence of
surfactants.

RESULTS

The observation of liposomes was confirmed by electron microscopic
observation (Fig.1).

The size of the vesicles was 80 nm + 20 nm. The dose-response
relationships of four surfactants with respect to the release of 5-6-C-F
are shown in Fig.2.

From the curves we obtained, we were able to determine the
"concentration minus logarithm" at which 50% of 5-6-C-F were released in
the presence of each surfactant (- log C50).
For each material the dose-response curves and the -log C50 values
obtained in different assays using new liposome preparations were quite
the same.
- The different values of - log C50 are presented in Table 2.
- Compared with the classification obtained by the in vivo test (DRAIZE
  test) we have simplified the findings and have summarized the in vitro
  and in vivo effects (table 3).

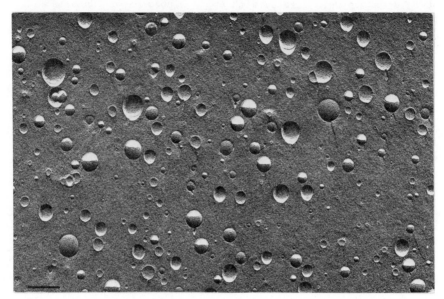

Fig.1. Freeze-fracture electron micrograph of soya PC (Epikuron 200 -
Lucas Meyer) and B-sitosterol liposomes (8/2 - M/M) - Bar = 0.2 um.
The liposome suspension was produced according to the "spray
drying" and "high pressure homogenization" processes (see 20 and
21).

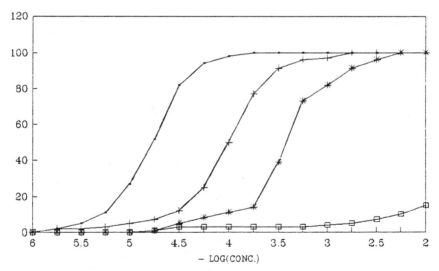

Fig.2. Dose-response relationship for the release of 5-6-C-F from
liposomes by a range of concentration (g/ml) of 4 surfactants.
(☐ : product B, * : product D, + : product G, ● : product J).
For more details see Materials and Methods.

TABLE 2

CHEMICAL NAME AND IN VITRO DATA (- log C50) OF THE DIFFERENT SURFACTANTS
TESTED

Chemical name	- log C50[a]
Alkylamino betaine	< 3.00
Polyoxypropylene-poly oxyethylene esters	< 3.00
Protein fatty acid condensation product	3.43
Sodium N.laurylimido-diproprionate	3.43
Sodium lauryl sulfate	3.80
Propylamido betaine	3.85
Na C12-18 alkyl sulfate	3.96
Na Lauryl ether sulfate	3.95
Na Imidazolin-dicarboxyl sulfate	3.99
Alkyl monoethanolamide	4.75

[a] : - log. of the surfactant concentration that releases 50% 5-6-C-F.

TABLE 3

COMPARISON AND CLASSIFICATION OF THE IN VITRO AND IN VIVO EFFECTS

In vitro (liposome test)	In vivo (Draize test)
- log C 50 < 3.00	Non irritant
3.00 ⩽ - log C 50 < 3.50	Very slightly irritant
3.50 ⩽ - log C 50 < 4.00	Slightly irritant
4.00 ⩽ - log C 50 < 4.50	Irritant
4.50 ⩽ - log C 50	Dramatically irritant

DISCUSSION

It seems reasonnable to assume that the irritant effect of particular chemicals may be triggered by damage to the cell membrane. This damage may be due to a destabilization or a complete loss of membrane integrity. The liposome, in terms of a phospholipid membrane model, is a good candidate to predict the in vivo effects of surfactants.

The in vivo model developed in this preliminary study demonstrates the following :

- When liposomes are "broken" by a low concentration of surface active agents, their products are very irritating in the in vivo test.
- On the contrary, when high concentrations of surfactant are necessary to "break" liposomes, these molecules are slightly irritant or non irritant in the DRAIZE test.
- In comparison with other alternative in vitro techniques, this method, therefore, is reproducible, simple and less costly.

ACKNOWLEDGMENTS

We would like to express particular thanks to Jacqueline Harnay, translator, for reviewing the English version and typing this manuscript.

REFERENCES

1. Bangham, A.D., Standish, M.M., and Watkins, J.C., 1965, J.Mol.Bio., 13:238.
2. Borenfreund, E., and Borrero, O., 1984, Cell Biol.Toxicol., 1:33.
3. Coxell, R.D., Sullivan, D.R., and Szuhaj, B.F., 1982, Surfactant Sci.Ser., 12:229.
4. Curri, S.B., Gezzi, A., Longhi, M.G., and Castelpietra, R., 1986, Fitoterapia, 57:217.
5. Draize, J.H., and Kellery, E.A., 1952, Drug Cosmet.Ind., 71:36.
6. Draize, J.H., 1959, Appraisal of the Safety of Chemicals in Foods, Drug and Cosmetics, p.46.
7. Fost, D.L., 1987, Ann. Meet. Soc. Cosm. Chem., 38:358.
8. Gregoriadis, G., 1984, Liposome Technology, ed. CRC Press Inc., Bocaraton, Florida.
9. Kato, S., Itagaki, H., Chiyoda, I., Hagino, S., Kobayashi, T., and Fujiyama, Y., 1988, Toxic in Vitro, 2:125.

10. Mitani, T., and Sugai, T., 1986, Skin Res., 28:703.
11. Motitschke, L., 1984, Parfümerie und Kosmetik, 65:177.
12. Noro, S.I., Ishii, F., and Haraguchi, Y., 1985, Yakugaku Zasshi, 105:634.
13. Olecniacz, W.S., 1976, U.S. Patent 3,957,971.
14. Orsinger, K., 1983, Seifen Ole Bette Waschse, 109:495.
15. Rebmann, H., 1977, Soap Perfum. Cosmet., 50:361.
16. Redziniak, G., and Meybeck, A., 1985, U.S. Patent 4,508,703.
17. Redziniak, G., and Meybeck, A., 1986, U.S. Patent 4,621,023.
18. Rovesti, P. and Curri, S.B., 1981, Riv.Ital. E.P.P.O.S., 63:23.
19. Schneider, M., 1985, Seifen Ole Bette Waschse, 111:16.

20. Shopsis, C., and Sathe, S., 1984, _Toxicology_, 29:195.
21. Sunamoto, J., Iwamoto, K., Imokawa, G., and Tsuchiya, S., 1987, _Chem. Pharm. Bull_., 35:2958.

PHOSPHOLIPID LIPOSOMES AS DRUG CARRIERS: PREPARATION AND PROPERTIES

Herbert Stricker

Institut für Pharmazeutische Technologie
und Biopharmazie der Universität Heidelberg, FRG

INTRODUCTION

Phospholipids such as Phosphatidylcholines have the tendency to form molecular bilayers and to minimize their surface by forming vesicles. Such liposomes have been tested with regard to their suitability as drug carriers. The number of drugs which has been entrapped into liposomes up till now is considerable. The main groups of interest today are: cytostatics, antibiotics, biological response modifiers and hormones. With some of these preparations clinical studies are now being carried out. Examples are: liposomes containing Amphotericin B, or Adriamycin, or lipophilic derivatives of Muramyldipeptide, etc. Many publications have been written about the properties of liposome-entrapped drugs and the biological effects which can be achieved with them[2,12]. From today's point of view it is clear that liposomes as a dosage form have significant biopharmaceutical advantages, but only in specific fields (e.g. parenteral depots, drug targeting via the reticuloendothelial system, RES, etc.)

During the development of liposome preparations different biopharmaceutical and technological problems have to be overcome. They arise in connection with:

- the preparation,
- storage stability[4],
- drug release[4],
- pharmacokinetics[5] and
- biological effects of the drug-containing liposomes.

The physicochemical properties of the liposomes significantly affect their pharmacokinetic properties, especially with regard to:

- biological degradation,
- the lack of membrane penetration,
- the half-life in blood after i.v. application (0.1 to 36h),
- accumulation in the RES,
- transport via lymphatics after subcutaneous application, and
- the depot effect after a local, i.g. an intratumoral application.

The behaviour of liposomes in the organism after parenteral application depends mainly on the lipid composition, the liposome structure, the ligands and the vesicle size.

Frequently used lipids - which have to satisfy the demand of pharmaceutical purity - are natural phosphatidylcholines, for example from lecithin (PC), hydrogenated PC from soybean lecithin (HPC) and synthetic lipids with different fatty acid chains such as dimyristoylphosphatidylcholine (DMPC). Further usual components are: charged phospholipids (like dipalmitoylphosphatidylserine, DPPS), cholesterol (CHol) and many other lipophilic substances (for examples see reference[1]). Important properties result from lipid composition, especially the phase transition temperature T_c and with that the membrane permeability, the membrane density, as well as the lipid charge and the membrane stability.

With regard to structure and the one differentiates uni- and multi-lamellar vesicles (UV, MLV). This structure determines the encapsulation ratio of the aqueous phase and thereby the entrapment efficiency of hydrophilic drugs. The vesicle size has the same significance, and is one of the important factors for the fate of the liposomes in the organism. Therefore it is important to strive for size fractions as narrow as possible. Between 20 and 2000 nm one differentiates small, large and very large liposomes (SUV, MLV and LMLV), respectively.

A large entrapment efficiency cannot however be realized in all cases. It depends also on the lipid solubility of the particular drug, which in turn affects the drug´s location within a liposome. Consequently hydrophilic drugs are entrapped in the inner aqueous phase of a liposome; hydrophilic, polar drugs also can be adsorptively bound on the surface of charged liposomes; and lipophilic drugs distribute mainly into the lipid phase. On the other hand amphiphilic and lyophobic drugs are difficult to entrap.

This article deals mainly with the production of drug-containing liposomes which can be performed in different ways. Decisive for the best method in each specific case are the required physicochemical and pharmacokinetic properties of the liposomes, the properties of the encapsulated drug and the batch size. In the following examples the preparation of liposomes on a larger scale under aseptic GMP conform conditions with a hydrophilic, a polar and a lipophilic drug is described.

PREPARATION AND PROPERTIES OF DRUG CONTAINING LIPOSOMES

The preparation of liposomes and the entrapment of drugs has to be performed in several steps (see Table 1), which are more closely defined in the following examples. Important liposome properties depend on the dispersion step, which can be done using various methods (see Table 1). It is typically found that no universal procedure exists. Each has its specific advantages and disadvantages: some are only suitable for the laboratory scale (for example the sonication method); some do not result in different size fractions or maximal entrapment volume; some are not suitable for certain lipid combinations; and some require very stable drugs.

Only the two first dispersion methods mentioned (Table 1) are dealt with in the following examples, because they have important advantages: they are suitable for larger batches, they allow antiseptic techniques, they are variable, and they yield, depending on the experimental conditions, special liposome properties.

TABLE 1

LIPOSOME PREPARATION, STEPS AND DISPERSION METHODS

Preparation steps	Dispersion methods
1. solvation	- homogenization
2. dispersion	(high pressure)
3. fractionation/	- extrusion (filter)
separation	- sonication
4. sterilization	- detergent dialysis
5. filling	- ethanol injection
	- reverse phase evaporation
	- ether infusion
	- freezing and thawing
	- calcium-induced fusion
	- pH-adjustment

EXAMPLE 1: ASEPTIC PRODUCTION OF MAB-MARKED, MTX CONTAINING LIPOSOMES

Methotrexate-sodium (MTX), a water soluble cytostatic drug, can be entrapped into liposomes with defined properties with high efficiency and under aseptic conditions as a one litre batch. As lipids HPC and CHol were used. Dipalmitoylphosphatidylethanolaminepyridyldithiopropionate (DPPE-PDP[9]) serves as a linking reagent for the covalent bounding of the antibody (monoclonal mouse-antimelanoma-antibody MAB, M 2.9.4., IgG_{2a}[9,10]) on the liposome surface.

The schema of Fig. 1 shows that the production is undertaken in seven steps: the hydration of the lipid film with MTX solution, the

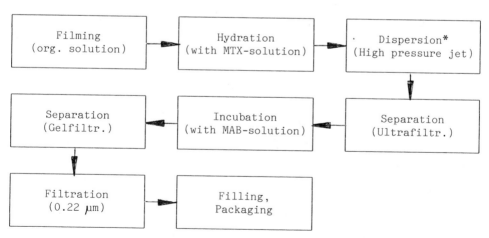

Fig. 1. Aseptic production of MAB-labelled Methotrexate containing liposomes made from HPC, CHol and DPPE-PDP

Batch size: 1 l (10 % lipid). Molar ratio: 5:5:1.
* jet 2; 2,000 psi; 35°C; n = 30 (= 3.5 l/h); N_2

117

dispersion with a high pressure homogenizer[6] and the separation by ultrafiltration from MTX which has not been entrapped (for example with the Minitan[R] system, Millipore). After incubation of the liposomes with the thiolated antibody the unbound protein is separated by gel filtration[9], the dispersion is then filtered through a 0.22 um membrane filter and filled into ampoules. A decisive step is the homogenization with an apparatus which is schematically shown in Fig. 2. Its construction corresponds to the common principle for the preparation of emulsions or for cell disruption. It is built out of two piston pumps which press the lipid dispersion with an ajustable pressure between 40 and 30,000 psi through a jet, made preferably from ceramic.

Variables for the process are: pressure, temperature, time of homogenization and jet construction. The two jets shown in Fig. 2 differ mainly in the lock system: jet 1 has a ball, jet 2 a half ball. This construction difference is of importance for the resulting liposome properties (Fig. 3): jet 2 results in larger vesicle diameters and larger encapsulation ratios than jet 1. A special advantage of this homogenizer is that the higher temperatures which are necessary in the case of lipids with high phase transition temperatures occur only during the short jet passage. The greatest part of the liposome dispersion can be held at essentially lower temperatures (e.g. 35°C). For temperature labile drugs this is of decisive significance. The chosen experimental conditions for the discussed lipsomes preparation are: 2,000 psi, 35°C, 3.5 1/h., N_2 and jet type 2.

The properties of liposomes produced in the described way are summarised in Table 2.

This product is being tested in animal studies with respect to its targeted effect on liver micrometastases.

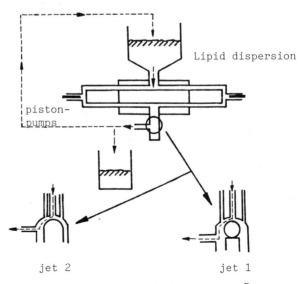

jet 2   jet 1

Fig. 2. High Pressure Homogenizer (Stansted[R], Stansted UK)

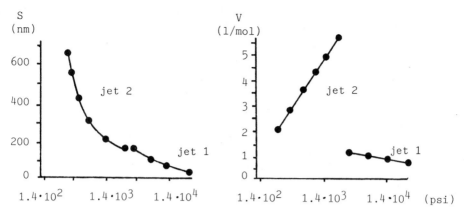

Fig. 3. Mean vesicle size S and encapsulation ratio V at
different pressure and with different jets

Conditions: 35°C, n = 30 (= 3.5 1/h), N$_2$
Lipid: HPC/CHol (1:1)

TABLE 2

PROPERTIES OF ANTIBODY-LABELLED, METHOTREXATE-
CONTAINING LIPOSOMES

mean vesicle size	: $\bar{x}$ = 190 nm (90 - 270)
encapsulation ratio	: 5 1/mol lipid
MTX	: 50 mg/g lipid
MAB	: 6 molecules/liposome
storage stability (6°C):	3 month

EXAMPLE 2: ASEPTIC PRODUCTION OF γ-INF AND MDP CONTAINING LIPOSOMES

The schema in Fig. 4 describes the series of process steps which
are necessary in order to entrap two immunostimulating peptides: the
water soluble muramyldipeptide (MDP) in the inner phase of negative
charged liposomes made from DMPC, CHol, DPPS and the positively charged
interferon γ (IFN) at the liposome surface. The lipids are first hydrated
with the muramyldipeptide solution and afterwards dispersed by shaking
under defined conditions, followed by a filter extrusion. The large
liposomes are separated from non-entrapped muramyldipeptide by centrifugation, followed by an incubation with the interferon solution. Because
of the positive charge of this protein (i.e.p. = 8.6) at pH 7.4 and
the negative charged liposomes (conditioned by DPPS), there occurs a
very quick, almost complete and strong protein adsorption with a loss
of antiviral activity of about 40 %. The very small amount of unbound
interferon is separated from the liposomes by centrifugation and the
liposome dispersion is filled into polyethylene ampoules. Because these
large vesicles cannot be sterile filtered or sterilized at 121°C they
are treated at 35,000 psi and 40°C. This new sterilization method ensures

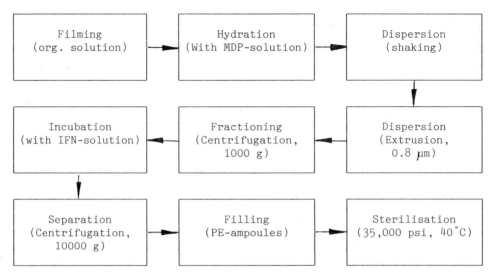

Fig. 4. Aseptic production of γ-IFN and Muramyldipeptide-containing liposomes made from DMPC, CHol and DPPS

Molar ratio: 5:2:3

the death of microorganisms, but does not distroy the liposomes and does not reduce interferon activity.

The product (Table 3) has a mean vesicle size of 400 nm and an entrapment efficiency of 2 mg muramyldipeptide + 1 mg interferon per gram lipid.

The interferon is so strongly absorbed that the storage stability at 6°C of this product is more then 4 weeks. This product is being tested on the activation of macrophages with respect to the type of entrapment, and a possible synergism between interferon and muramyldipeptide.

TABLE 3

PROPERTIES OF γ-IFN AND MDP-CONTAINING LIPOSOMES

mean vesicle size	: $\bar{x}$ = 400 nm (320 - 480)
MDP	: 2 mg/g lipid
γ-IFN	: 1 mg/g lipid (2 10^4 U/g)
IFN-adsorption	: 0.03 (d^{-1}) > k_{des} << k_{ads}

EXAMPLE 3: ASEPTIC PRODUCTION OF LOMUSTINE CONTAINING LIPOSOMES

This example deals with the aseptic production of liposomes made from DMPC, CHol and the lipophile cytostatic lomustine. This drug is very sensitive to hydrolysis and requires, therefore, a dry liposome preparation. Fig. 5 shows that there are 4 preparation steps which require special attention of times and temperatures. After brief shaking of the dry preparation with water pH 5, liposomes are formed with encapsulated lomustine and a mean vesicle size of 600 nm. The chemical stability of lomustine in this formulation is increased remarkably (Table 4). Such

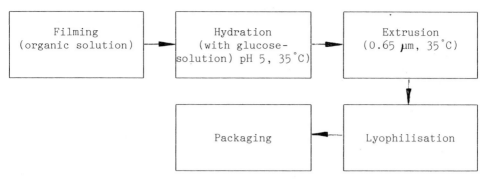

Fig. 5. Aseptic production of a dry preparation of lomustine containing
LMLV-liposomes made from DMPC and CHol

Step 2 to 4: max. 1h.

large liposomes remain for longer times within the tumour[8,11], after
intratumoral application. This product is therefore being tested in animal
studies as a potential local chemotherapeutic agent for the treatment
of inoperable brain tumours.

TABLE 4

PROPERTIES OF LOMUSTINE CONTAINING LIPOSOMES
(after shaking with H_2O, pH 5)

mean vesicle size	: $\bar{x}$ = 600 nm (450 – 650)
lomustine	: 10 mg/g lipid/10 ml
chemical stability	
(serum, 37°C)	: $t_{50} > 4h$

SUMMARY

The preparation of liposomes on a larger scale and under aseptic conditions containing either a hydrophilic, a polar or a lipophilic drug
(Methotrexate, Y-Interferon, Muramyldipeptide, Lomustine) has been described. Corresponding process parameters (such as the homogenization
pressure) have been optimized for each separate step of the production.
The properties of the resulting products, including vesicle size and
structure, encapsulated ratio, entrapment efficiency and stability on
storage have been discussed. Intended animal experiments with respect
to possible advantages of the liposome preparations for cancer chemotherapy have been considered.

Literature

1. Arndt, D. and Fichtner, I., 1986, "Liposomen", Akademie Verlag Berlin
2. Gregoriadis, G., 1984, "Liposome Technology", CRC Press, Boca Raton
3. Ostro, M.J., 1987, "Liposomes", Marcel Dekker, Inc New York, Basel
4. Kibat, P. and Stricker, H., 1986, Storage of Dispersions of Soybean
   Lecithin Liposomes, Pharm. Ind. 38:1184-1189

5. Kibat, P. and Stricker H., 1988, Elimination of Liposome-Encapsulated Substances Following Intravenous Administration, Arzneim.Forsch./Drug Res., 38:1472-1478
6. Mentrup, E. and Stricker, H., 1990, Herstellung von Liposomen mit einem Hochdrucksterilisator, Pharm. Ind. in press
7. Mentrup, E., Butz, P., Stricker, H. and Ludwig, H., 1988, High Pressure Sterilisation of Liposomes, Pharm.Ind. 50:363-366
8. Mentrup, E., Stricker, H., Wowra, B., Zeller, W.J. and Sturm, V., 1989, In vitro Release and Depot Effect after Intratumoral Application of Metrizamide-containing Liposomes, Arzneim.Forsch./Drug Res., 39:421-423
9. Weckenmann, H.P., Matzku, S. and Stricker, H., 1988, Optimization of the Covalent Binding of Monoclonal Antibodies to Liposomes, Arzneim.Forsch./Drug Res., 38:1556-1563
10. Weckenmann, H.P., Matzku, S., Stricker, H. and Sinn, H., 1989, Stability in Biological Fluids and Clearance of Immunoliposomes, Arzneim.Forsch./ Drug Res., 39:415-420
11. Wowra, B., Mentrup, E., Zeller, W.J., Stricker, H. and Sturm, V., 1988, CT-Kinetik intratumoraler Liposomendepots, Onkologie, 11:81-84
12. Yagi, K., 1986, Medical application of liposomes, Japan Scientific Societies Press, Tokyo

BIOLOGICAL BEHAVIOUR OF LIPOSOMES

Gregory Gregoriadis

Centre for Drug Delivery Research
The School of Pharmacy, University of London
London WC1N 1AX

INTRODUCTION

Many drugs in therapeutic, diagnostic and preventive medicine would be more effective if they were to act selectively, where they are needed (4). Barring a few exceptions (e.g. antibiotics interfering with metabolic pathways in bacteria), however, drug selectivity is often poor. For example, in cancer chemotherapy cytostatic agents also damage normal rapidly proliferating cells. Additional obstacles to effective drug action include inaccessibility of targets (e.g. many antimicrobial agents are unable to reach microbe-infested intracellular sites), drug instability in the biological milieu, premature drug loss through excretion and allergic reactions to some drugs.

Accumulating information on the three dimensional structure of drugs and on quantitative structure-activity relationships is already contributing to better drug design and the optimisation of the interaction with relevant molecular targets. Unfortunately, there are problems, such as those encountered within the biological milieu interposed between the site of drug administration and the target, which are unlikely to disappear, at least in the foreseeable future. In this respect, progress in such diverse areas as monoclonal antibodies, the chemistry of polymers and colloids, and the understanding of ligand-receptor interactions and related intracellular events, has contributed to the rapid progress of an alternative approach to conferring selectivity on drugs, namely targeting. The concept is based on the use of carriers to deliver drugs to the intended site of action. Carriers can do so because of an inherent or acquired ability to interact selectively with respective cells. Thus, antibodies bind with exquisite specificity to cell surface antigens, glycoproteins interact through specific terminal groups with receptors, and colloid microspheres (e.g. liposomes) are endocytosed by a variety of cells. Injected drug-loaded carriers are expected, once in the vicinity of the target, to associate with it on contact. Through a variety of scenarios, the drug is subsequently freed to act. Such considerations have formed, during the last two decades, the basis for the study of carrier behaviour in vivo and the control of such behaviour. This is especially true for liposomes for which a wealth of relevant data has been already amassed (9).

LIPOSOMES: STRUCTURAL CHARACTERISTICS

Liposomes (phospholipid vesicles) were originally described (1) as a suitable tool for the study of membrane biophysics. They consist of one or more concentric phospholipid bilayers alternating with aqueous spaces and are chiefly characterised by structural versatility. A variety of phospholipids with an array of gel-liquid crystalline transition temperatures (Tc) can, in association with sterols, ensure a wide range of membrane fluidity which, in turn, controls membrane permeability to entrapped solutes, bilayer stability and behaviour in vivo. Further, a complement of amphiphiles incorporated into the lipid structure will impose a negative or positive surface charge as required (8). A great number of water or lipid soluble drugs, including antitumour and antimicrobial agents, enzymes, peptides, hormones, antigens, genetic material, etc., have been already incorporated either in the aqueous or the lipid phase of liposomes. In addition, targeting ligands, for instance antibodies, certain glycoproteins, neoglycoproteins and glycolipids can also be covalently or otherwise linked to the outer bilayer (8). In short, it is now possible to design liposomes to satisfy particular needs in a variety of applications ranging from biochemical and immunological assay kits and diagnostic reagents to therapeutic preparations for enteral and parenteral use and vaccines (9). In this brief contribution, I will summarize some of the major findings on the interactions of liposomes with the biological milieu which have helped towards applying the system both experimentally and clinically.

FATE OF LIPOSOMES IN VIVO

Soon after the introduction of liposomes in the early seventies (7) as a drug delivery system, it was shown (7) that liposome behaviour in vivo will vary according to the type of vesicles used, the drug that is carried, the site of administration, and the animal species, as well as its physiological state.

Behaviour in Blood

Following intravenous injection, drug-containing "conventional" liposomes, i.e. vesicles which are made solely of phospholipids, interact almost immediately with at least two different groups of plasma proteins (7). The first group includes opsonins which adsorb onto the surface of the vesicles and render them, together with their drug contents, prey to the phagocytic cells of the reticuloendothelil system (RES). The second group, lipoproteins (HDL) attack vesicles and, in concert with a protein exhibiting phosphatidylcholine transfer activity, remove phospholipid molecules from the bilayer structure. This leads to the disintegration of the vesicles (or to formation of pores on their surface), following which, entrapped solutes are released into the circulation. At the

124

same time vesicles, at various stages of destabilization and with some of their solute contents still entrapped, are rapidly removed by the RES, probably through recognition of adsorbed opsonins. The rate of clearance of liposomes that have preserved their vesicular structure is controlled by their size with larger vesicles being cleared more rapidly than smaller ones. Surface charge of liposomes also influences the rate of clearance, with negatively charged liposomes being removed at a faster rate than neutral (uncharged) ones.

Liposomes with a positive surface charge have, in all studies so far, exhibited rates of clearance similar to those seen with neutral liposomes. However, recent work from this laboratory (15) indicates that the role of a positively charged stearylamine, used in all previous studies, is lost from liposomes (which, therefore, become uncharged), probably by binding to proteins. Under conditions where the retention of the charged lipid in serum is facilitated by the composition of the vesicles (e.g. densely packed bilayers made of excess cholesterol and distearoyl phosphatidylcholine; DSPC), liposomes are cleared from the circulation at rates similar to those seen when the surface charge is negative. In addition to the theoretical implications of this finding in the understanding of mechanisms responsible for liposome clearance and uptake by tissues, it will now be possible to predict or explain the effect of liposome-incorporated drugs or ligands, expressing a net positive surface charge, on liposomal behaviour in vivo.

Loss of liposomal stability in the circulating blood and ensuing premature release of contents are incompatible with the effective use of the system in delivering drugs to tissues quantitatively. It was reasoned (7) that packing the relatively loose phospholipid bilayers (usually made of egg phosphatidylcholine; PC) through the addition of cholesterol or the replacement of PC with saturated phospholipids which exhibit Tcs higher than 37°C and are therefore "solid" at the body's temperature, would curtail the ability of HDL to remove phospholipids from the vesicles. This was confirmed in appropriate experiments (5): liposomes, regardless of their size, when composed of PC and increasing amounts of cholesterol lost correspondingly decreasing amounts of their PC component and of water-soluble markers both after exposure to blood at 37°C or in the circulation of injected animals. Similarly, there was very little loss of phospholipid and entrapped marker when liposomes were made of DSPC (Tc = 54°C) only. In addition, for similar liposomes made of equimolar DSPC and cholesterol, bilayer stability remained unchanged even after incubation with blood for over 48 hours. Indeed, a relationship between stability of liposomes and their rate of clearance (i.e. the more stable the vesicles the longer their half-life) was observed. This, as in the case of bilayer stability, was true for vesicles of all sizes and, for small unilamellar vesicles (SUV) composed of equimolar DSPC and cholesterol (40-80nm diameter), a half-life of 20 h was recorded in mice. A hypothesis was thus formulated (7) to explain such a relationship: vesicles with a "loose" bilayer donate some of their phospholipid to HDL. This is followed by the formation of "gaps" leading to the escape of entrapped solutes and also to the adsorption of opsonins, the latter facilitating vesicle uptake by the RES. On the other hand,

125

vesicles with packed bilayers are not or are only minimally affected by HDL, there is no escape of solutes, and opsonins are unable to adsorb in sufficient amounts so as to mediate rapid vesicle removal by the RES.

Progress in the control of liposome stability and presence in the circulation have had important ramifications in targeting the system to cells other than those (i.e. RES) which normally intercept it. For instance, liposomes coated with antibodies or other ligands and destined to interact with specific cells, must remain in the circulating blood for periods of time long enough for them to seek and interact with their target effectively. Long half lives are also a prerequisite for other uses, a good example being haemosomes (liposome carrying haemoglobin as a blood surrogate) (9). Generally, it is advantageous to employ large liposomes for drug delivery as a large aqueous phase will accommodate greater quantities of drugs, regardless of molecular size. However, even the most stable large liposomes exhibit brief half-lives, although prolonged clearance rates (about 10 h in mice) have been recently claimed for vesicles larger (70-120 nm) than those mentioned above, incorporating gangliosides (5).

Uptake by Tissues

As already discussed, removal of the majority of liposomes from the blood is carried out by the RES. Uptake of the vesicles by the fixed macrophages (and probably by the circulating monocytes), is effected though endocytosis following which vesicles end up in the lysosomal apparatus of the cells. Within lysosomes, vesicles are disintegrated by phospholipases and freed drugs can then act within the organelles or, if of small enough size, diffuse through the lysosomal membranes to reach other cell compartments. Because of their size (about 25nm minimum diameter), intravenously injected liposomes cannot undergo transcapillary passage in normal animals. However, such passage has been reported (9) in inflamed areas in the body where vessels become leaky. Whilst large liposomes preferentially localize in the macrophages of the liver and spleen, small vesicles can pass through the fenestrae in the liver to reach the hepatic parenchymal cells by which they are endocytosed. In addition, small liposomes with long half-lives will accumulate extensively in the bone marrow macrophages (9).

Engulfment of liposomes by phagocytic cells will also occur following administration by routes other than the intravenous. For example, although after intramuscular or subcutaneous injection the majority of liposomes either reach the blood circulation (and eventually the RES) via the lymphatics (small vesicles) or disintegrate locally (large vesicles), some of them end up in the lymph nodes draining the injected site (9). Circulation through the lymphatic system also appears to account for the recovery of a proportion of liposomes injected intraperitoneally, in the RES and the thoracic, renal, lumbar and intestinal lymph nodes. Indeed, in terms of uptake per unit weight of tissue, lymph nodes take up a greater proportion of liposomes than do the liver and spleen.

Attempts (2) have been made to resolve the question as to whether or not orally administered liposomes protect entrapped agents in, and facilitate their uptake by the gut. Several laboratories (2,9) had reported, in experiments with liposomal insulin, a reduction in blood glucose levels in diabetic rats. More recent work, however, suggests that partial recovery of insulin and a number of non-absorbable agents given orally in the liposome form, in the blood circulation of treated animals, occurs through the absorption of phospholipid-bile salt-agent complexes rather than the absorption of intact vesicles (9).

TARGETING OF LIPOSOMES

Interaction of drug-containing liposomes with cells which do not normally take them up effectively, has been facilitated by antibodies and other cell-specific ligands covalently or hydrophobically linked to the outer bilayer of liposomes. It has been demonstrated repeatedly in vitro that polyclonal or monoclonal antibodies raised against a repertoire of cell surface antigens mediate the association, and subsequent incorporation, of the drug-containing liposomal moiety (to which such antibodies are linked), into the respective cells (7). In vivo targeting of liposomes, however, has proved a much more challenging proposition (17), especially when mediated via antibodies the Fc portion of which binds to its receptors on the macrophage, thus accelerating removal of the carrier by the RES. This problem can be circumvented by the use of the antigen-recognizing Fab portions of the immunoglobulin molecule as a ligand or by taking advantage of the already long half-lives of small, stable vesicles. Such complications do not occur when using certain glycoprotein and glycolipid ligands which associate exclusively with their receptors in vivo. Asialoglycoproteins, for instance, bind quantitatively to the galactose receptor on hepatic parenchymal cells, and together with mannose-terminating glycoproteins and neoglycoproteins (i.e. proteins such as albumin coupled to sugars) have been already linked to liposomes and shown to mediate targeting to cells expressing the relevant receptors (6).

The use of ligand-coated liposomes to transport drugs intracellularly (as opposed to the direct use of ligands) has several advantages: (a) much larger quantities of drugs can be delivered by a single vesicle bearing one molecule of ligand than with the same molecule of ligand as such; (b) covalent attachment of excess drug to ligand often results in the masking of its receptor-recognizing portion of ligand, its denaturation and altered pharmacokinetics. This is not the case with liposomes where the drug is contained within the bilayers and cannot interfere in any way with the function of the externally attached ligand; (c) whereas appropriately manufactured liposomes (with or without ligands) will retain their drug content quantitatively in the presence of blood, some of the bonds in drug-ligand complexes can be vulnerable to the action of plasma hydrolases; and (d) in contrast to drug-ligand complexes for which the mechanism of intracellular drug release from the carrier will vary according to the bond, the mode of drug release is identical for all liposomal preparations, namely vesicle disruption in the lysosomes.

IMPLICATIONS IN MEDICINE

Successful use of liposomal drugs in the treatment of prevention of disease in experimental animals and in clinical trials suggests that clinical applications may be forthcoming (12). In this respect, the first and obvious prerequisite is that a liposomal drug preparation designed to treat a particular disease has significant advantages over the conventional use of the therapeutic agent. These may include a lower dosage, reduction of drug waste, and improved access to the target. Novel drug toxicities as a result of carrier-induced altered pharmacokinetics should, in addition, be minimal.

Cancer Chemotherapy

The use of liposome-entrapped cytostatic drugs in the treatment of cancer has been considered in spite of the obvious difficulties of vesicle access to areas where tumours reside. However, indirect evidence from in vivo work suggests that certain tumours take up drugs entrapped in small liposomes. Whether this could be explained on the basis of higher endocytic activity of some tumour cells combined with increased local permeability of adjacent capillaries, drug diffusion from circulating vesicles followed by preferential drug localization in the tumour because of increased blood flow in that area, or indeed, as a result of migration of monocytes (together with engulfed liposomes) to tumours, remains an open question. A role for liposomal drugs in cancer chemotherapy appears more convincing in work aimed at reducing toxicity while at the same time maintaining the tumouricidal effect of the drug. Studies with liposomes containing anthracycline cytostatics have clearly shown reduction of cardiotoxicity and dermal toxicity and prolonged survival of tumour-bearing animals compared to controls receiving the free drug (14). It has been suggested (9) that liposomal drug taken up by the RES tissues is slowly released to penetrate neighbouring malignant cells and exert its effect. In this respect, there have been already promising results with at least two of the ongoing related clinical trials (12). An alternative approach uses liposomes containing macrophage activation agents which transform macrophages to a tumouricidal state and eradicate metastases in experimental animals (12).

Antimicrobial Therapy

A role for site-specific drug delivery in the treatment of microbial disease is envisaged mostly because of the inability of many agents to enter infected intracellular sites effectively. A large variety of microorganisms reside in the liver and spleen, especially their RES component, and liposomes with their propensity to localize in these tissues are the obvious choice of carrier. Many workers have shown that liposomes are superior to the free agents not only in terms of distribution to the appropriate intracellular sites but also in being therapeutically effective. Liposomes have also been applied in the treatment of fungal diseases (e.g. in immunosuppressed patients). Amphotericin B used for the treatment of such diseases (e.g. candidiasis), acts by binding to the ergosterol of fungi membranes thus creating channels through which vital molecules leak from the cells which die as a result. Because the drug also binds to the cholesterol of mammalian cells, it induces toxicity. Several reports have shown that disseminated candidiasis in animals can be treated successfully with amphotericin B incorporated into liposomes composed of saturated phospholipids, and similarly encouraging observations were made in patients with fungal disease (12). The benificial effect of liposomal amphotericin B has been attributed to its considerably lower toxicity to mammalian cells. More recently, work with liposome-mediated direct killing of microbes is now being supplemented with attempts to deliver (via liposomes) immunostimulating agents such as muramyl dipeptide and its derivatives, in order to activate macrophages to a microbiocidal state (12).

Storage Diseases

Two groups of storage diseases are candidates for treatment with liposome-entrapped agents (9). The first group includes inherited metabolic disorders such as lysosomal storage diseases in which deficiency of a hydrolase in the lysosomes leads to the accumulation of the substrate, normally hydrolysed by the enzyme, in the organelles. This leads, in turn, to tissue enlargement and malfunction. The second group concerns metal storage diseases in which iron, copper, zinc, plutonium and other metals accumulate in tissues in a variety of circumstances.

A large number of intravenous injections of liposome-entrapped human glucocerebroside β-glucosidase into a Gaucher's disease patient over a period of five years, reduced the size of her liver (11). Correction of the malfunctioning gene, or its replacement when absent, would be the ideal approach to the management of enzyme deficiencies including lysosomal storage diseases. In this respect, targeted gene delivery via liposomes, already demonstrated in vitro (8), is an attractive possibility.

With regard to the treatment of metal storage diseases, targeted delivery of chelators could improve their efficacy. Taking into consideration all major prerequisistes for an ideal chelator, i.e. retention of its structural integrity in blood, avoidance of interaction with other, irrelevant metals, and accessibility to the intracellularly stored metal, usually within phagocytic cells, it would appear that liposomes are the vehicles of choice. As early as 1973, liposomes were tested as a carrier of chelating agents in plutonium-poisoned mice and subsequently, by the same group (13) and others (9) in animal models loaded experimentally with such metals as iron, aluminium, ytterbium and cadmium. According to most related studies, the effect of liposome-entrapped chelator in removing deposited metals from tissues was greater than that of the free agent.

Immunopotentiation

One of the areas in medicine where the prospect of applying liposomes seems particularly realistic, is immunopotentiation, with particular emphasis on vaccines. Recently, developments in recombinant DNA techniques, elucidation of the immunological structure of proteins, and the ways by which cells and mediators interact to induce immune responses, have led to a new generation of vaccines. Thus subunit vaccines (e.g. hepatitis B surface antigen) have been produced through gene cloning, and a number of peptides which mimic very small regions of protein on the outer coat of viruses and are capable of eliciting virus neutralizing antibodies, have been synthesized. Subunit and peptide vaccines are, however, often only weakly active or non-immunogenic in the absence of immunological adjuvants (10). The latter are a variety of substances of unrelated structure which induce immune responses to antigens by activating macrophages to release mediators such as interleukin 1. This in turn stimulates the proliferation of lymphocytes leading to humoural and or cell-mediated immunity. Another (indirect) mechanism by which antigen-incorporating adjuvants may act, is slow degradation at the site of injection, thus facilitating uptake of the antigen by the regional lymph nodes. Unfortunately, many of the adjuvants available, including aluminium salts, which are the only adjuvant licensed for use in man and, in additon, does not produce cell mediated immunity, can induce side reactions (10).

Liposomes interact avidly with macrophages, persist at the site of injection, and it is perhaps not surprising that they potentiate strong immune responses to entrapped antigens. The immunoadjuvant action of liposomes was originally established (10) for diphtheria toxoid and later confirmed with a large number of antigens relevant to human and veterinary immunization (10). They include tetanus toxoid, Streptococcus pneumoniae serotype 3, Salmonella typhimurium lipopoly-saccharide, cholera toxin, influenza virus subunits, hepatitis B surface antigen, Epstein-Barr virus gp 340 protein, synthetic peptides of foot-and-mouth disease virus and rat spermatozoal polypeptide fractions. In some of the studies, protection of animal models was achieved by immunization with antigen-containing liposomes. At least one of these preparations is presently undergoing clinical trials (9,10). A major advantage of liposomes over other adjuvants is their variability in structural characteristics and mode of antigen accommodation, suggesting versatility in immunoadjuvant action and vaccine design (3). Targeted adjuvanticity has been demonstrated (6), for instance, with liposomes coated with a mannosylated ligand, the latter facilitating binding to antigen presenting cells expressing the mannose receptor. In addition, cytokines such as interleukin-2 co-entrapped with the antigen (tetanus toxoid) in the same liposomes also improve further liposomal adjuvanticity (16).

CONCLUSIONS

Twenty years of research have established liposomes as a promising drug carrier. Liposomes as such are limited in tissue selectivity because of their inability to undergo transcapillary passage. However, their scope can be broadened by targeting to a variety of (accessible) cells by the use of ligands linked to the outer bilayer surface. A wide range of applications are envisageed in areas as diverse as antimicrobial and cancer therapy (especially through the activation of microbiocidal and tumouricical macrophages, respectively), storage diseases and vaccines. Other applicaitons, based on properties which are peculiar to liposomes and related more to optimal drug release or action than to direct targeting, are blood surrogates and topical and oral therapy (9).

REFERENCES

1. Bangham, A.D., Hill, M.W., and Miller, N.G.A., 1974,
    Preparation and use of liposomes as models of
    biological membranes, Meth. Membr. Biol. 1:1-68.

2. Chiang, C.M., and Weiner, N., 1987, Gastrointestinal
    uptake of liposomes: In vivo studies, Int. J.
    Pharmaceutics 40:143-150.

3. Davis, D., and Gregoriadis, G., 1987, Liposomes as
    adjuvants with immunopurified tetanus toxoid:
    Influence of liposomal characteristics, Immunology,
    61:229-234.

4. de Duve, C., 1979, Forward in: "Drug Carriers in Biology
    and Medicine," G. Gregoriadis, ed., Academic Press,
    London, pp. IX-XI.

5. Gabizon, A., and Papahadjopoulus, D., 1988, Liposome
    formulations with prolonged circulation time in blood
    and enhanced uptake by tumours, Proc. Nat. Acad.
    Sci., USA, 85:6949-6953.

6. Garcon, N., Gregoriadis, G., Taylor, M., and Summerfield,
    J., 1988, Targeted immunoadjuvant action of tetanus
    toxoid-containing liposomes coated with mannosylated
    albumin, Immunology, 64:743-745.

7. Gregoriadis, G., 1983, Targeting of Drugs with molecules,
    cells and liposomes, Trends Pharm Sci., 4:304-307.

8. Gregoriadis, G. (ed.), 1984, "Liposome Technology", vols
    1-3, CRC Press Inc., Boca Raton.

9. Gregoriadis, G. (ed.), 1988, "Liposomes as Drug Carriers:
    Recent Trends and Progress", Wiley, Chichester .

10. Gregoriadis, G., 1990, Immunological adjuvants: A role
    for liposomes, Immunology Today, 11:89-97.

11. Gregoriadis, G., Neerunjun, D.E., Meade, T.W., and Bull,
    G.M., 1980, Experiences after long-term treatment of
    an adult Gaucher's disease patient with liposomal
    glucocerebroside:β-glucosidase in: "Enzyme Therapy
    in Genetic Diseases," by J. Desnick, Ed., Alan R.
    Liss Inc., New York, pp. 383-392.

12. Lopez-Berestein, G., and Fidler, I. (eds.), 1989,
    "Liposomes in the Therapy of Infectious Diseases and
    Cancer", Alan R. Liss, Inc., New York.

13. Rahman, Y.E., Rosenthal, M.W., and Cerny, E.A., 1973,
    Intracellular plutonium removal by liposome-
    encapsulated chelating agent, Science, 180:300-302.

14. Sells, R.A., Gilmore, I.T., Owen, R.E., New, R.R.C., and
    Stringer, R.E., 1987, Reduction in doxorubicin
    toxicity following liposomal delivery, Cancer
    Treatment Reports, 14:383-387.

15. Tan, L., and Gregoriadis, G., 1989, The effect of positive surface charge of liposomes on their clearance from blood and its relation to vesicle lipid composition Biochem. Soc. Trans., 17:690-691.

16. Tan, L., and Gregoriadis, G., 1989, The effect of interleukin-2 on the immunoadjuvant action of liposomes, Biochem. Soc. Trans., 17:693-694.

17. Wolff, B., and Gregoriadis, G., 1984, The use of monoclonal antibody Thy_1 IgG_1 for the targeting of liposomes to AKR-A cells in vitro and in vivo, Biochem. Biophys. Acta., 802:259-273.

THERAPEUTIC ASPECTS OF LIPOSOMES

F. Puisieux, G. Barratt, L. Roblot-Treupel J. Delattre*,
P. Couvreur and J.Ph. Devissaguet

Laboratoire de Pharmacie Galénique et de Biopharmacie,
URA CNRS 1218, Université de Paris XI, Châteany-Malabry
Cédex, France and *Laboratoire de Biochimie, Université de
Paris V, Paris, France

Abbreviations used: ADM, adriamycin; AMB, amphotericin B; Ara-C, cytosine arabinoside; ATP, adenosine triphosphate; Chol, cholesterol; DCP, dicetylphosphate; DMPC, dimyristoylphosphatidylcholine; DMPG, dimyristoylphosphatidylglycerol; DOPC, dioleoylphosphatidylcholine; DOPE, dioleoylphosphatidylethanolamine; DPPC, dipalmitoylphosphatidylcholine; DTPA, diethylene triamine pentaacetic acid; MDP, muramyldipeptide; MLV, multilamellar vesicles; MTP-PE, muramyltripeptide-phosphatidylethanolamine; OA, oleic acid; PC, phosphatidylcholine; PS, phosphatidylserine; rIFN, recombinant human interferon; RES, reticuloendothelial system; REV, reverse phase vesicles; SA, stearylamine; SOPE, DN-succinyldioleoylphosphatidylethanolamine; SUV, small unilamellar vesicles; T_c, phase transition temperature.

INTRODUCTION

The fate of a drug after administration *in vivo* is determined by a combination of several processes: distribution and elimination when given intravenously and absorption, distribution and elimination when an extravascular route is used.   Regardless of the mechanisms involved, each of these processes depends mainly on the physico-chemical properties of the drug and therefore, for the most part, on its chemical structure.

During the last few decades, research workers have been trying to develop systems which would allow them to control the fate of drugs within the patient by means of the form administered (Fig. 1).   Using forms known as controlled release systems, pharmacists have succeeded, at least in part, in governing the first process, i.e. drug absorption.   The aim of the current research was to control, if possible, the second process - drug distribution within the organism - by the use of carriers.

Phospholipids
Edited by I. Hanin and G. Pepeu
Plenum Press, New York, 1990

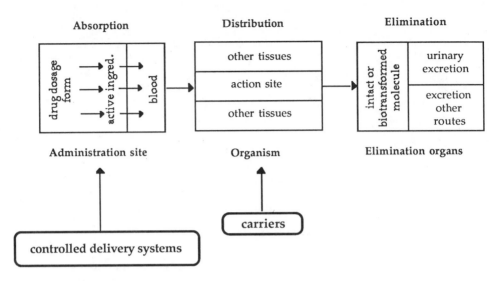

Figure 1. Fate of drugs after extravascular administration *in vivo*

As shown in Table 1, the carriers available at present may be divided into three main groups: first - second - and third generation carriers (80).

TABLE 1
PRINCIPLE CARRIER SYSTEMS UNDER STUDY (80)

Generation	Size	Definition	Examples
FIRST	>1μm	SYSTEMS able to release drug at the target site but necessitating a particular type of administration	MICROCAPSULES and MICROSPHERES for chemoembolization
SECOND	<1μm	CARRIERS able to transport a drug to the target site, which can be given by a general route	Passive carriers: ERYTHROCYTES, LIPOSOMES, NANOSPHERES Active carriers: temperature- or pH-sensitive LIPOSOMES magneticNANOSPHERES
THIRD	<1 μm	CARRIERS able to recognise the target specifically	MONOCLONAL ANTIBODIES, 2nd generation carriers targeted with MAbs

The so-called "first-generation" carriers are not true carriers but rather systems capable of delivering the active substance specifically to the intended target, as shown in the table.   However, in order to do this, they have to be implanted as closely as possible to the site of action.   Microcapsules and microspheres for chemoembolization, studied in our laboratory, for over 10 years (12, 13), belong to this group.

In contrast "second-generation" carriers are true carriers (usually of colloidal size) and are capable of not only releasing an active product at the intended target but also of carrying it there.   This group includes so-called passive carriers such as liposomes, nanocapsules and nanospheres and certain "active" carriers such as temperature - sensitive liposomes and magnetic nanospheres .

The carriers referred to as "third-generation" are also true carriers and are capable of specific recognition of the target.   For example, monoclonal antibodies belong to this group, as do certain second-generation carriers (liposomes, nanocapsules, nanospheres) piloted by monoclonal antibodies.

Liposomes began life as membrane models for biologists and were first suggested as drug carriers by Gregoriadis *et al.* in 1971, for the encapsulation of enzymes (38).   As a result of this pioneering work, many groups took up similar research and the potential of liposomes in many varied fields of therapeutics rapidly came under study.   Since then, in spite of some undeniable disadvantages, liposomes remain the second-generation carrier to have been studied the most (36, 59, 69, 79).

This chapter, which will only deal with their therapeutic potential, is divided into two parts.   The first is a short analysis of the initial studies, carried out mainly in the 1970s and early 1980s, in order to give an idea of the possible uses of liposomes in various areas of therapeutics.   The second part is devoted to current research.

EARLY RESEARCH

A wide range of groups (78) quickly embarked on studies in various fields of therapeutics, following the initial work of Gregoriadis *et al.* (38).   These studies appear today to have been mainly based on two elements: on one hand the vesicular structure of liposomes, and on the other the fact that they seemed to be able to interact with the target cells.   Taking into account the vesicular nature of these carriers some workers attempted to use liposomes either to protect certain labile products (enzymes, insulin) or, more infrequently, to protect the organism from the toxic effects of some substances.   Considering their possible interactions with cells, other research groups tried to encapsulate a number of products (e.g. cancer chemotherapeutic agents, chelating agents), with the principal aim of improving their entry into cells.

As the following examples demonstrate, the studies described above did not take sufficient account of the processes involved and the numerous and considerable obstacles encountered by a drug carrier, between its site of administration and the site of action of the encapsulated drug.   As a consequence these studies did not always achieve the desired results.

The work on the encapsulation of enzymes, as stated above, was initiated by Gregoriadis *et al.* in 1971.   Between 1971 and 1982, about twenty different enzymes were studied (78) including, notably, certain enzymes whose deficiency leads to various metabolic diseases (35). In the case of metabolic diseases the effects sought after by encapsulation were, above all, improved stability of the enzymes *in vivo*, reduced antigenicity and better penetration into the target cells (i.e. the cells in which the substrate tends to accumulate). On the whole, the results were negative. In particular, the antigenicity of the enzymes was increased by encapsulation in liposomes.

The first studies on the encapsulation of antitumoral drugs were also carried out by Gregoriadis *et al.* (34) and, during the next 10 years, about 25 other products were encapsulated in liposomes (78). The initial objective, based at least in part on an alleged increased endocytic activity in malignant cells, was to improve specificity. This hope was not fulfilled; but other interesting effects were often obtained, in particular a reduction in the toxicity and a slowing-down of the apparent rate of elimination.

The encapsulation of insulin was studied for the first time in 1976 by Dapergolas *et al.* (23). After certain authors claimed that encapsulated insulin was protected and could be given orally, many different liposome compositions were tested. Once again, the greater part of the results were negative. Today, the instability of conventional liposomes in digestive juices is accepted and only a few studies are now being carried out on their use in the oral administration of insulin (24).

During the 1970s and early 1980s a large number of other drugs were encapsulated in liposomes (78). Interesting, even spectacular, results were sometimes obtained. In this way, the ability of liposomes to promote intracellular penetration (chelating agents, antimonials, antibiotics, immunomodulators), to prolong a systemic effect (chelating agents, immunomodulators) or a local effect (anti-inflammatory steroids by intraarticular injection), and to reduce side-effects (antimony drugs, intraarticular anti-inflammatory steroids) has been unequivocably demonstrated.

CURRENT RESEARCH

The work which is now being carried out is certainly based on more solid foundations. It has benefited from not only the foregoing research on the possible therapeutic uses of liposomes but also from more fundamental work, especially that concerning the stability of the vesicles in the presence of biological fluids and their fate *in vivo* (passage across certain physiological barriers, distribution in the body, fate within the cell).

One of the most important studies must be that of Poste, published in 1983 (76). By showing that liposomes are incapable of penetrating the walls of most blood capillaries (the exception being those with a discontinuous endothelium), Poste shed new light on the research on the therapeutic applications of liposomes and all other colloidal drug carriers (nanocapsules, nanospheres).

Subsequent research has taken more account of anatomical, physiological and pathological information and is now, for the most part, based on the fate of liposomes *in vivo*. This is true for all types of liposomes, whether they be "simple", "targetable" or "targeted", and for all routes of administration (intravenous, interstitial, intracavity, oral and other routes).

1   SIMPLE LIPOSOMES

The liposomes referred to as "simple" in this article are all liposomes formed from one or more phospholipid species with the possible addition of a sterol, usually cholesterol, and one or more charged lipids.

1.1   Intravenous Route

The fate of simple liposomes *in vivo* after i.v. administration is characterized by a purely passive distribution depending only on their physico-chemical properties.

Many authors have studied this distribution including Eichler *et al.* (27), Fruhling *et al.* (32), Poste (76), Scherphof *et al.* (86), Senior *et al.* (90), Senior and Gregoriadis (91), Spanjer *et al.* (93), Vidal *et al.* (102) and Waser *et al.* (104).   The subject also has been extensively reviewed (4, 11, 14, 26, 37, 42, 60, 76).

Insofar as the therapeutic potential is concerned, three major factors have to be considered.

a)   The inability of simple liposomes to pass through the walls of most blood capillaries.

Blood capillaries may be classified into three groups according to the architecture of the endothelial lining and the underlying subendothelial basement membrane (basal lamina): discontinuous or so-called sinusoidal capillaries; fenestrated capillaries and continuous capillaries (76).

Discontinuous (sinusoidal) capillaries, found only in the liver, spleen and bone marrow, are thin-walled vessels.   Their endothelium has relatively large gaps which may be as wide as several hundred nanometers in diameter.   Moreover, in most species, the sinusoidal capillaries of the liver lack a basement membrane.

Fenestrated capillaries are present in most exocrine and endocrine glands, in the gastro-intestinal tract and in renal glomeruli and peritubular capillaries.   Their endothelium is interrupted by fenestrae (30-80 nm in diameter) but these do not represent simple openings but are spanned by a thin membranous diaphragm (4-6 nm). Moreover, the subendothelial basal lamina is continuous.

Continuous capillaries are the type which occur in most tissues and organs: skeletal, smooth and cardiac muscle, connective tissue, the central nervous system, exocrine pancreas, gonads and lung.   In these capillaries the endothelium is in the form of a continuous lining in which adjacent endothelial cells adhere *via* tight junctions.   In addition, continuous capillaries typically possess an uninterrupted subendothelial basement membrane.

In 1983 Poste studied the extravasation of liposomes of different sizes (MLV, SUV) from different kinds of capillaries.   All the results tended to indicate that small liposomes could probably penetrate the walls of vessels with discontinuous endothelia, but that they were unable to pass through continuous or even fenestrated endothelia.

This observation is particularly important because it forces us to consider tissues and organs as belonging to two main groups:

- those which are accessible to liposomes; i.e. those served by vessels with a discontinuous endothelium. These are the liver, spleen and bone marrow.

- those which are inaccessible to liposomes because they are served by vessels with fenestrated or discontinuous endothelia. This is the case for most of the major organ systems of the body.

b) The rapid uptake of liposomes by phagocytic reticuloendothelial (RE) cells.

This uptake takes place, after opsonization, by endocytosis; the most active cells are the macrophages lining the sinusoids of the liver (Kupffer cells) and the spleen. In addition to this uptake by fixed phagocytes of the RES, intravenously injected liposomes are also phagocytosed by circulating monocytes (76).

Many studies have aimed to reduce the uptake of circulating liposomes by the mononuclear phagocyte system. Some workers have attempted to do this by modifying the composition of the liposomal membranes (5, 36, 37, 89, 90), whereas others have tried a "blockade" of the phagocytic cells by pretreating with saturating doses of liposomes or other inert particles (71). Some interesting results have been obtained by coating the liposomes with polysaccharides (98). However, neither of the approaches described above has been totally successful.

c) The pronounced tendency of liposomes to concentrate encapsulated substances in the liver and spleen, and also to bring about an increase in the circulating concentration and half-life of these substances while reducing their apparent elimination (70, 107).

As an example, Figure 2 shows the retention, as found by Poste (76) of liposome-derived (^{14}C)-inulin in the liver of mice at different times after intravenous injection of various doses of multilamellar liposomes (PS : PC : Chol, 3/3/4) containing encapsulated (^{14}C)-inulin. Regardless of the dose injected and the time after administration, the radioactivity in the liver always represented about 80 % of the total radioactivity in the body at the given time-point.

The second illustration, Figure 3, shows the distribution of liposomes within the liver as observed by Storm (96). The smallest liposomes seem to be able to pass through the gaps in the sinusoids but larger liposomes are rapidly endocytosed by the Kupffer cells.

Figure 4 shows the different processes which occur as a result of the endocytosis of liposomes: inclusion inside a phagosome, fusion of the latter with a lysosome, digestion of the liposomal wall by lysosomal enzymes and passage of the active ingredient into the cytoplasm of the cell (44). The last process obviously depends on whether the active substance is sufficiently stable in the intra-lysosomal environment and whether it is able to pass through the phagosomal membrane.

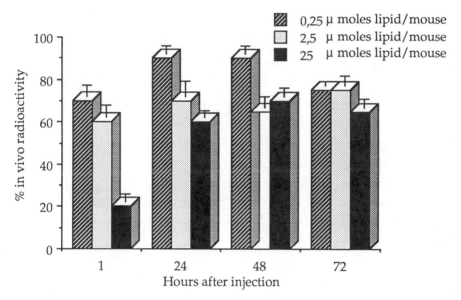

Figure 2.   Retention of liposomes derived [^{14}C]-inulin in the liver at different times after intravenous injection of different doses of MLV liposomes (PS : PC : Chol, 3/3/4) containing encapsulated [^{14}C]-inulin.   Liposomes were suspended in 0.2 mL PBS and injected into the tail vein of unanaesthetized C57 BL/6 mice: (After Poste; 76).   Reprinted with permission.

The major therapeutic possibilities for simple liposomes given by the intravenous route are summarized in Figure 5.   In accordance with the fate *in vivo* described above, these can be categorized into three main groups.

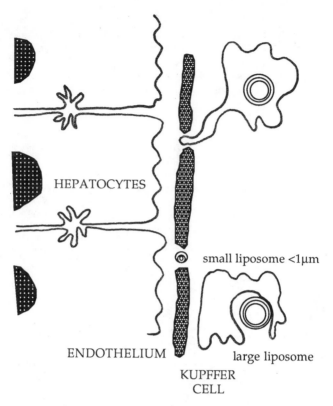

Figure 3.   Intrahepatic uptake of liposomes (After Storm; 96:p25).
Reprinted with permission.

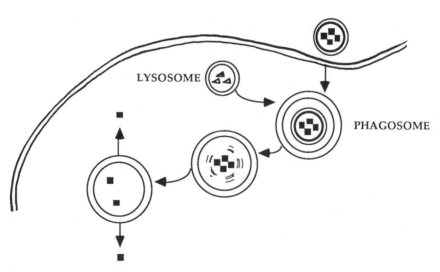

Figure 4.   Endocytosis of liposomes (After Juliano; 44).   Reprinted with
permission

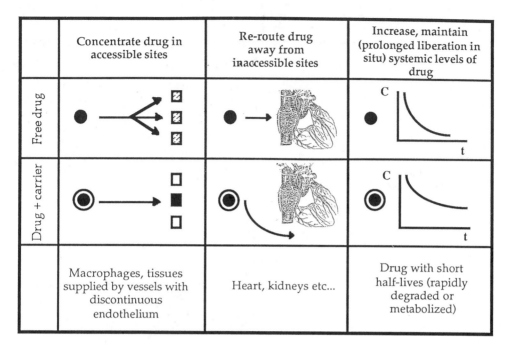

	Concentrate drug in accessible sites	Re-route drug away from inaccessible sites	Increase, maintain (prolonged liberation in situ) systemic levels of drug
Free drug			
Drug + carrier			
	Macrophages, tissues supplied by vessels with discontinuous endothelium	Heart, kidneys etc...	Drug with short half-lives (rapidly degraded or metabolized)

Figure 5. Possible therapeutic uses of liposomes by the intravenous route (After Puisieux et Roblot-Treupel; 80). Reprinted with permission.

Concentration (targeting) of drugs to sites accessible to liposomes (passive targeting). This is possible for:

- tissues supplied by vessels with a discontinuous endothelium, such as the liver parenchyma, where it might be interesting to target certain chemotherapeutic agents in cancer; and

- above all, macrophages, to which it would be interesting to deliver several types of substance: antiparasitic drugs in the case of intracellular parasitosis; antibacterial, antifungal and antiviral agents to treat intracellular infections; chelating agents in the event of intracellular accumulation of certain metals; immunomodulators in order to stimulate macrophages in their role in the non-specific defence of the organism.

As examples of the use of entrapped antiparasitic agents we should cite the work of Alving *et al.* (6), Black *et al.* (15) and New *et al.* (67), who have clearly demonstrated the advantages of the encapsulation of antimony derivatives for the treatment of leishmaniasis. The liposome-mediated delivery of these products permits their concentration in mononuclear phagocytes, where most of the parasites are found, and thus to increase their efficacy dramatically. Liposome-entrapped antimonials injected by the intracardiac route daily for four days had an activity 800 times higher than that of the free drug as judged by the concentration necessary to achieve 50 % inhibition of parasite growth in the liver.

A large number of articles, recently reviewed by Coune (21) and Svenson et al. (97), could be cited to illustrate the encapsulation of anti-infectious agents. From among these we have chosen the work of Bakker-Woudenberg et al. (7). They showed a definite improvement in the efficacy of ampicillin after entrapment in MLV (Chol : PC : PS, 5/4/1). In an experimental infection caused by L. monocytogenes, a 90-fold enhancement of the therapeutic activity of the antibiotic was found in mice. Substantial amounts of liposomal ampicillin were recovered from isolated Kupffer cells, the target cells for L. monocytogenes, after intravenous inoculation (7).

The advantages of encapsulating chelating agents withing liposomes have been extensively studied between 1973 and 1979, in particular by Rahman's group (83). An example is the work of Blank et al. (16), who investigated the efficacy of DTPA salts entrapped in liposomes (MLV, PC : Chol : DCP, 7/2/1) for removing colloidal Ytterbium-169 from rat tissues. The results obtained by a number of authors combine to demonstrate that encapsulation usually increases the efficacy of chelating agents by concentrating them at the level of the target cell, and by prolonging their action due to a reduction in their clearance from the plasma.

Initiated by Poste et al. (77), studies of the entrapment of immunomodulators have given rise to a great deal of research activity during the last few years.

Since 1979-1981, Poste, in collaboration with Fidler and his colleagues (30, 31) showed that systemic administration of MDP (muramyldipeptide, the minimum structure conserving adjuvant activity after chemical or enzymatic degradation of bacterial peptidoglycan), which activates macrophages to render them cytotoxic towards tumour cells, when encapsulated in liposomes, is highly effective in augmenting macrophage-mediated destruction of tumour cells in vitro and of established lung metastases in vivo.

For example, Table 2 clearly demonstrates the positive effect of encapsulation on the efficiency of MDP. Free MDP, at doses around 100 μg per mouse, has practically no effect on the development of lung-metastases in C57 BL/6 mice inoculated intravenously with cells from the B16-BL6 melanoma line. When encapsulated, MDP is effective after administration of doses 40 times lower.

MDP, a hydrosoluble compound, possesses several disadvantages: low encapsulation ratio and rapid leakage after encapsulation. For these reasons, workers have gradually orientated themselves towards the encapsulation of lipophilic derivatives: muramyltripeptide phosphatidyl-ethanolamine (MTP-PE; 29); 6-0-stearoyl MDP; 58); muramyldipeptide-glyceryldipalmitate (MDP-GDP; 74); MDP-L-alanylcholesterol (MTP-Chol; 75). In addition, over the past few years, several groups have investigated the activating effect of a lymphokine, recombinant human γ-interferon, both alone (51) and in combination with MDP (85) or MTP-PE (92).

One illustration of this is shown in Figure 6, which is taken from the work of Sone et al. (92), using MTP-PE. Their results clearly demonstrate both the

TABLE 2
TREATMENT OF SPONTANEOUS B16-BL6 METASTASES IN C57 BL/6
MICE BY INJECTION OF MLV CONTAINING MDP
(AFTER FIDLER *et al.*; 31)

| Treatment group | Mice with metastases no. total | Pulmonary metastases | | Mice without visible metastases % |
		Median no	Range no	
DB saline control mice	24/28	49	0-107	14
Free MDP (100 µg)	8/10	55	0-94	20
"Empty" MLV and free MDP (2.5 µg)	23/26	68	0-308	12
MLV-encapsulated MDP (2.5 µg)	7/27	0	0-7	74

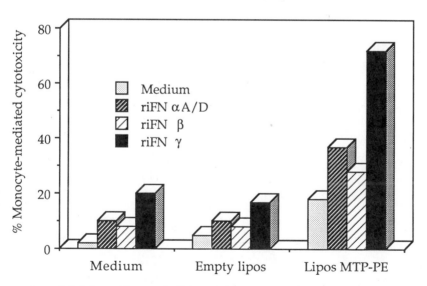

Figure 6.   Additive effects of riFN-αA/D and riFN-β with liposomes MTP-
PE in human monocyte activation (After Sone *et al.*; 92).
Reprinted with permission.

advantages of encapsulation in augmenting the activating properties of MTP-PE (MLV, PC : PS, 7/3) and the synergy which may be obtained with γ-interferon.

Liposomes containing biological response modifiers (BRM) are also highly effective in stimulating macrophage-mediated host resistance to microbial infections (76). Furthermore, evidence has been obtained demonstrating synergy in the action of liposome-encapsulated MDP or MDP derivatives and anti-infectious agents in the therapy of viral (51), fungal (28) and protozoan infections.

Re-routing of drugs away from sites particularly sensitive to their side-effects

This is possible for all sites supplied by vessels with continuous or fenestrated epithelia. This approach is very interesting, for example, in diverting from the heart and the kidneys certain drugs which are particularly toxic to these organs. Much work has been devoted to cisplatinum (73, 84, 94) and especially to doxorubicin (33, 40, 81, 82, 88, 101).

Examples are shown in Figure 7 and Table 3, which are taken from the work of Rahman *et al.*, (81) and Herman *et al.* (40), respectively. From these illustrations it is evident that the encapsulation of doxorubicin in positively charged cardiolipin liposomes leads to a reduction in the concentration in the heart, and at the same time decreased cardiotoxicity.

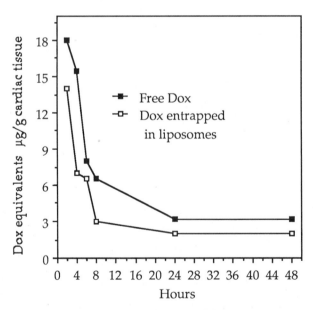

Figure 7.   Cardiac doxorubicin levels after mice received either free doxorubicin or doxorubicin entrapped in liposomes (After Rahman *et al.*; 81).   Reprinted with permission.

TABLE 3

COMPARISON OF THE INCIDENCE AND SEVERITY OF
CARDIOMYOPATHY IN BEAGLES TREATED CHRONICALLY WITH
EITHER FREE OR LIPOSOMAL-ENCAPSULATED DOXORUBICIN
(After Herman et al.; 40)[a]

Groups	Cardiomyopathy score					
	0	1	2	3	4	>2[b]
Free doxorubicin	0	0	1	1	3	5/5
Liposomal doxorubicin	5	0	0	0	0	0/5
Liposomes	2	0	0	0	0	0/2
0.9% NaCl solution	3	0	0	0	0	0/3

a   Dogs were given 7 i.v injections of free doxorubicin (1.75 mg/kg), liposomal doxorubicin (1.75 mg/kg), doxorubicin-free liposomes, or 0.9% NaCl solution at 3 weeks intervals and were sacrificed within 1 week after the last injection.

b   Numerator number of animals with a cardiomyopathy score of 2 or more; dénominator, the number of animals examined

Increased and maintained circulating concentrations of certain drugs

A reduction in the apparent rate of elimination as a result of entrapment within liposomes has been shown for a large number of drugs, particularly cancer chemotherapeutic agents (70, 107). As a dramatic example, we would cite work carried out in our own laboratory, in conjunction with that of J. Delattre (25) and that of P. Rossignol by Laham et al. (52,53) on the encapsulation of ATP.

Based on biochemical evidence, ATP has been suggested as a potential agonist in the treatment of some severe hypoergic cerebral syndromes. However, this hypothesis has never been substantiated experimentally in vivo, probably because of the poor passage of ATP through the blood brain barrier, due to its highly hydrophilic character. Furthermore, the lack of activity of ATP in brain resuscitation could also be explained by its rapid degradation in vivo.

Laham et al. (52) administered ATP intracerebroventricularly to rats submitted to episodes of cerebral ischaemia and lowering of systemic blood pressure by clamping the carotid arteries. They observed that ATP greatly increased the number of ischaemic episodes tolerated before brain electrical silence and death only when it was encapsulated in liposomes.

Later, Laham et al. (53) tested the same preparations by the intracarotid route and found a similar protective effect of liposome-entrapped ATP.

In order to explain these results Delattre et al. (25) studied plasma concentrations of ATP in the rat during ischaemic episodes after administration of either free ATP, or ATP encapsulated in reverse phase vesicles (REV, PC : Chol : sulphatide, 7/2/1). The results, in Figure 8, clearly show that the concentrations are significantly increased only when encapsulated ATP is given. The beneficial effect obtained during cerebral ischaemia could therefore be explained, at least in part, by prolonged liberation, *in situ,* of ATP from the liposomes.

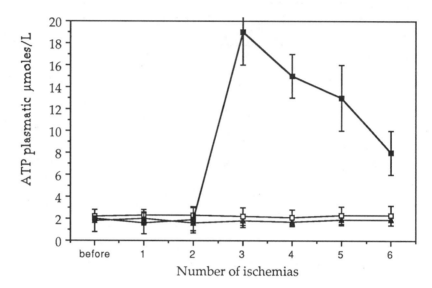

Figure 8.   Plasma concentration of ATP (μmoles/L) in rats subjected to ischaemic episodes (mean of 6 rats) ▲ non treated ischaemic control rats. ☐ ischaemic rats given an intracarotid injection of free ATP between the 2nd and the 3rd ischaemic episodes. ■ ischaemic rats given intracarotid injection of encapsulated ATP between the 2nd and the 3rd ischaemic episodes. (After Delattre *et al.*; 25). Reprinted with permission.

Although they are not so easily related to the fate of liposomes *in vivo*, some other interesting results obtained after i.v. administration deserve to be mentioned. One of the most promising concerns the use of liposomes in the treatment of systemic fungal infections with amphotericin (55).

Systemic fungal infections are most often seen in people whose resistance is depressed by disease or by immunosuppressant drugs. Indeed, these infections can cause disability and death in victims of acquired immune deficiency syndrome (AIDS) and in cancer patients undergoing chemotherapy. *Candida,spp* and *Aspergillus,spp* are the most common pathogens, and the treatment essentially consists of the parenteral administration of amphotericin B. Unfortunately, currently available preparations of amphotericin have significant toxicity. The systemic use of the anti-fungal is associated with severe acute reactions consisting mainly of fever and chills. Nephrotoxicity and central nervous system side-effects often result from its chronic use.

In mice, many studies have shown that liposomal amphotericin B cures systemic fungal infections more effectively than the free drug, largely because the entrapped drug is less toxic and higher doses can be given (45, 51, 72, 100). In human patients suffering from systemic fungal diseases more striking results have been obtained. Lopez-Berestein et al. (56) treated 12 patients with haematological malignancies complicated by fungal infections with liposomal amphotericin B. Nine patients were granulocytopenic and the three others were immunodepressed. All had previously failed to respond to therapy with conventional antifungal agents including amphotericin B (Fungizone). Doses of 0.8-1.0 mg/kg of liposomal amphotericin B (DMPC : DMPG 7/3) were administered intravenously every 24-72 h. Three patients had a complete remission and five a partial remission. Moreover, the side-effects were considerably reduced.

Sculier et al. (87) have also performed a pilot study with liposomal amphotericin B (PC : Chol : SA, 4/3/1) in cancer patients with fungal infections. Once again, liposomal amphotericin B was found to have a better therapeutic index than Fungizone® because of better tolerance and, possibly, increased antifungal activity which could be related to higher serum concentrations.

For example, Figure 9 summarizes some of the results of Juliano and Lopez-Berestein (45) obtained in mice. The curves clearly demonstrate the beneficial effect of encapsulation on the therapeutic index of amphotericin B.

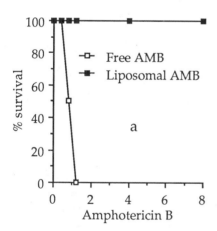 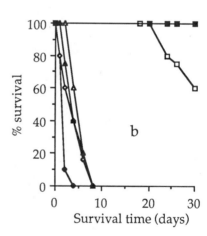

Figure 9.   Toxicity and therapeutic efficacy of liposomal amphotericin B
a) Toxicity: mice were injected i.v. with various doses of free AMB ❑ or liposomal AMB ■
b) Therapy: mice were infected with Candida albicans; 2 days after infection they were treated i.v. with: ♦ saline (control); ◊ empty liposomes; Δ free AMB 0.8 mg.kg^{-1}; ▲ liposomal AMB 0.8 mg.kg^{-1}; ❑ liposomal AMB 2.0 mg.kg^{-1}; ■ liposomal AMB 4.0 mg.kg^{-1}. (After Juliano and Lopez Berenstein; 45). Reprinted with permission.

<u>Interstitial (i.m., s.c.) and intracavity (i.p., intra-articular,</u>
       <u>intrapleural) routes</u>

The encapsulation of drugs within liposomes usually has two main consequences:

-    a slowing-down of their transfer to the circulation, corresponding
     to a slow release from the site of administration; and

-    an increase in their lymphatic drainage.

Figure 10, taken from Weinstein (108) illustrates the reasons for the increased lymphatic drainage.   The main one is the difference in structure between the blood capillaries and the lymphatic vessels.   Liposomes are largely prevented from entering blood capillaries by the basement membrane and the endothelial cell layer; instead they are transported into lymphatic capillaries which have large clefts between endothelial cells and little or no continuous basement membrane (108).

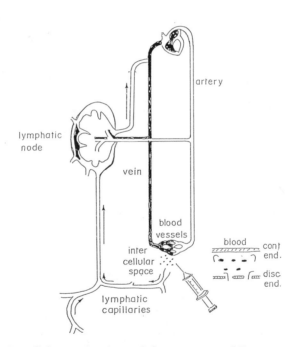

Figure 10.   Schematic view of the passage of liposomes from an
             interstitial injection site into the lymphatics
             (After Weinstein; 108).   Reprinted with permission.

The possible advantages of liposomes are summarized in Figure 11.   In accordance with the fate *in vivo,* they can be categorized under three main headings.

Increase and maintenance in circulating concentrations of certain drugs after interstitial administration (s.c., i.m.).

An example is the work of Stevenson and colleagues with insulin in diabetic dogs (70, 95).   Figure 12 shows that after subcutaneous administration the increase in blood insulin and hypoglycaemia were much more prolonged when insulin was entrapped in MLV (PC : Chol : DCP, 10/2/1).

Increased and maintained local concentration of certain drugs after intracavity administration (i.p., intrapleural, intra-articular).

The work of Kim et al. (47) may be cited as an example.   They administered liposomes containing Ara-C intraperitoneally with a view to treating certain cancers which tend to develop in this cavity, such as ovarian carcinoma and some mesotheliomas.   Liposomal entrapment prolonged the half-life of Ara-C in the peritoneal cavity 79-fold.   The use of empty liposomes, in conjunction with liposomal Ara-C, resulted in a 619-fold increase in the drug's peritoneal half-life, as compared to that of the free drug.

Lymphatic targeting of some drugs after interstitial or intracavity administration.

One example from among those described by Hirano and Hunt (41) is the work of Khato et al. (46) who injected liposomes containing melphalan subcutaneously into rats bearing lymphatic metastases.   By concentrating the drug in the lymphatics, liposomes were able to enhance the activity of the drug.

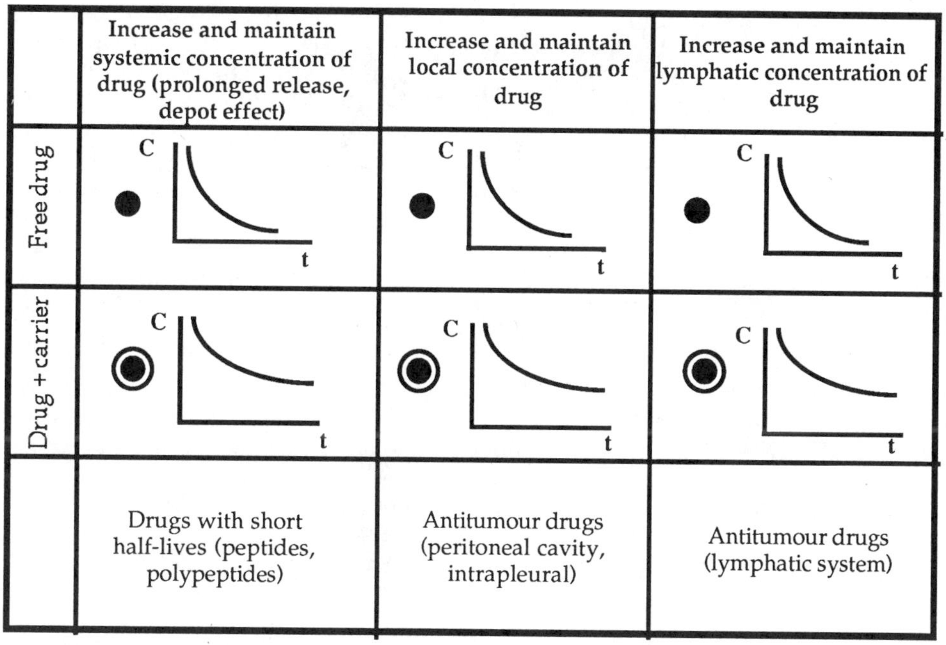

Figure 11.   Possible uses of liposomes by interstitial and intracavity routes
(After Puisieux and Roblot-Treupel; 80).
Reprinted with permission.

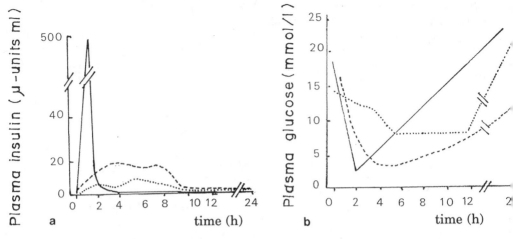

Figure 12. Effect of subcutaneous injection of 10 units of free and
liposomally entrapped insulin on plasma insulin (a) and
glucose (b) concentrations in diabetic dogs. ——— free insulin;
------ insulin entrapped in negatively charged MLV;
······· as ------ but treated with trypsin to remove insulin
associated with the surface of the liposomes (After Patel; 70).
Reprinted with permission.

1.3   The Oral Route

After the early successes obtained by the oral route with substances such as
insulin (23) and Factor VIII (39), the fate of liposomes after oral
administration *in vivo* has become the object of much attention (26, 48, 105).
Several groups have undertaken systematic studies, in the course of which
they have particularly focused on the effects of pH, bile salts, pancreatic
lipase and phospholipases.   There appear to be differences between various
liposome compositions but, one the whole, all authors have come to the
conclusion that liposomes are rapidly destroyed in digestive juices (64, 103).

Despite some recent results describing a favourable effect of liposomal
encapsulation on the intestinal absorption of a few drugs (24, 43),
conventional liposomes now seem to hold little potential for oral drug
administration.

This is illustrated in Figure 13, which presents some results obtained in our
laboratory with insulin.   When given intragastrically, insulin entrapped
within MLV (DPPC : Chol, 7/2) did not bring about any hypoglycaemia (106).

In this area, the results obtained with liposomes are in marked contrast to
those recently obtained by our group (22) using small polymeric vesicles,
nanocapsules (3).   To all appearances stable in the intestinal environment,
nanocapsules seem to be able:

-   to protect the digestive tract from the side-effects of certain drugs
    (non-steroidal anti-inflammatory agents); and

-   to protect some drugs during their passage through the digestive
    tract, to promote their absorption and to prolong their action to a
    greater or lesser extent (peptides, polypeptides, proteins).

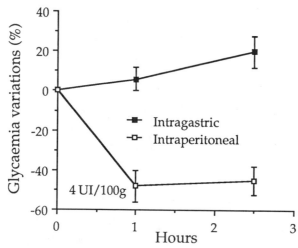

Figure 13.   Percentage change in glycaemia after administration of MLV
(DPPC:Chol, 7/2) containing insulin intragastrically (■) and
intraperitoneally □ to normal rats (4 UI/100g)
(After Weingarten *et al.*; 106).   Reprinted with permission.

1.4   Other Routes (topical application)

One way of targeting a drug to an organ is topical application.   Organs which
are accessible to this type of administration include the skin, eyes, lungs and
certain other body cavities.

It is not always certain, however, that a good therapeutic effect will be
obtained.   In some cases the drug applied to the surface may not penetrate
the organ and may not reach the site of action within that organ; in other
instances the drug may be absorbed into the blood and lymphatic systems and
cause a systemic rather than the expected local effect.

With this problem in mind, a number of studies have been undertaken in
order to assess the usefulness of liposomal encapsulation of drugs to be
applied topically (reviewed by Mezei, (61)).   Among the more recent, we
would mention, for the cutaneous route, those of Knepp *et al.* (50) and
Wohlrab and Lasch (109) and, for ophtalmic applications, those of Morimoto
et al., (63) and Taniguchi *et al.* (99).   In addition, Mihalko *et al.* (62) have
reviewed the use of liposomes by the pulmonary route.

The combined results indicate that liposomes applied topically can act, to use
the expression carried by Mezei (61), as excellent "localizers".   They seem to
be able:

-       to increase and maintain the local concentration of the drug and to
        prolong its action;

-       sometimes, to change the distribution of the drug (within the skin
        or eye) and thus permit access to alternative sites (2, 61); and

to reduce the passage of the drug into the circulation and thereby reduce its systemic side-effects.   The main reason for this effect is the inability of liposomes to penetrate the capillary endothelium, as mentioned above.

As an example, Figure 14 shows some results described by Patel (70), obtained after the administration of ^{3}H-methotrexate onto the skin of nude mice.   The retention of the entrapped drug in the skin was 2-3 times that of the free form.

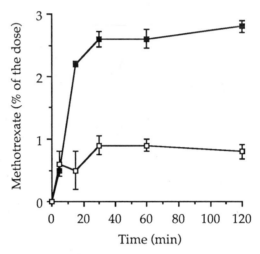

Figure 14.   Uptake of liposomally entrapped (■) and free (□) [^{3}H]-methotrexate by the skin of nude mice, after topical application (After Patel; 70).   Reprinted with permission.

Another application worthy of note is that liposomes given by the intratracheal route may be used to deliver phospholipids in the treatment of Respiratory Distress Syndrome.   Several laboratories, including our own (17) are working in this area and are trying to use liposomes as a way of administering, in aqueous form, to premature infants the phospholipids which are lacking in their alveoli.   In this case, the optimum liposome formulation is one that can be rapidly adsorbed at the liquid-air interface and can form a layer which reduces the surface tension sufficiently.

2   TARGETABLE LIPOSOMES

In this chapter "targetable" refers to liposomes which have been designed either to be guided to a particular target (magnetic liposomes) or to deliver their contents more or less specifically at a site raised to a particular temperature (temperature-sensitive liposomes) or having a specific pH (pH-sensitive liposomes).

Magnetic liposomes (15 mg Fe_3O_4 per μmol of lipid) containing an aqueous phase marker-(^{3}H)-inulin-have been prepared recently.   Experiments were carried out *in vivo* with rats bearing the Yoshida sarcoma implanted in the food pad. With the aid of an external magnet, a very small proportion of the liposomes, but significantly more than the control, was trapped in the tumour tissue (49).

The first studies of temperature-sensitive liposomes were carried out by Yatvin et al. (112). The principle was based on the fact that liposomes release their contents most rapidly at the phase transition temperature (T_c) of the chosen phospholipid mixture. The strategy is therefore:

a) to prepare liposomes with a T_c slightly above physiological temperatures (e.g. 42°C); and

b) to raise the target tissue (e.g. a tumour) to this temperature.

This approach has been tested, particularly with liposomes containing methotrexate and cis-platinum (10, 107, 108). As an illustration, Table 4 presents some results obtained by Yatvin et al. (111) with cis-platinum in the mouse sarcoma 180 model. The uptake of cis-platinum by tumours heated to the T_c was almost 4-fold greater when the drug was associated with liposomes than when it was administered in the free form.

The appeal of temperature-sensitive liposomes is that it is not necessary for them to leave the circulation in order to deliver drug to extravascular target; the major disadvantage is that the location of the tumour needs to be known. Hence this approach is not the answer to the major problem of cancer: widespread metastases.

The use of pH-sensitive liposomes as a drug delivery system was suggested by the observation that pathological tissues often have an ambient pH considerably lower than that of normal tissues (65). This is the case for primary tumours, metastases and sites of inflammation and infection, which all seem to have a reduced pH (< 7) in the local environment.

The general strategy used in the construction of pH-sensitive liposomes has been to incorporate lipids containing pH-sensitive groups into liposomal membranes. The first lipid used was N-palmitoyl-homocysteine (110), with which the mechanism of release might be the formation of a thiolactone ring at acidic pH (20).

TABLE 4
ORGAN RADIOACTIVITY 4 HOURS POST-INJECTION. ORGAN RADIOACTIVITY IS PRESENTED AS PERCENTAGE OF TOTAL INJECTED DOSE (After Yatvin et al., 111)

	Free PDD	Unilamellar liposomes	Unilamellar liposomes +42°C	Multilamellar liposomes
Tumor	2.4±0.5[a]	4.4±0.7	8.2±0.5	2.1±0.6
Liver	9.9±2.0	17.8±0.7	19.0±3.0	55.4±0.6
Spleen	0.5±0.2	0.8±0.2	0.8±0.2	2.4±0.4
Kidney	3.6±0.4	4.1±0.2	4.4±0.4	1.3±0.3

a: mean±SD (n=6)

Table 5 illustrates the results obtained by Nayar and colleagues (65, 66) concerning the activation, *in vitro*, of macrophages for cytotoxicity against tumour cells by liposomes containing MDP and recombinant γ-interferon. No activation was obtained with MLV made from the pH-sensitive phospholipid SOPE (DN-succinyldioleoylphosphatidylethanolamine), which release their internal contents preferentially at pH 7.4. The effective MLV were those made from DOPE : SOPE (7/3) which release their contents at pH 4. According to the authors, pure SOPE MLV were probably ineffective because the immunomodulators leaked from the liposomes (at the neutral pH of the media) before they were internalized by the macrophages.

TABLE 5

IN VITRO ACTIVATION OF MACROPHAGES TO THE TUMORICIDAL
STATE BY LIPOSOMES CONTAINING rIFN-γ AND MDP
(After Nayar *et al.*; 65).

Treatment of macrophages	Radioactivity in viable cells on day 3
None, BL16-tumor cells alone	2869±130
Untreated (medium)	2117±166
Free IFN-γ (25U)	254± 46 (88%)[a]
SOPE liposomes	
MLV containing HBSS	2017±154
MLV containing IFN-γ (6U) and MDP (0.2 ng)	2506±143
DOPE/SOPE liposomes (7 : 3)	
MLV containing HBSS	1622±104
MLV containing IFN-γ (6U) and MDP (0.2 ng)	773±224 (63%)[a]

a: percentage cytotoxicity

3. TARGETED LIPOSOMES

The targeted liposomes described in this section are, like all "third-generation" carriers, systems linked to a ligand which is capable of recognizing the target cell. This distinguishes them from the "targetable" liposomes discussed in the previous section, where the targeting agent is external and not directly associated with the particles.

Like all "third-generation" drug delivery systems, targeted liposomes were conceived to avoid excessive uptake by phagocytic cells of the RES, to widen the range of sites accessible to the liposomes and to further increase the specificity of drug delivery to the target cells.

Many ligands are being studied at the moment (9, 107, 108). The class which has attracted the most attention are monoclonal antibodies. The methods now being used to attach them to liposomes can be categorized into three groups (19).

- methods based on the introduction of a hydrophobic group into the antibody, followed by the incorporation of this derivative into the phospholipid bilayer of the liposome;

- methods in which a covalent link is formed between the antibody and a phospholipid (usually phosphatidylethanolamine) by means of a bifunctionnal spacer molecule; and

- a related method developed by Leserman *et al.* (54) in which staphylococcal protein A is covalently linked to the liposomes as above. The liposomes are then targeted by the addition of target-specific antibody which binds to the protein A through its Fc portion. This method has the advantage that a single liposome preparation can be used for different targets by changing the associated specific antibody.

The fate of targeted liposomes *in vivo* has not been studied as extensively as for simple liposomes, but it is certainly very similar.

By the intravenous route, targeted liposomes, like simple liposomes seem to be:

- incapable of crossing the walls of most blood vessels; and

- rapidly taken up by the cells of the RES.

As far as the interstitial and intracavity routes are concerned, as for simple liposomes, targeted liposomes appear to promote the lymphatic drainage of the substances which they carry.

These considerations place obvious contraints on the therapeutic potential of targeted liposomes. These liposomes could be used to concentrate a drug within the targets recognized by the attached ligand only if this target is accessible to the carrier after administration by the chosen route.

At the moment, most work has been concerned with cancer therapy, in which susceptible target cells would be intravascular (leukemia) intracavity (intraperitoneal, intrapleural cancers) or intralymphatic (metastases). Other possible targets might be T lymphocytes (113), which could be usefully destroyed when an immunosuppression is required (organ transplants and maybe some viral or autoimmune diseases) (114). Furthermore, targeted liposomes could be used to stimulate certain subpopulations of defensive cells more specifically than with simple liposomes and for targeting to vascular endothelial cells (76).

One example (Figure 15) is the results of Onuma *et al.* (68) concerning the cytotoxicity of adriamycin toward bovine leukemia cells *in vitro*. Increased cytotoxicity is clearly obtained when the anti-tumour drug is entrapped in liposomes conjugated with monoclonal antibody against a specific tumour-associated antigen.

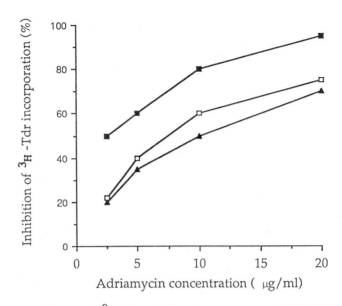

Figure 15.   Inhibition of [³H]-thymidine incorporation of BLSC-KU 1 cells by C 14 3 -conjugated liposomes containing ADM.   Cells were treated with various concentrations of ADM in C 14 3 conjugated liposomes (■), ADM in normal mouse IgG-conjugated liposomes (□) and ADM in unconjugated liposomes (▲) (After Onuma *et al.*; 68).   Reprinted with permission.

A second illustration is given in Figure 16, taken from Collins *et al.* (18) and shows the cytotoxicity of Ara-C towards L929 cells *in vitro*.   The curves compare the effects of free Ara-C, and Ara-C encapsulated within antibody-targeted liposomes (anti-H2K) either pH-insensitive (DOPC), or pH-sensitive (DOPE : OA 8/2).   In this example, the results are doubly interesting, since they show:

-   the increased toxicity due to the encapsulation of Ara-C in conventional immunoliposomes (with respect to the free drug); and

-   the particular advantage of pH-sensitive immunoliposomes.   This seems to be related to the instability of Ara-C in the lysosomal environment. pH-sensitive immunoliposomes seem to preserve its activity by releasing their contents in an acidified prelysosomal compartment such as the endosome.

Despite the promising nature of the results described above, the possible applications of targeted liposomes, and especially immunoliposomes, are limited by a number of problems.   Over and above their inability to penetrate the capillary endothelium and their rapid uptake by the RES one could also

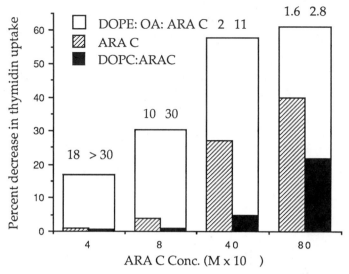

Figure 16.   Cytotoxicity of immunoliposomes-encapsulated Ara-C.   L 929
cells were incubated with pH-sensitive immunoliposomes
(DOPE : OA : Ara-C), pH insensitive immunoliposomes
(DOPC : Ara-C) both containing Ara-C or free drug (Ara-C)
(After Collins *et al.*; 8).   Reprinted with permission.

mention: the random nature of their uptake after binding to target cells (8);
the antigenic heterogeneity of tumour cells; antibody-induced antigenic
modulation of antigen-positive cells; the presence of cross-reactive isotopes on
normal cells, and possible adverse immune reactions resulting from the
formation of anti-idiotypic antibodies against the administered antibodies
(76).

CONCLUSION

The concept of the "magic bullet", first formulated by Ehrlich in 1906,
specifically directed towards the therapeutic target, has inspired intensive
research during the last twenty years.   The object of the initial studies,
liposomes, have remained the drug delivery system which has attracted the
most attention.   Early research often met with failure because certain
anatomical, pathological and physiological considerations were not
sufficiently taken into account.   Nevertheless, the work continued, in a
remarkabe multidisciplinary effort and the possibilities for liposomes, as well
as for other colloidal carriers (nanocapsules, nanospheres), gradually came
to be defined more precisely.   In addition, new approaches have been
developed, including some which have not yet been mentioned in this
chapter: preparation of polymerizable liposomes more stable than ordinary
liposomes; coating of liposomes to render them more stable and maybe to
alter their distribution; the concept of "carrier-dependent drugs", which

should lead to more critical consideration of the nature and properties of the drugs to be delivered in this way and more recently the concept of "stealth" liposomes.

Other promising applications of liposomes and other colloidal vectors can be found in transfusion (artificial red cells) and in a number of areas more or less related to therapeutics: pharmacotechnology, where liposomes might help to resolve some technical problems, medical imaging and diagnosis, radioprotection, genetic engineering and biotechnology, vaccination, cosmetology and bioreagents in various fields of Industry.

ACKNOWLEDGEMENTS

The authors would like to thank Sophie Parussolo for rapid and efficient preparation of the camera-ready copy and Claude Puisieux for preparing the figures.

REFERENCES

1   ADINOLFI L.E., BONVENTRE P.F., VAN DER PAS M. and EPPSTEIN D.A., 1985, Synergistic effect of glucanthine and a liposome-encapsulated muramyl dipeptide analog in therapy of experimental visceral leishmaniasis, Infect. Immun., 48, 409-416.

2   AHMED I. and PATTON T.P., 1986, Selective intraocular delivery of liposome encapsulated inulin via the non-corneal absorption route, Inter. J. Pharm., 14, 163-167.

3   AL KHOURI FALLOUH N., ROBLOT-TREUPEL L., FESSI H., DEVISSAGUET J.P. and PUISIEUX F., 1986, Development of a new process for the manufacture of polyisobutylcyanoacrylate nanocapsules, Int. J. Pharm., 28 : 125-132.

4   ALLEN T.M., 1988, Interaction of liposomes and other drug carriers with the mononuclear phagocyte system, in "Liposomes as Drug Carriers. Recent Trends and Progress", G. Gregoriadis, Ed., J. Wiley and Sons, Chichester, New York, Brisbane, Toronto, Singapore, 37-50.

5   ALLEN T. and CHONN A., 1987, Large unilamellar liposomes with low uptake into the reticuloendothelial system, FEBS Lett., 223 : 42-46.

6   ALVING C.R., STECK E.A., CHAPMAN W.L., WAITS V.B., HENDRICKS L.D., SWARTZ G.M. and HANDSON W.L., 1978, Therapy of leishmaniasis, superior efficacies of liposome-encapsulated drugs, Proc. Natl. Acad. Sci., USA, 75 : 2959-2963.

7   BAKKER-WOUDENBERG I.A.J.M., LOKERSE A.F., VINK VAN DEN BERG J.C. and ROERDINK F.H., 1988, Liposome-encapsulated ampicillin against Listeria monocytogenes in vivo and in vitro, Infection, 16, Suppl. 2 : S165-S174.

8   BARBET J., MACHY P. and LESERMAN L.D., 1985, Pilotage des liposomes, in "Les liposomes, Applications Thérapeutiques", F. Puisieux et J. Delattre, Eds., Lavoisier Tec. et Doc., Paris, 189-204.

9   BARRATT G., TENU J.P., YAPO A. and PETIT J.P., 1986, Preparation and characterization of liposomes containing mannosylated phospholipids capable of targeting drugs to macrophages, Biochim. Biophys. Acta, 862 : 153-164.

10 BASSETT J.B., TACKER J.R., ANDERSON R.U. and BOSTWICK D., 1988, Treatment of experimental bladder cancer with hyperthermia and phase transition liposomes containing methotrexate, J. Urol., 139 : 634-636.

11 BENITA S. et DEVISSAGUET J.P., 1985, Pharmacocinétique des liposomes et des principes actifs encapsulés dans les liposomes, in "Les Liposomes, Applications Thérapeutiques", F. Puisieux and J. Delattre, Eds., Lavoisier, Tec. et Doc., Paris,173-188.

12 BENOIT J.P., 1985, Microencapsulation et chimio-embolisation, in "Formes Pharmaceutiques Nouvelles. Aspects Technologique, Biopharmaceutique et Médical", P. Buri, F. Puisieux, E. Doelker and J.P. Benoit, Eds., Lavoisier, Tec. et Doc., Paris-613-656.

13 BENOIT J.P. and F. PUISIEUX, 1986, Microcapsules and microspheres for embolization and chemoembolization, in "Polymeric Nanoparticles and Microspheres", P. Guiot and P. Couvreur, Eds., CRC Press, Boca Raton, 137-174.

14 BIENVENUE A. and PHILIPPOT J., 1985, Interaction des liposomes avec les cellules, in "Les Liposomes, Applications Thérapeutiques", F. Puisieux and J. Delattre, Eds., Lavoisier, Tec. et Doc., Paris,147-171.

15 BLACK C.D.V., WATSON G.J. and WARD R.J., 1977, The use of pentostam liposomes in the chemotherapy of experimental leishmaniasis, Trans. R. Soc. Trop. Med. Hyg., 71 : 550-554.

16 BLANK M.L., BYRD B.L., CRESS T.A., WASBURN L.C. and SYNDER F., 1984, Liposomal preparations of calcium or zinc-DTPA have a high efficacy for removing colloidal ytterbium-169 from rat tissues, Toxicology, 30 : 275-282.

17 BONTE F., TAUPIN C. and PUISIEUX F., 1986, Liposomes de surfactant artificiel et transition de phase, Rev. Mal. Resp., 3 : 129-132.

18 COLLINS D., NORLEY S., ROUSE B. and HUANG L., 1988, Liposomes as carriers for antitumor and antiviral drugs : pH sensitive immunoliposomes and sustained release immunoliposomes, in "Liposomes as Drug Carriers. Recent Trends and Progress", G. Gregoriadis, Ed., J. Wiley and Sons, Chichester, New York, Brisbane, Toronto, Singapore, 761-770.

19 CONNOR J., SULLIVAN S. and HUANG L., 1985, Monoclonal antibodies and liposomes, Pharmac. Ther., 28 : 341-365.

20 CONNOR J., YATVIN M.B. and HUANG L., 1984, pH-sensitive liposomes : acid induced liposome fusion, Proc. Natl. Acad. Sci., USA, 81 : 1715-1718.

21 COUNE A., 1988, Liposomes as drug delivery systems in the treatment of infectious diseases. Potential applications and clinical experience, Infection, 16 : 141-147.

22 DAMGE C., MICHEL C., APRAHAMIAN M. and COUVREUR P., 1988, New approach for oral administration of insulin with polyalkylcyanoacrylate nanocapsules as drug carrier, Diabetes, 37 : 246-251.

23 DAPERGOLAS G., NEERUNJUN E.D. and GREGORIADIS G., 1976, Penetration of target areas in the rat by liposome-associated bleomycin, glucose-oxidase and insulin, FEBS Lett., 63 : 235-240.

24 DAS N., BASU M.K. and DAS M.K., 1988, Oral application of insulin encapsulated liposomes, Biochem. Inter., 16 : 983-989.

25 DELATTRE J., LAHAM A., CLAPERON N., CHAPAT S., COUVREUR P., PUISIEUX F. and ROSSIGNOL P., 1989, Nouvelles perspectives offertes par l'administration de liposomes d'adénosine-triphosphate (ATP) dans l'ischémie cérébrale expérimentale, STP Pharma., 5, 92-98.

26   DELATTRE J. et VASSON M.P., 1985, Stabilité des liposomes dans les liquides biologiques. Effets de certains constituants, in "Les Liposomes, Applications Thérapeutiques", F. Puisieux and J. Delattre, Eds., Lavoisier, Tec. et Doc., Paris115-145.

27   EICHLER H.G., SENIOR J., STADLER A., GASIC S., PFUNDNER P. and GREGORIADIS G., 1988, Kinetics and disposition of fluorescein-labelled liposomes in healthy human subjects, Eur. J. Clin. Pharmacol., 34 : 475-479.

28   EPPSTEIN D.A., VAN DER PAS M.A., FRASER-SMITH E.B., KURAHARA C.G., FELGNER P.L., MATTHEWS T.R., WATERS R.V., VENUTI M.C., JONES G.H., MEHTA R. and LOPEZ-BERESTEIN G., 1986, Liposome-encapsulated muramyldipeptide analogue enhances non-specific host immunity, Int. J. Immunother., II, (2) : 115-126.

29   FIDLER I.J., 1986, Optimization and limitations of systemic treatment of murine melanoma metastases with liposomes containing muramyl tripeptide phosphatidylethanolamine, Cancer Immunol. Immunother., 21 : 169-173.

30   FIDLER I.J., RAZ A., FOGLER W.E., KIRSH R., BUGELSKI P. and POSTE G., 1980, Design of liposomes to improve delivery of macrophage-augmenting agents to alveolar macrophages, Cancer Res., 40 : 4460-4464.

31   FIDLER I.J., SONE S., FOGLER W.E. and BARNES Z.L., 1981, Eradication of spontaneous metastases and activation of alveolar macrophages by intravenous injection of liposomes containing muramyldipeptide, Proc. Natl. Acad. Sci., USA, 78 : 1680-1681.

32   FRUHLING J., COUNE A., GHANEM G., SCULIER J.P., VERBIS A., BRASSINE C., LADURON C. and HILDEBRAND J., 1984, Distribution in man of ^{111}In-labelled liposomes containing a water-insolubles antimitotic agent, Nucl. Med. Commun, 5 : 205-208.

33   GABIZON A.A. and BARENHOLZ Y., 1988, Adriamycin containing liposomes in cancer chemotherapy, in "Liposomes as Drug Carriers. Recent Trends and Progress", G. Gregoriadis, Ed., J. Wiley and Sons, Chichester, New York, Brisbane, Toronto, Singapore, 365-379.

34   GREGORIADIS G., 1973, Drug entrapment in liposomes, FEBS Lett., 36: 292-296.

35   GREGORIADIS G., 1978, Liposomes in the therapy of lysosomal storage diseases, Nature, 275 : 695-696.

36   GREGORIADIS G., 1988, Liposomes as Drug Carriers. Recent Trends and Progress, J. Willey and Sons, Chichester, New York, Brisbane, Toronto, Singapore.

37   GREGORIADIS G., 1988, Fate of injected liposomes : observations on entrapped solute retention, vesicle clearance and tissue distribution in vivo, in "Liposomes as Drug Carriers. Recent Trends and Progress", G. Gregoriadis, Ed., J. Wiley and Sons, Chichester, New York, Brisbane, Toronto, Singapore, 3-18.

38   GREGORIADIS G., LEATHWOOD P.D. and RYMAN B.E., 1971, Enzyme entrapment in liposomes, FEBS Lett., 14 : 95-99.

39   HEMKER H.C., MULLER A.D., HERMENS W.T. and ZWAAL R.F.A., 1980, Oral treatment of haemophilia A by gastrointestinal absorption of factor VIII entrapped in liposomes, Lancet, 12 : 70-71.

40   HERMAN E.H., RAHMAN A., FERRANS V.J., VICK J.A. and SCHEIN P.S., 1983, Prevention of chronic doxorubicin cardio-toxicity in beagles by liposomial encapsulation, Cancer Res., 43 : 5427-5432.

41 HIRANO K. and HUNT A., 1985, Lymphatic transport of liposome-encapsulated agents : effects of liposome size following intraperitoneal administration, J. Pharm. Sci., 74, 915-921.

42 HWANG K.J. and BEAUMIER P.L., 1988, Disposition of liposomes *in vivo*, in "Liposomes as Drug Carriers. Recent Trends and Progress", G. Gregoriadis, Ed., Wiley and Sons, Chichester, New York, Brisbane, Toronto, Singapore, 19-35.

43 JASKIEROWICZ D., GENISSEL F., ROMAN V., BERLEUR F. and FATOME M., 1985, Oral administration of liposome-entrapped cysteamine and the distribution pattern in blood, liver and spleen, Int. J. Radiat. Biol., 47: (6), 615-619.

44 JULIANO R.L., 1981, Liposomes as a drug delivery system, Trends Pharmacol., Sci. Feb., 39-41.

45 JULIANO R.L. and LOPEZ-BERESTEIN G., 1985, New lives for old drugs : liposomal drug delivery systems reduce the toxicity but not the potency of certain chemotherapeutic agents, Pharm. Inter., July : 164-167.

46 KHATO J., DEL CAMPO A.A. and SIEBER S.M., 1983, Carrier activity of sonicated small liposomes containing melphalan to regional lymph nodes of rats, Pharmacol., 26 : 230-240.

47 KIM S., KIM D.J. and HOWELL S.B., 1987, Modulation of the peritoneal clearance of liposomal cytosine arabinoside by blank liposomes, Cancer Chemother. Pharmacol., 19, 307-310.

48 KIMURA T., 1988, Transmucosal passage of liposomal drugs, in "Liposomes as Drug Carriers. Recent Trends and Progress", G. Gregoriadis, Ed., J. Wiley and Sons, Chichester, New York, Brisbane, Toronto, Singapore, 635-647.

49 KIWADA H., SATO J., YAMADA S. and KATO Y., 1986, Feasibility of magnetic liposomes as a targeting device for drugs, Chem. Pharm. Bull., 34 : 4253-4258.

50 KNEPP V.M., HINZ R.S., SZOKA F.C. and GUY R.H., 1988, Controlled drug release from a novel liposomal delivery system. I. Investigation of transdermal potential, J. Control. Release, 5 : 211-221.

51 KOFF W.C. and FIDLER I.J., 1984, Human monocytes activated by immunomodulators in liposomes lyses Herpes-virus-infected but not normal cells, Science, 224 : 1007-1009.

52 LAHAM A., CLAPERON N., DURUSSEL J.J., FATTAL E., DELATTRE J., PUISIEUX F. COUVREUR P. and ROSSIGNOL P., 1987, Liposomally entrapped ATP : improved efficiency against experimental brain ischemia in the rat, Life Sci., 40 : 2011-2016.

53 LAHAM A., CLAPERON N., DURUSSEL J.J., FATTAL E., DELATTRE J., PUISIEUX F., COUVREUR P. and ROSSIGNOL P., 1988, Intracarotidal administration of liposomally-entrapped ATP : improved efficiency against experimental brain ischemia, Pharmacol. Research Comm., 20 : 699-705.

54 LESERMAN L.D., BARBET J., KOURILSKY F. and WEINSTEIN J., 1980, Targeting to cells of fluorescent liposomes covalently coupled with monoclonal antibody or protein A, Nature, 288 : 602-604.

55 LOPEZ-BERESTEIN G., 1988, Liposomal amphotericin B in antimicrobial therapy, in "Liposomes as Drug Carriers. Recent Trends and Progress", G. Gregoriadis, Ed., J. Wiley and Sons, Chichester, New York, Brisbane, Toronto, Singapore, 345-352.

56 LOPEZ-BERESTEIN G., FAINSTAIN V., HOPFER R., MEHTA K., SULLIVAN M.P., KEATING M., ROSENBLUM M.G., MEHTA R., LUNA M., HERSH E.M., REUBEN J., JULIANO R.L. and BODEY G.P., 1985, Liposomal Amphotericin B for the treatment of systemic fungal infections in patients with cancer : a preliminary study, J. Infect. Dis., 151 : 704-710.

57 LOPEZ-BERESTEIN G., HOFFER R.L., MEHTA R., MEHTA K., HERSH E.M. and JULIANO R.L., 1984, Prophylaxis of *Candida albicans* infection in neutropenic mice with liposome-encapsulated Amphotericin B, Antimicrob. Agents. Chemother., 25 : 366-367.

58 LOPEZ-BERESTEIN G., MEHTA K., MEHTA R., JULIANO R.L. and HERSH E.M., 1983, The activation of human monocytes by liposome-encapsulated muramyldipeptide analogues, J. Immunol., 130 : 1500-1502.

59 MACHY P. and LESERMAN L., 1987, Les liposomes en Biologie Cellulaire et Pharmacologie. Les Editions Inserm, J. Libbey Eurotext Ltd., Paris.

60 MARGOLIS L.B., 1988, Cell interactions with solid and fluid liposomes *in vitro* : lessons for "liposomologists" and cell biologists, in "Liposomes as Drug Carriers. Recent Trends and Progress", G. Gregoriadis, Ed., J. Wiley and Sons, Chichester, New York, Brisbane, Toronto, Singapore, 75-92.

61 MEZEI M., 1988, Liposomes in the topical application od drugs : a review, in "Liposomes as Drug Carriers. Recent Trends and Progress", G. Gregoriadis, Ed., J. Wiley and Sons, Chichester, New York, Brisbane, Toronto, Singapore, 663-677.

62 MIHALKO P.J., SCHREIER H. and ABRA R.M., 1988, Liposomes : a pulmonary perspective, in "Liposomes as Drug Carriers. Recent Trends and Progress", G. Gregoriadis, Ed., J. Wiley and Sons, Chichester, New York, Brisbane, Toronto, Singapore, 679-694.

63 MORIMOTO K., NAKAMURA T., NAKAI T. and MORISAKA K., 1988, Effects of liposomes on permeability of various compounds through rabbit corneas, Arch. Int. Pharmacodyn., 293 : 7-13.

64 NAGATA M., YOTSUYANAGI T. and IKEDA K., 1988, A two-step model of disintegration kinetics of liposomes in bile salts, Chem. Pharm. Bull., 36 : 1508-1513.

65 NAYAR R., FIDLER I.J. and SCHROIT A.J., 1988, Potential applications of pH-sensitive liposomes as drug delivery systems, In "Liposomes as Drug Carriers. Recent Trends and Progress", G. Gregoriadis, Ed., J. Wiley and Sons, Chichester, New York, Brisbane, Toronto, Singapore, 771-782.

66 NAYAR R. and SCHROIT A.J., 1985, Generation of pH-sensitive liposomes. Use of large unilamellar vesicles containing N-succinyldioleoylphosphatidylethanolamine, Biochemistry, 24 : 5967-5971.

67 NEW R.R.C., CHANCE M.L., THOMAS S.C. and PETERS W., 1978, Antileishmanial activity of antimonials entrapped in liposomes, Nature, 272 : 556-561.

68 ONUMA M., ODAWARA T., WATARAI S., AIDA Y. OCHIAI K., SYUTO B., MATSUMOTO K., YASUDA T., FUJIMOTO Y., IZAWA H. and KAWAKAMI Y., 1986, Antitumor effect of adriamycin entrapped in liposomes conjugated with monoclonal antibody against tumor-associated antigen of bovine leukemia cells, Cancer Res. (Gann), 77, 1161-1167.

69 OSTRO M.J., 1987, Liposomes from Biophysics to Therapeutics, M. Dekker Inc., New York, Basel.

70   PATEL H.M., 1985, Liposomes as a controlled-release system, <u>Biochem. Soc. Trans.</u>, 13 : 513-516.

71   PATEL K.R., LI M.P. and BALDESCHWIELER J.D., 1983, Suppression of liver uptake of liposomes by dextran sulfate 500, <u>Proc. Natl. Acad. Sci. USA</u>, 80 : 6518-6522.

72   PAYNE N.I., COSGROVE R.F., GREEN A.P. and LIU L., 1987, *In vivo* studies of Amphotericin B liposomes derived from protoliposomes : effect of formulation on toxicity and tissue disposition of the drug in mice, <u>J. Pharm. Pharmacol.</u>, 39 : 24-28.

73   PEREZ-SOLER R., KHOKHAR A.R. and LOPEZ-BERESTEIN G., 1988, Development of lipophilic cisplatin analogs encapsulated in liposomes, in "Liposomes as Drug Carriers. Recent Trends and Progress", G. Gregoriadis, Ed., J. Wiley and Sons, Chichester, New York, Brisbane, Toronto, Singapore, 401-417.

74   PHILLIPS N.C., MORAS M.L., CHEDID L., LEFRANCIER P. and BERNARD J.M., 1985, Activation of alveolar macrophage tumoricidal activity and eradication of experimental metastases by freeze-dried liposomes containing a new lipophilic muramyldipeptide derivative, <u>Cancer Res.</u>, 45 : 128-134.

75   PHILLIPS N.C., MORAS M.L., CHEDID L., PETIT J.F., TENU J.P., LEDERER E., BERNARD J.M. and LEFRANCIER P., 1985, Activation of macrophage cytostatic and cytotoxic activity *in vitro* by liposomes containing a new lipophilic muramyl peptide derivative, MDP-L-Alanyl-Cholesterol (MTP-CHOL), <u>J. Biol. Response Mod.</u>, 4 : 464-474.

76   POSTE G., 1983, Liposome Targeting *in vivo* : problems and opportunities, <u>Biol. Cell.</u>, 47 : 19-38.

77   POSTE G., KIRSH R., FOGLER W. and FIDLER I.J., 1979, Activation of tumoricidal properties in mouse macrophages by lymphokines encapsulated in liposomes, <u>Cancer Res.</u>, 39 : 881-892.

78   PUISIEUX F., 1985, Les liposomes et leur emploi en thérapeutique. Bilan des recherches réalisées, in "Les Liposomes. Applications Thérapeutiques", F. Puisieux and J. Delattre, Eds., Lavoisier, Tec. et Doc., Paris, 205-253.

79   PUISIEUX F. and DELATTRE J., 1985, Les Liposomes. Applications Thérapeutiques, Lavoisier, Tec. et Doc., Paris.

80   PUISIEUX F. et ROBLOT-TREUPEL L., 1989, Vectorisation et vecteurs de médicaments, 1989, <u>STP Pharma</u>, 1989, 5, 107-113.

81   RAHMAN A., GANJEI A. and NEEFE J.R., 1986, Comparative immunotoxicity of free doxorubicin and doxorobucin encapsulated in cardiolipin lipsomes, <u>Cancer Chemother. Pharmacol.</u>, 16 : 28-34.

82   RAHMAN A. and SCHEIN P.S., 1988, Use of liposomes in cancer chemotherapy, in "Liposomes as Drug Carriers. Recent Trends and Progress", G. Gregoriadis, Ed. J. Wiley and Sons, Chichester, New York, Brisbane, Toronto, Singapore, 381-400.

83   RAHMAN Y.E. and WRIGHT B.J., 1975, Liposomes containing chelating agents. Cellular penetration and a possible mechanism of metal removal, <u>J. Cell. Biol.</u>, 65 : 112-117.

84   RESZKA R., FICHTNER I., NISSEN E., ARNOT D. and LADHOFF A.M., 1987, Preparation, characterization, therapeutic efficacy and toxicity of liposomes containing the antitumour drug cis-dichlorodiamine-platinum (II), <u>J. Microencaps.</u>, 4 : 201-212.

85 SAIKI I., SONE S., FOGLER W.E., KLEINERMAN E.S., LOPEZ-BERESTEIN G. and FIDLER I.J., 1985, Synergism between human recombinant γ-interferon and muramyldipeptide encapsulated in liposomes for activation of antitumor properties in human blood monocytes, Cancer Res., 46 : 6188-6193.

86 SCHERPHOF G., VAN LEEUVEN B., WILSCHUT J. and DAMEN J., 1983, Exchange of phosphatidylcholine between small unilamellar liposomes and human plasma high-density lipoprotein involves exclusively the phospholipid in the outer monolayer of the liposomal membrane, Biochim. Biophys. Acta, 732 : 595-599.

87 SCULIER J.P., COUNE A., MEUNIER F., BRASSINNE C., LADURON C., HOLLAERT C., COLLETTE N., HEYMANS C. and KLASTERSKY J., 1988, Pilot study of Amphotericin B entrapped in sonicated liposomes in cancer patients with fungal infections, Eur. J. Cancer Clin. Oncol., 24: 527-528.

88 SELLS R.A., GILMORE I.T., OWEN R.R., NEW R.R.C. and STRINGER R.E., 1987, Reduction in doxorubicin toxicity following liposomal delivery, Cancer Treat. Rev., 14 : 383-387.

89 SENIOR J., 1987, Fate and behaviour of liposomes in vivo : a review of controlling factors. CRC Crit. Rev. Therapeutic Drug Carrier Systems, 3: 123-125.

90 SENIOR J., CRAWLEY J.C.W. and GREGORIADIS G., 1985, Tissue distribution of liposomes exhibiting long half-lives in the circulation after intravenous injection, Biochim. Biophys. Acta, 839 : 1-8.

91 SENIOR J. and GREGORIADIS G., 1984, Role of lipoproteins in stability and clearance of liposomes administered to mice, Biochem. Soc. Trans., 12 : 339-340.

92 SONE S., TANDON P., UTSUGI T., OGAWARA M., SHIMIZU E., NH A. and OGURA T., 1986, Synergism of recombinant human interferon gamma with liposome-encapsulated muramyltripeptide in activation of the tumoricidal properties of human monocytes, Int. J. Cancer, 38 : 495-500.

93 SPANJER H.H., VAN GALEN M., ROERDINK F.H., REGTS J. and SCHERPHOF G.L., 1986, Intrahepatic distribution of small unilamellar liposomes as a function of liposomal lipid composition, Biochim. Biophys. Acta, 863 : 224-230.

94 STEERENBERG P.A., STORM G., DE GROOT G., CLAESSEN A., BERGERS J.J., FRANKEN M.A.M., VAN HOESEL Q.G.C.M., WUBS K.L. and DE JONG W.H., 1988, Liposomes as drug carrier system for cis-diamminedichloroplatinum (II). II. Antitumor activity in vivo, induction of drug resistance, nephrotoxicity and Pt distribution, Cancer Chemother. Pharmacol., 21 : 299-307.

95 STEVENSEN R.W., PATEL H.M., PARSONS J.A. and RYMAN B., 1982, Prolonged hypoglycemic effect in diabetic dogs due to subcutaneous administration of insulin in liposomes, Diabetes, 31 : 506-511.

96 STORM G., 1987, Liposomes as delivery system for doxorubicin in cancer chemotherapy. Thesis Rijksuniversity of UTRECHT.

97 SVENSON C.E., POPESCU M.C. and GINSBERG R.S., 1988, Preparation and use of liposomes in the treatment of microbial infections, Crit. Rev. Microbiol., 15 : S1-S31.

98 TAKADA M., YUZURIHA K., KATAYAMA K., IWAMOTO K. and SUNAMOTO J., 1984, Increased lung uptake of liposomes coated with polysaccharides, Biochim. Biophys. Acta, 802 : 237-244.

99   TANIGUCHI K., ITAKURA K., YAMAZAWA N., MORISAKI K., HAYASHI S. and YAMADA Y., 1988, Efficacy of a liposome preparation of antiinflammatory steroid as an ocular drug-delivery system, J. Pharmacol. Dyn., 11 : 39-46.

100 TREMBLEY C., BARZA M., FIORE C. and SZOKA F., 1984, Efficacy of liposome-intercalated Amphotericin B in the treatment of systemic candidiasis in mice, Antimicrob. Agents and Chemother., 26 : 170-173.

101 VAN HOESEL Q.G.C.M., STEERENBERG P.A., CROMMELIN D.J.A., VAN DIJK A., VAN OORT W., KLEIN S., DOUZE J.M.C., DE WILDT D.J. and HILLEN F.C., Reduced cardiotoxicity and nephrotoxicity with preservation of antitumor activity of doxorubicin entrapped in stable liposomes in the Lou/M WS1 rat, Cancer Res., 44 : 3698-3705.

102 VIDAL M., BIENVENUE A. SAINTE-MARIE J. and PHILIPPOT J., 1984, The influence of the internal content of negatively charged liposomes on their interaction with high-density lipoprotein, Eur. J. Biochem., 138 : 399-405.

103 WALDE P., SUNAMOTO J. and O'CONNOR C.J., 1987, The mechanism of liposomal damage by taurocholate, Biochim. Biophys. Acta, 905 : 30-38.

104 WASER P.G., MULLER U., KREUTER J., BERGER S., MUNZ K., KAISER E. and PFLUGER B., 1987, Localization of colloidal particles (liposomes, hexylcyanoacrylate nanoparticles and albumin nanoparticles) by histology and autoradiography in mice, Inter. J. Pharm., 39 : 213-227.

105 WEINER N. and CHIANG C.M., 1988, Gastrointestinal uptake of liposomes, in "Liposomes as Drug Carriers. Recent Trends and Progress", G. Gregoriadis, Ed., J. Wiley and Sons, Chichester, New York, Brisbane, Toronto, Singapore, 599-607.

106 WEINGARTEN C., MOUFTI A., DESJEUX J.F., LUONG T.T., DURAND G., DEVISSAGUET J.P. and PUISIEUX F., 1981, Oral ingestion of insulin liposomes : effects of the administration route, Life Sci., 28 : 2747-2752.

107 WEINSTEIN J.N., 1984, Liposomes as drug carriers in cancer therapy, Cancer Treat. Rep., 68 : 127-135.

108 WEINSTEIN J.N., 1987, Liposomes in the diagnosis and treatment of cancer, in "Liposomes from Biophysics to Therapeutics", M.J. Ostro, Ed., M. Dekker, Inc., New York, Basel, 277-338.

109 WOHLRAB W. and LASCH J., 1987, Penetration kinetics of liposomal hydrocortisone in human skin, Dermatologica, 174 : 18-22.

110 YATVIN M.B., KREUTZ W., HORWITZ B.A. and SHINITZKY M., 1980, pH-sensitive liposomes : possible clinical implications, Science, 210 : 1253-1255.

111 YATVIN M.B., MUHLENSIEPEN H., PORSCHEN W., WEINSTEIN J.N. and FEINENDEKEN L.E., 1981, Selective delivery of liposome-associated cis-dichlorodiammineplatinum (II) by heat and its influence on tumor drug uptake and growth, Cancer Res., 41 : 1602-1607.

112 YATVIN M.B., WEINSTEIN J.N., DENNIS W.H. and BLUMENTHAL R., 1978, Design of liposomes for enhanced local release of drugs by hyperthermia, Science, 202 : 1290-1293.

113 YEMUL S.S., BERGER C., ESTABROOK A., SUAREZ S., EDELSON R. and BAYLEY H., 1987, Selective killing of T lymphocytes by phototoxic liposomes, Proc. Natl. Acad. Sci., USA, 84 : 246-250

114 ZAGURI D., 1980, Les cellules tueuses, La Recherche, 114 : 928-937.

THE EFFECT OF PHOSPHOLIPIDS AND LIPOTROPIC FACTORS ON XENOBIOTIC TOXICITY

G. Caderni

Department of Preclinical and Clinical Pharmacology
University of Florence
Viale G.B. Morgagni, 65
Florence, ITALY

INTRODUCTION

Although fat is an essential component of the diet, epidemiological and experimental studies indicate that a high intake of fat increases the risk of developing colon cancer (2,34,25). To explain this correlation it has been suggested that dietary fat damages the intestinal mucosa leading to an increased proliferation of the colonic cells (11,33). Hyperproliferation has been associated with an increased risk of developing colon cancer both in experimental animals and humans (20).

The phospholipid phosphatidylcholine and the lipotropic nutrients choline and methionine have important effects on the transport of dietary fats in the body. We thought it of interest, therefore, to study in mice whether these nutrients could affect the toxicity of dietary fat in the colon.

In this paper we summarize briefly the role of phosphatidylcholine and choline in the transport of dietary fat from the intestine and in the mobilization of fat from the liver. We also describe some experiments on the effect of phosphatidylcholine, choline and methionine on the hyperproliferation induced by dietary fat in the intestinal mucosa of C57Bl/6J mice.

Role of phosphatidylcholine and choline in the transport and mobilization of dietary fat

Dietary fat consists mainly of triglycerides (TG), although phospholipids and cholesterol are also present in the diet (31). In the lumen of the small intestine, dietary lipids are partially hydrolyzed and absorbed by the intestinal mucosa cells. In the enterocytes the hydrolytic products of fats are then resynthetized into TGs, combined with cholesterol, phospholipids and apoproteins to form intestinal lipoproteins, chilomicrons and very low density lipoproteins (VLDL) that are then released into the lymphatic channels (29).

Several studies have demonstrated that the absorption and transport of dietary fat into the lymph require phosphatidylcholine, or its biosynthetic precursor choline in the intestinal lumen (30,31,24,29). Luminal phosphatidylcholine in fact provides both phospholipids for the maintenance and functioning of enterocyte membrane, and the phospholipid coat of the newly formed intestinal lipoproteins (31). Increased phosphatidylcholine synthesis has been demonstrated in rat jejunum during fat absorption (27), and it has also been shown that luminal phosphatidylcholine or choline stimulate mucosal protein biosynthesis (24).

Other studies have pointed out the role played by phosphatidylcholine in the transport of dietary fat from the enterocytes into the lymph (31,4,29,22). It has been reported that in bile-diverted rats infused with triolein via an intraduodenal tube, the lymphatic output of triglycerides was greatly enhanced when the infusate was supplemented with phosphatidylcholine (30). An impairment in fat transportation through the lymph has also been reported in choline-deficient rats (29), and it has been estimated that over 70% of the phospholipid coat of chilomicrons is composed of phosphatidyl choline (30).

Besides their role in the transport of dietary fat from enterocytes, phosphatidylcholine and choline also have important effects on the mobilization of dietary fat from the liver (lipotropic action). The lipotropic effect of phosphatidylcholine and choline has been known since the 1930s when it was shown that these nutrients prevent the deposition of fat observed in the liver of normal rats fed a low-choline, high-fat diet (5,6).

Choline, besides being a precursor of the phospholipids phosphatidylcholine and sphingomyelin, also serves as a methyl donor in transmethylation reactions and in the synthesis of the neurotransmitter acetylcholine. Choline-deficiency produces, in fact, various pathological lesions in different organs of the body in the experimental animal (1, 35). With regard to the lipotropic action of choline, several studies have suggested that the accumulation of fat in the liver observed during choline-deficiency is mediated by a diminished synthesis of phosphatidylcholine (19). This decreased formation of phosphatidylcholine leads to alterations in the membranes of hepatocytes and to an impairment of the synthesis and secretion of lipoproteins in the plasma, thus affecting the transport of hepatic TGs from the liver (21,23). Phosphatidylcholine is synthetized mainly from choline (in the form of cytidine diphosphate-choline), and 1,2- diacylglycerol (18); a second pathway for the synthesis of phosphatidylcholine involves the sequential methylation of phosphatidylethanolamine with methyl groups given by S-adenosylmethionine (15). Therefore the metabolism of choline and methionine are strictly correlated, and methionine, together with other nutrients involved in transmethylation reactions (such as Vitamin B12 and folic acid), affect the lipotropic action of choline (19,1).

Toxicity of dietary fat to the intestinal mucosa; correlation with the promotion of colonic carcinogenesis by dietary fat

Epidemiological and experimental studies indicate that diet is an important etiological factor in the development of colon cancer (2,25,34). In particular epidemiological data from different populations suggest that the incidence of colon cancer is positively correlated with the intake of dietary fat (2,34). Studies in the experimental animal also support a role of dietary fat in colon cancer development indicating that high-fat diets act as a promoter of chemically induced carcinogenesis (10,25).

To explain the correlation between fat consumption and colon cancer it has been suggested, on the basis of experimental studies, that fat and/or bile acids damage the intestinal mucosa leading to a compensatory increase in cell proliferation (11,33). Increased proliferative activity in the colonic mucosa is associated in fact with an high risk of developing colon cancer in humans (20); moreover, studies in experimental animals suggest that the proliferative rate of the colonic mucosa varies the sensitivity of the colon to chemically induced carcinogenesis (3,16).

Several studies have investigated the effect of dietary lipids on colonic proliferation (7,32). It has been demonstrated that corn oil or beef tallow given as a bolus to mice cause a loss of surface epithelial cells in the colon and a compensatory increase in proliferative activity (7). Similar effects of oral boluses of fat have also been observed in the human colon (28). Free fatty acids and bile acids that increase in the intestinal lumen during fat consumption, when given intrarectally to mice, damage the intestinal mucosa and increase the proliferative activity of the colon (32,33).

In contrast with the marked increase in cell proliferation induced by a bolus of corn oil, the effect of dietary lipids when mixed in a balanced diet and fed chronically to mice is much more limited (9,17,12). This phenomenon suggested to us that some dietary component may interfere with the toxic effect of fat and, given and the role played by phosphatidylcholine, choline and methionine on the absorption and transport of dietary fat in the intestine, we decided to study whether hyperproliferation caused by dietary fat in mice might be affected, besides other dietary components, by phosphatidylcholine, and the lipotropic nutrients choline and methionine.

RESULTS AND DISCUSSION

The first indication that dietary components might counteract the burst of proliferative activity induced by a bolus of corn oil in colonic mucosa came from experiments in which C57Bl/6J female mice were gavaged with 0.3 ml of corn oil or with an aqueous slurry of the high fat diet described in Table I which contained the same amount of oil.

169

TABLE I

COMPOSITION OF THE HIGH-FAT DIET USED IN THE BOLUS EXPERIMENTS*.

Corn Oil	Casein	Sucrose	Starch	Cellulose	Mineral Mix #	Vitamin Mix @	CH	MT
30.4	25	12.7	19	6	4.4	1.37	0.3	0.4

* g/100 g diet, based on AIN-76 Diet (26). #AIN mineral mix. @AIN Vitamin Mix. CH=Choline bitartrate. MT=DL-Methionine.

The proliferative activity was measured by colchicine arrest 18 hrs after the treatment, and by counting mitotic figures in histological sections of the colon as described previously (14). The results indicated that while the bolus of corn oil alone produced a peak of proliferative activity , the high fat diet containing the same amount of corn oil did not (Table II).

TABLE II

PROLIFERATIVE ACTIVITY OF MICE COLONIC MUCOSA AFTER TREATMENT WITH ORAL BOLUSES OF DIFFERENT COMPOSITION.

Bolus	Mitotic figures/crypt#
H_2O	0.87 +/- 0.17
Corn oil	2.13 +/- 0.30 **
High-fat diet	0.79 +/- 0.06

\# Values are means +/- SE. n=10 (number of mice/group).
** $p < 0.01$ compared with the H_2O and diet treated animals.

The experimental diet contained the macronutrients casein, sucrose, starch and cellulose as well as vitamins, minerals, choline and methionine. To find out whether a particular component was responsible for the protective effect of the diet, we treated mice with a bolus of corn oil (0.4 ml) mixed with different macronutrients and minerals at the relative proportions in which they were present in the 30% fat diet (Table I), and cell proliferation was measured as before. The results of these experiments indicated (Table III), that while sugar and casein did not affect the proliferative activity induced by corn oil, the starch, cellulose and minerals were able to lower colonic proliferation.

TABLE III

PROLIFERATIVE ACTIVITY IN COLONS OF MICE AFTER TREATMENT WITH ORAL BOLUSES OF DIFFERENT COMPOSITION.

	Mitotic figures/crypt @
H_2O	0.33 +/- 0.07
Corn oil	1.30 +/- 0.24
Corn oil + Sucrose	1.45 +/-0.25
Corn oil + Casein	1.26 +/-0.31
Corn oil + Starch	0.91 +/-0.23
Corn oil + Cellulose	0.55 +/- 0.10 *
Corn oil + Mineral Mix	0.63 +/- 0.16 *

@ Values are means +/- SE n=20.* $p < 0.05$ as compared with the corn oil group. Modified from Caderni et al.(14).

Given the importance of phosphatidylcholine in the absorption of dietary fat in the intestinal lumen, we investigated whether this phospholipid, and its biosynthetic precursors in the diet,choline and methione, might also affect the proliferative response induced in the colon by corn oil. Choline and methionine are incorporated in semi-synthetic diets to provide, among other functions, sufficient amounts of phosphatidylcholine. Mice were treated with a bolus of corn oil mixed with methionine or choline in their relative proportions of the high-fat diet in Table I, and the proliferative activity measured as described before. The results indicate that both methionine and choline did not affect significantly the hyperproliferation caused by corn oil (Table IV).

TABLE IV

PROLIFERATIVE ACTIVITY IN COLON OF MICE AFTER TREATMENT WITH ORAL BOLUSES OF DIFFERENT COMPOSITION.

	Mitotic figures/ crypt @
H_2O	0.24 +/- 0.06
Corn oil	1.36 +/- 0.27
Corn oil + Methionine	1.58 +/- 0.44
Corn oil + Choline	1.91 +/- 0.41

@ Values are means +/- SE. n=10.

Similarly, when different amounts of phosphatidyl-choline were administerd together with a bolus of corn oil, the proliferative activity observed was not significantly affected (Table V).

TABLE V

PROLIFERATIVE ACTIVITY IN MICE COLONIC MUCOSA AFTER TREATMENT WITH ORAL BOLUSES WITH DIFFERENT COMPOSITION.

	Mitotic figures/crypt @
Corn oil	1.81 +/- 0.44
Corn oil + 0.5% Phosphatidylcholine	1.18 +/- 0.23
Corn oil + 1% Phosphatidylcholine	2.55 +/- 0.41
Corn oil + 10% Phosphatidylcholine	1.65 +/- 0.39

@ Values are means +/- SE. n=10.

Since the response of the colonic epithelium to dietary components could be different in an acute exposure (such as with the bolus treatments) compared with chronic exposure, we decided to test, in chronic feeding studies, the significance of the results obtained with the bolus treatments. Therefore we devised an experimental model in which mice were fed for a week with high fat-diets lacking different components (see Table VI for the composition of these diets). At the end of this period mice were sacrified and the proliferative activity measured as described before.

TABLE VI

COMPOSITION OF THE DIETS USED FOR THE CHRONIC FEEDING EXPERIMENTS@.

	Oil*	Cas	Suc	Vit	Sta	Cel	Min	CH	MT
Diet A	30	25	43	1.4	-	-	-	-	-
Diet A+Min, CH and MT	30	25	38	1.4	-	-	4.4	0.3	0.4
Diet A+Sta and Cel	30	25	18	1.4	19	6	-	-	-
Complete diet	30	25	13	1.4	19	6	4.4	0.3	0.4

@g/100 g of diet. * Symbols are: Oil=Corn oil, Cas=caseine, Suc=sucrose, Vit=AIN Vitamin Mix, Sta=starch, Cel=cellulose, Min=AIN Mineral Mix modified to provide 0.1% calcium in the diet (8), CH=choline bitartrate, MT=DL-methionine.

The results of these experiments (Table VII) showed that the "deficient" diet A caused a significant increase in cell proliferation when compared to a complete diet. However, when the "deficient" diet A was supplemented with starch and cellulose or with minerals, choline and methionine we observed a significant decrease in cell proliferation.

TABLE VII

PROLIFERATIVE ACTIVITY OF MICE COLONIC MUCOSA FED FOR 1 WEEK WITH THE FOLLOWING DIETS.

	Mitotic figures/crypt @	
Diet A	1.27+/-0.12	n=48
Diet A + min, choline and methionine	0.92+/-0.14 *	n=34
Diet A + starch and cellulose	1.00+/-0.09 *	n=27
Complete diet	0.60+/-0.06 **	n=46

@ Values are means +/- SE. n= number of animals. * $p<0.05$ as compared to diet A, **$p<0.01$ as compared to diet A.

In order to clarify the relative role of minerals, choline and methionine in lowering the proliferative activity caused by the "deficient" diet A, mice were fed for a week with Diet A supplemented with Mineral mix or with methionine and choline. The results of this experiment showed that both minerals and the lipotropic nutrients choline and methionine were able to lower the proliferative activity of the "deficient" diet A (Table VIII).

TABLE VIII

PROLIFERATIVE ACTIVITY OF COLON MUCOSA OF MICE FED FOR 1 WEEK WITH DIFFERENT DIETS.

	Mitotic figures/crypt
Diet A	1.65+/-0.35
Diet A + Mineral Mix	0.66+/-0.10 **
Diet A + choline and methionine	0.75+/-0.11 **
Complete diet	0.64+/-0.14 **

* * $p<0.01$ as compared to diet A. Values are means +/- SE.

In conclusion, we have studied the effect of different nutrients on fat toxicity. The protective effect of starch, cellulose and minerals has been discussed elsewhere and will be further investigated by us (13, 14). The experiments on phosphatidylcholine and its precursors in the diet have indicated that phosphatidylcholine, choline and methionine do not affect corn oil toxicity when these components are administerd together with a bolus of corn oil. However, when diets deficient in choline and methionine and high in fat were supplemented with methionine and choline, we observed a normalization of cell proliferation, which suggests a protective effect of these nutrients on colonic mucosa.

Diets lacking choline and methionine, or low in these components, produce a series of alterations in animals (19). In particular, a diminished synthesis of phosphatidylcholine has been reported during choline-deficiency (21). Although we have no data on phospholipid biosynthesis in the mice fed the "deficient" diet, or on chronic feeding with "deficient" diets supplemented with phosphatidylcholine, it is possible to speculate that the protective effect observed with the supplementation of choline and methionine, might be due to an adequate synthesis of phosphatidylcholine. This would lead to an increased absorption of fat in the small intestine and to a diminution of the amount of fat reaching the colon.It is also possible that variations in the supply of phosphatidyl-choline might affect the enterocyte membranes. As mentioned in the introduction, decreased fat absorption in the small intestine and alterations in the structure of enterocyte cell membranes have been described in rats fed choline-deficient diets (29).

It should be noted, however, that neither choline and methionine, nor phosphatidylcholine, at least at the dosages used in our experiments, affect the hyperproliferation induced by a bolus of corn oil in the colonic mucosa. The differing effects of phosphatidylcholine, choline and methionine on the toxicity of dietary lipids suggest that the mechanism underlying this phenomenon is complex and needs further clarification.

ACKNOWLEDGEMENTS: This work was supported in part by a Grant from "Regione Toscana". The author thanks Dr. W. R. Bruce, Director of the Ludwig Institute for Cancer Research in Toronto, Canada, where part of the the experiments were performed.

REFERENCES

1. Appel, J.,A., and Briggs, G.,M., 1980, Choline, in: "Modern nutrition in health and disease", Goodhart, R.S., and Shils M.S., eds., Lea and Febiger, Philadelphia, pp282-286.
2. Armstrong,B., and Doll,R.,1975, Environmental factors and cancer incidence and mortality in different countries, with special reference to dietary practices,J. Cancer 15, 617-631.
3. Barthold, S.,W., and Beck, D.,1980, Modification of early dimethylhydrazine carcinogenesis by colonic mucosa hyperplasia, Cancer Res., 40, 4451-4455.
4. Bennet Clark, S., 1978, Chylomicron composition during duodenal triglycerides and lecithin infusion, Am. J. Physiol. 235, E183-E190.

5. Best, C.,H.,Hershey, J.,M., and Huntsman, M, E, 1932, The effect of lecithine on fat deposition in the liver of normal rat, J. Physiol., 75, 56-66.

6. Best,C., H., and Huntsman, M., L., 1932, The effect of the components of lecithine upon deposition of fat in the liver, J. Physiol., 75, 405-412.

7. Bird, R., P., Medline, A., Furrer, R., and Bruce W., R., 1985, Toxicity of orally administerd fat to the colonic epithelium of mice, Carcinogenesis 6, 1063-1066.

8. Bird, R., P., Schneider, R., Stamp, D., and Bruce W., R., 1986, Effect of dietary calcium and cholic acid on the proliferative indices of murine colonic epithelium, Carcinogenesis 7, 1657-1661.

9. Bird, R.P., and Stamp, D., 1986, Effect of high fat diet on the proliferative indices of murine colonic epithelium, Cancer Letters, 31, 61-67.

10. Bull, A.,W., Soullier, B.,K., Wilson, P.,S., Van Hayden, M., and Nigro, N.,D., 1979, Promotion of azoxymethane-induced intestinal cancer by high-fat diet in rats, Cancer Res., 39, 4956-4959.

11. Bull, A., W., Marnett, L., S., Dame, E., J., and Nigro, N.,D.,1983, Stimulation of deoxythymidine incorporation in the colon of rats treated intrarectally with bile acids and fats, Carcinogenesis, 4, 207-210.

12. Caderni, G., 1988, Dietary components affect the proliferative activity of colonic epithelium in mice, in: "Nutritional and toxicological aspects of food processing", R. Walker, and E. Quattrucci eds. Taylor and Francis London, New York and Philadelphia, pp293-298.

13. Caderni,G., Stuart,E.W., and Bruce W.,R., 1988, Dietary factors affecting the proliferation of epithelial cells in the mouse colon, Nutr. Cancer, 11, 147- 153.

14. Caderni, G., Bianchini, F., Dolara, P., and Kriebel, D., Proliferative activity in the colon of the mouse and its modulation by dietary starch, fat, and cellulose, Cancer Res., 49, 1655-1659.

15. Chan, M., 1984, Choline and carnitine, in: "Handbook of Vitamins", Nutritional, Biochemical and Clinical Aspects, L., J., Machlin Ed. Marcel Dekker Inc., New York and Basel, pp 549-570.

16. Deshner, E., E., Long, F., C., Hakissian, M., and Hermann, S., L., 1983, Differential susceptibility of AKR, C57Bl/6J, and CF1 mice to 1,2-dimethylhydrazine-induced colonic tumor formation predicted by proliferative characteristics of colonic epithelial cells, J.N.C.I., 70, 279- 282.

17. Jacobs, L., R., 1983, Effect of short-term dietary fat on cell growth in rat gastrointestinal mucosa and pancreas, Am. J. Clin. Nutr., 37, 361-367.

18. Kennedy, E., P., and Wiess, S., B., 1956, The function of coenzymes in the biosynthesis of phospholipids, J. Biol. Chem., 222, 193-214.

19. Kuksis, A., and Mookerja, S., 1984, Choline, in: "Present knowledge in nutrition", Nutrition Foundation, Inc., New York, NY, pp 383-399.

20. Lipkin, M., 1988, Biomarkers of increased susceptibility to gastrointestinal cancer: new application to studies of cancer prevention in human subjects, Cancer Res.,48, 235-245.

21. Lombardi, B., 1971, Effects of choline deficiency on rat epatocytes, Fed., Proc. 30, 139-142.
22. Mansbach II, C., M., and Arnold, A., 1986, Steady-state kinetic analysis of triacylglycerol delivery into mesenteric lymph, Am. J. Physiol., 251, G263-G269.
23. Mookerjea, S., 1971, Action of choline in lipoprotein metabolism, Fed. Proc., 30, 143-150.
24. O'Doherty, P., J., A., Kakis, G., and Kuksis, A., 1973, Role of luminal lecithin in intestinal fat absorption, Lipids 8, 249-255.
25. Reddy, B., S., 1983, Tumor promotion in carcinogenesis, in:"Mechanisms of tumor promotion", Slaga, T., S., ed., CRC Press, Boca Raton FL, pp 107-109.
26. Report of the American Institute of Nutrition. Ad hoc Committee on Standard nutritional values", 1977, J. Nutr. , 107, 1340-1348.
27. Shaikh, N., A., and Kuksis, A., 1982, Further evidence for enhanced phospholipid synthesis by rat jejunal villus cells during fat absorption, Can. J. Biochem. Cell. Biol., 61, 370-377.
28. Stadler, J., Stern, H., S., Yeung, K., S., Mc Guire, V., Furrer, R., Marcon, N., and Bruce, W., R., 1988, Effect of high fat consumption on cell proliferation activity of human colorectal mucosa and soluble fecal bile acids, Gut, 29, 1326-1331.
29. Takahashi, Y., and Mizunuma, T., 1984, Cytochemistry of fat absorption, Int. Rev. Cytol., 89, 115-136.
30. Tso, P., Kendrick, H., Balint, J., A., and Simmonds, W., J., 1981, Role of biliary phosphatidylcholine in the absorption and transport of dietary triolein in the rat, Gastroenterology, 80, 60-65.
31. Tso, P., 1985, Gastrointestinal digestion and absorption of lipid, Adv. Lipid Res., 21, 143-186.
32. Wargovich, M., J., Eng V., W., S., Newmark, H., L., and Bruce R., W., 1983, Calcium ameliorates the toxic effect of deoxycholic acid on colonic epithelium, Carcinogenesis, 4, 1205-1207.
33. Wargovich, M., J., Eng, V., W., S., and Newmark, H., L., 1984, Calcium inhibits the damaging and compensatory proliferative effects of fatty acids on mouse colon epithelium, Cancer Lett., 23, 253-258.
34. Willet, W., C., and MacMahon B., 1984, Diet and Cancer - an overview, New En. J. Med., 310, 697-703.
35. Zeisel, S., H., 1981, Dietary choline: biochemistry physiology and pharmacology, Ann. Rev. Nutr. 1, 95-121.

PHOSPHOLIPIDS IN ONCOLOGY

Dieter Arndt

Academy of Sciences of German Democratic Republic
Central Institute of Cancer Research
GDR - 1115 Berlin-Buch

INTRODUCTION

Phospholipids are, besides glycolipids, cholesterol and proteins, the main components of biological membranes. Membrane phospholipids are possibly involved in differences between normal cells and tumor cells. Furthermore, phospholipids in the cell membrane may be plausible targets for anticancer agents. A few phospholipids, especially etherlipids, are very promising anticancer agents and, finally, phospholipids are important as the main bilayer component for the encapsulation of anticancer drugs in liposomes.

PHOSPHOLIPIDS AND MEMBRANE MODIFICATION IN TUMOR CELLS

It is generally recognized that the search for structrural or metabolic differences between tumor and normal cells has not revealed substantial differences. This assertion is also valid for phospholipids as the essential structural elements of the cell membrane. But there are differences in lipid composition and structural order of the lipid domain of anticancer agent-resistant cancer cells when compared to the parent line [3]. P 388 murine leukemia cells resistant against the anticancer drug doxorubicin, contain relatively less phosphatidylcholine and more sphingomyelin than drug-sensitive cells. Also the lipid domain of the plasma membrane in drug-resistant cells is structurelly more ordered than the sensitive line [25]. Further, there are differences between tumor and normal cells with regard to membrane physical properties and certain cellular functions, including carrier-mediated transport, ion channels, and eicosanoid production [31]. In many cases these alterations in membrane properties are related to changes in membrane proteins. In other words, the lipid fluidity may be of essential importance for protein activity. Based on experimental evidence and thermodynamic arguments, it has been asserted that, upon increase in the membrane lipid microviscosity, the protein-lipid interaction will decrease and the equilibrium position of the protein will be shifted more towards the aqueous domains. Decrease in the lipid microviscosity will result in the converse shift in equilibrium position [29]. Membrane modification of tumor cells by fatty acids or by alkylglycerols may also enhance the sensitivity to hyperthermia and doxorubicin or the transport of anticancer drugs across the blood-brain-barrier [3,31].

MEMBRANES AS TARGETS FOR THERAPEUTIC AND TOXIC ANTICANCER DRUG ACTION

It is generally accepted that DNA is the presumed intracellular target of some of the more clinically important anticancer agents, including doxorubicin, bleomycin, cisplatin and cyclophosphamide. However, it has also been reported that doxorubicin could be cytotoxic without entering the cell. In those experiments, murine cancer cells (L 1210) were exposed to large, insoluble polymeric agarose beads to which doxorubicin had been covalently attached. Since the support was larger than the cells, the drug could not penetrate to the cytoplasm or nucleus. The absence of free doxorubicin was demonstrated by high-performance liquid chromatography [33]. It is possible that immoblized doxorubicin could lead to the formation of toxic activation oxygen species followed by membrane rigidification.

Also for other anticancer drugs, e.g. alkylating agents, there is evidence showing that this class of drugs may have effects at the level of the cell surface [14]. But also the side-effects of anticancer drugs are connected with drug-membrane interactions. Anthracycline glycosides, especially doxorubicin, display toxic side effects against a large variety of cells. Their cardiotoxicity is, however, very specific and sets a limit to the total dose that may be given. For these drugs it was demonstrated that they form stable complexes with acidic phospholipids, and more specifically with cardiolipin [1].

In these studies two essential interactions are responsible for the complex stabilization: a) an electrostatic interaction between the protonated amino groups of the sugar residues of the drug and ionized phosphate residues of the lipid; and b) a hydrophobic interaction between the fatty acid chains of the lipid and hydrophobic parts of the drug [2].

PHOSPHOLIPIDS AS ANTICANCER AGENTS

Eukaryotic cells contain a family of genes termed cellular oncogenes or proto-oncogenes, which are thought to regulate normal cell growth and development. In certain abnormal circumstances, activation of these genes causes tumors and leukemias in animals. Possible mechanisms of activation of cellular oncogenes include: point mutation, amplification, activation by internal rearrangement and recombinant events. Most, but not all, oncogenes code for growth factor receptors or elements of the growth factor signal transduction pathways [15]. Many parts of these pathways are phospholipid-dependent, especially phospholipid-sensitive Ca^{++}-dependent protein kinase C [22]. Protein kinase C is also a target for a new class of anticancer agents: ether lipids. Other studies with ether lipids, especially with 1-0-octadecyl-2-0-methyl-sn-glycero-3-phosphocholine, suggest that the antitumor action of these agents may be further attributed to generation of cytotoxic macrophages, direct cytotoxicity, diminished activity of the alkyl cleavage enzymes present in tumors, and selective cell membrane interaction [30]. We could also observe these effects with a series of halogen-containing alkylglycerolipid analogs [5].

In a search for the minimal structural requirement for alkyl-lysophospholipid action, the antitumor activity of alkylphosphocholines was found [34]. Best therapeutic results were obtained with nitrosourea-induced mamma tumors in rats. In murine P 388 models with hexadecyl-, octadecyl- and eicosanyl phosphocholine in different schedules we did not observe any antitumor activity [35].

An interesting observation of specific cytotoxicity of a natural phospholipid against tumor cells was made by the Alving group [1]. Liposomes containing plant phosphatidylinositol (PI) exhibited potent killing activity against virtually all cultured malignant cell lines tested, but did not kill normal cells. Maybe this effect is connected with the higher content of linoleic acid in plant PI in comparison to animal PI and a higher inherent phospholipase A activity in malignant cells than in nonmalignant cells[1].

LIPOSOMES AS DRUG CARRIERS IN CANCER CHEMOTHERAPY

Phospholipids play an important role in oncology as the bilayer-forming component in liposomes used for the encapsulation of anticancer drugs, not only in experimental cancer chemotherapy but also in clinical trials. The rationale for this use is that liposomal encapsulation alters the pharmacological parameters of entrapped antineoplastics (for reviews see [2,13]). Initially, these small particles were thought to be the ideal vehicle to target cancer cells, as a result of the supposed higher endocytic activity of tumor cells in comparison to normal cells. But now it has been shown that, in general, no differences in endocytic or pinocytic capacity between tumor and normal cells occurrs with liposomes. Furthermore, liposomes are too large to pass efficiently across epithelial and endothelial cell layers, or across intact basement membranes[23].

On the other hand, it has been demonstrated that there are quantitative differences in macromolecular leakage from the vasculature of tumor and normal tissue, reflecting a leaky structure of the neovasculature in the tumor area[17]. Considering this, the rationales for the use of liposomes as carriers for anticancer drugs are the following:

1. Sequestration of the liposome as a particle
2. Prolonged circulation
3. Protection of liposome contents against the host
4. Protection of the host against liposome contents
5. Fixation to an anatomic compartment
6. Possibility of targeting (passive, compartmental, ligand-mediated, and physical)

To date, nearly all classes of anticancer drugs have been encapsulated in liposomes. Within the different groups of antineoplastics the following results of liposomal encapsulation have been obtained[2,8].

1. Alkylating agents, nitrosoureas, cisplatin

Entrapment shows here only small advantages, for example circumvention of solubility problems, reduction of the rapid breakdown of nitrosoureas in vivo. One reason for the unsatisfactory results seems to be the high sensitivity of these drugs to hydrolysis.

2. Antimetabilites

Most results concern here the encapsulation of cytosine arabinoside or methotrexate, resulting in a depot effect and enhanced activity, but also increased toxicity in the case of methotrexate. The very low encapsulation efficency of 5-fluoruracil has been overcome by esterification of the OH-bonds in the sugar part of the drug molecule with longchain fatty acids. The drug is lipophilic now and quantitatively binds to the phospholipid bilayer of the liposomes.

3. Anthracyclines

Numerous publications describe the encapsulation of one of the most powerful anticancer drugs: doxorubicin (adriamycin). They agree that liposomal entrapment of this anthracycline leads to reduction of toxicity, especially heart and bone marrow damage, while the antitumor activity remains constant.

4. Miscellaneous anticancer agent

Methyl-GAG, methylglyoxal-bis-guanylhydrazone, is a polyamine analog and inhibitor of polyamine synthesis. Different phase II trials in patients with head and neck cancer make methyl-GAG a drug of interest for future exploration in patients with these tumors. Unfortunately the toxicity of the drug, especially in a daily dosage schedule, is extraordinary. Liposomal encapsulation of methyl-GAG in reverse phase evaporation vesicles leads to increased antitumor activity and reduced hypoglycemia, the main toxicity of the drug [3].

5. Biological response modifiers, cytokines

Many hormone-like mediators are secreted in the course of the immune, inflammatory reactions and blood formation response. Such mediators are generally termed cytokines. These mediators function as intracellular signals that regulate growth, differentiation and function of cells. Cytokines are generally synthetized and secreted as glycoproteins with molecular weights ranging from 6,000 to 60,000 daltons. Cytokines are extremely potent compounds which act at 10^{-10} to 10^{-15} mol/l concentration to stimulate target cell functions [9].

Important cytokines are the interferons (IF-α, β, γ), the interleukins (IL 1, IL2), the tumor necrosis factor (TNF) and the macrophage activation factor (MAF). Noteworthy in this connection are also foreign substances which stimulate macrophages to become cytotoxic against microorganisms, parasites or cancer. Macrophages are readily activated to a microbicidal state subsequent to interaction with microorganisms or their products, e.g., endotoxins, certain bacterial cell wall skeletons and components of bacterial cell walls such as muramyldipeptide (MDP), and muramyltripeptide (MTP). The in vivo effects of these substances are poor, because these drugs are rapidly cleared from the body after parenteral administration. These problems in activating macrophage tumoricidal properties can be overcome by liposomal-entrapped immunomodulators (MDP, MTP, MAF, IF-γ)[8]. Cytokines and muramyl-peptides encapsulated in liposomes are highly effective in activating antitumor or antiviral properties in different animal and human macrophages in vitro and in vivo [9]. The entrapment of soluble MDP in the aqueous space of liposomes poses problems in terms of its efficient incorporation. For this reason, the lipophilic analog of MDP (muramyl tripeptide phosphatidylethanolamine, MTP-PE), which could be inserted directly into liposome membrane bilayers, was developed.

Numerous clinical trials with liposomal MTP-PE are now being started in Europe and the USA.

CLINICAL STUDIES WITH LIPOSOMAL ANTICANCER DRUGS

Table 1 summarizes recent, as well as current clinical trials with liposomal drugs in cancer patients. These drug preparations were obtained in large scale, and were sterile, pyrogen-free and stable.

The preliminary data from these studies indicate that the different types of liposomes entrapping various anticancer drugs are well tolerated by cancer patients. Relatively high amounts of lipids (up to 400 ml containing 8 g of lipids) were administered without impairment. No manifestations of late toxicity were observed when patients were followed up for more than one year. These studies indicate that liposomes may be used safely as carriers for administration of different anticancer drugs to humans.

The first phase II study in recurrent measurable breast cancer is presenty under way [32]. Response rate and toxicities, particularly cardiac, will be assessed in this trial. Preliminary results indicate a complete resolution of pleural effusion and skin metastasis, as well as complete regression of a supraclaricular lymph node.

Further investigation is necessary, in order to ensure the gain in life prolongation and life quality for cancer patients by liposomal drugs, in comparison to the free substance.

TABLE 1

LIPOSOMES IN CANCER PATIENTS

Drug	Liposome type (Size)	Lipid composition (Source)	Reference
MDP	MLV	PC, PS, CHOL	21
MTP-PE	MLV (< 10 μm)	PC, PS (synthetic)	6, 16, 18
NSC 251635	SUV (30-110 nm)	PC, CHOL, SA (egg yolk)	26
Amphotericin B	SUV	PC, CHOL, SA (egg yolk)	27
Amphotericin B	MLV (0,5-2 μm)	DMPC, DMPG (commercial)	19
^{99m}Tc	MLV (0,2-2 μm)	DMPC, DMPG (commercial)	20
Bleomycin	REV	DPPC, DPPA, CHOL (commercial)	10
Doxorubicin	SUV (40-180 nm)	PC, PG, CHOL (commercial)	11, 28
Doxorubicin		CL containing	24, 32

Abbrevations: CHOL, cholesterol; CL , cardiolipin; DMPC, dimyristoyl-phosphatidylcholine; DMPG, dimyristoylphosphatidylglycerol; DPPA, dipalmitoylphosphatidic acid; DPPC, dipalmitoyl-phosphatidylcholine; MDP, muramyldipeptide; MLV, multi-lamellar vesicles; MTP-PE, muramyltripeptide-phosphatidyl-ethanolamine; NSC 251635, water insoluble cytostatic agent; PC, phosphatidylcholine; PS, phosphatidylserine; SA, stearylamine; SUV, small unilamellar vesicles.

REFERENCES

1. Alving, C.R., 1987, Liposome techniques in cell biology, Nature 330: 189-190.
2. Arndt, D., and Fichtner, I., 1986, "Liposomen – Darstellung, Eigenschaften, Anwendung", Akademie-Verlag, Berlin.
3. Arndt, D., Hülsmann, W., and Schwabe, K., 1986, Resistenz gegenüber antineoplastischen Substanzen, in: "Allgemeine und spezielle Tumor-chemotherapie", St. Tanneberger, ed., Akademie-Verlag, Berlin.

4. Berger, M.R., Schmaehl, D., Muschiol, C., Unger, C., and Eibl, H., 1986, Alkylphosphocholines as anticancer agents, J. Cancer Res. Clin. Oncol., 111: S24.
5. Brachwitz, H., Langen, P., Arndt, D., and Fichtner, I., 1987, Cytostatic activity of synthetic O-alkylglycerolipids, Lipids 22: 897-903.
6. Creaven, P.J., Brenner, D.E., Cowens, J.W., Huben, R., Karakousis, C., Han, T., Dadey, B., Andrejcio, K., and Cushman, M.K., 1989, Initial clinical trial of muramyl tripeptide derivative (MTP-PE) encapsulated in liposomes: An interim report, in: "Liposomes in the therapy of infections diseases and Cancer", G. Lopez-Berestein, I.J., Fidler, Eds., Alan R. Liss, Inc., New York.
7. Eibl, H., 1984, Phospholipide als funktionelle Bausteine biologischer Membranen, Angew. Chem. 96: 247-262.
8. Fichtner, I., and Arndt, D., 1989, Stand und Perspektiven der Liposomenforschung, Pharmazie 44: 752-757.
9. Fidler, I.J., and Schroit, A.S., 1988, Recognition and destruction of neoplastic cells by activated macrophages: discrimination of altered self, Biochim. Biophys. Acta 948: 151-173.
10. Firth, G.P., Firth, M., McKeran, R., Rees, J., Walter, P., Uttley, D., and Marks, V., 1988, Application of radioimmunoassy to monitor treatment of human cerebral gliomas with bleomycin entrapped within liposomes, J. Clin. Pathol. 41: 38-45.
11. Gabizon, A., Sulkes, N., Peretz, T., Druckmann, S., and Barenholz, Y., 1989, Liposome-associated doxorubicin: Preclinical pharmacology and exploratory clinical phase 1,2, in: "Liposomes in the therapy of infections diseases and cancer", G. Lopez-Berestein, I.J. Fidler, eds., Alan R. Liss, Inc., New York.
12. Goormaghtigh, E., and Ruysschaert, J.M., 1984, Anthracycline glycoside membrane interactions, Biochim. Biophys. Acta 779: 271-288.
13. Gregoriadis, G., ed., 1988, Liposomes as drug carriers - Recent trends and progress, J. Wiley, Chichester.
14. Grunicke, H., Doppler, W., Hofmann, J., Lindner, H., Maly, K., Oberhuber, H., Ringsdorf, H., and Roberts, J.J., 1986, Plasma membrane as target of alkylating agents, Advances in Enzyme Regulation 24: 247-261.
15. Gustin, A.S., Leaf, E.B., Shipley, G.D., and Moses, H.L., 1986, Growth factors and cancer, Cancer Res. 46: 1015-1029.
16. Hanagan, J.R., Trunet, P., LeSher, D., Andrejcio, K., and Frost, H., 1989, Phase I development of CGP 19835A lipid (MTP-PE encapsulated in liposomes), in: "Liposomes in the therapy of infections diseases and cancer", G. Lopez-Berestein, I.J. Fidler, eds., Alan R. Liss, Inc., New York.
17. Heuser, L.S., and Miller, F.N., 1986, Differential macromolecular leakage from the vasculature of tumors, Cancer 57: 461-464.
18. Kleinermann, E.S., and Hudson, M.M., 1989, Liposome therapy: Anovel approach to the reatment of childhood osteosarcoma, in: "Liposomes in the therapy of infections diseases and cancer", G. Lopez-Berestein, I.J. Fidler, eds., Alan R. Liss, Inc., New York.
19. Lopez-Berestein, G., Bodey, G.P., Frankel, L.S., and Mehta, K., 1987, Treatment of hepatosplenic candidiasis with liposomal-amphotericin B, J. Clin. Oncol. 5: 310-317.
20. Lopez-Berestein, G., Kasi, L., Rosenblum, M.G., Haynie, T., Jahns, M., Glenn, H., Mehta, R., Mavligit, G.M., and Hersh, E.N., 1984, Clinical pharmacology of Tc-labeled liposomes in patients with cancer, Cancer Res. 44: 375-378.
21. Morere, J.F., Israel, L., Puisieux, F., Breau, J., Durieux, J., and Sponton, E., 1986, Phase I trial of systemic and regional muramyldipeptide (MDP) liposomes in cancer patients with advanced neoplasms, Proc. ASCO 5: 220.

22. Nishizuka, Y., 1986, Studies and perspectives of protein kinase C, Science 233: 305–312.
23. Poste, G., 1986, Pathogenesis of metastatic disease: Implications for current therapy and for the development of new therapeutic strategies, Cancer Treatm. Rep. 70: 183–199.
24. Rahman, A., Roh, J.-K., and Treat, J., 1989, Preclinical and clinical pharmacology of doxorubicin entrapped in cardiolipin liposomes, in: "Liposomes in the therapy of infections diseases and cancer", G. Lopez-Berestein, I.J. Fidler, Eds., Alan R. Liss, Inc. New York.
25. Ramu, A., Glaubiger, D., and Weintraub, H., 1984, Differences in lipid composition of doxorubicin-sensitive and -resistant P 388 cells, Cancer Treatm. Rep. 68: 637–641.
26. Sculier, J.P., Coune, A., Brassinne, C., Laduron, C., Atassi, G., Ruyschaert, J.M., and Frühling, J., 1986, 1986, Intravenous infusion of high doses of liposome containing NSC 251635, a water-insoluble cytostatic agent. Apilot study with pharmacokinetic data, J. Clin. Oncol. 4: 789–797.
27. Sculier, J.P., Coune, A., Meunier, F., Brassinne, C., Laduron, C., Houlaert, C., Collette, N., Heymans, C., and Klastersky, J., 1988, Pilot study of amphotericin B entrapped in sonicated liposomes in cancer patients with fungal infections, Eur. J. Cancer Clin. Oncol. 24: 527–538.
28. Sells, R.A., Gilmore, I.T., Owen, R.R., New, R.R.C., and Stringer, R.E., 1987, Reduction in doxorubicin toxicity following liposomal delivery, Cancer Treatm. Rev. 14: 383.
29. Shinitzky, M., 1979, Passive modulation of membrane proteins, in: "Physical and chemical aspects of cell surface events and cellular regulation", R. Blumenthal, C. de Lisi, eds., Elsevier/North Holland, Amsterdam.
30. Special Issue, 1987, Ether lipids in oncology, Lipids 22: No 11.
31. Spector, A.A., and Burns, C.P., 1987, Biological and therapeutical potential of membrane lipid modification in tumors, Cancer Res. 47: 4529–4537.
32. Treat, J., Greenspan, A.R., and Rahman, A., 1989, Liposome encapsulated doxorubicin preliminary results of phase I and phase II trials, in: "Liposomes in the therapy of infections diseases and cancer", G. Lopez-Berestein, I.J. Fidler, eds., Alan R. Liss, Inc., New York.
33. Tritton, T.R., and Yee, G., 1982, The anticancer agent adriamycin can be actively cytotoxic without entering cells, Science 217: 248–250.
34. Zeisig, R., Arndt, D., and Brachwitz, H., 1990, Etherlipide-Synthese und tumortherapeutische Verwendbarkeit, Pharmazie 45: in press.

PHOSPHOLIPIDS IN PULMONARY FUNCTIONS

Burkhard Lachmann

Dept. of Anesthesiology, Erasmus University
Rotterdam, The Netherlands

HISTORY

In 1929, von Neergaard wrote "It may be possible that
the surface tension of the alveoli is diminished by
concentration of surface-active substances against other
physiologic solutions" (14). He was referring to the
existence of a surface film in the alveoli and his assumption
was based on the following observations. Neergaard measured
pressure-volume diagrams from human and animal lungs, first
filling them with air and then with liquid. The surprising
result was that the pressure necessary for filling the lung
with liquid was only half the pressure necessary for filling
the lung with air (Fig. 1). His explanation of this
remarkable difference was based on the assumption that in
each alveolus there must be a barrier between air and fluid
(such as in the wall of a soap-bubble) with a tendency to
diminish its size according to the law of Laplace. The amount
of retraction pressure for the lung is larger than the
retraction force of the elastic fibers. By filling the
alveoli with liquid, the air-liquid barrier is replaced by a
liquid-to-liquid barrier without any surface tension. The
retraction pressure, measurable in the fluid-filled lung, is
therefore equal to the retraction pressure of the elastic
fibres. The same procedure for determining the influence of
surface tension on the overall retraction of the lung was
later reported by other scientists, and is now a well-
established method.

In 1955 Pattle showed that bubbles resulting from lung
edema, as well as bubbles squeezed from a lung cut, are very
stable (11). Pattle assumed that the walls of these bubbles
consist of surface active materials and that their stability
depended on the quantity and quality of the surfactant
phospholipids. In 1957 Clements (2) was the first to
investigate lung extracts in a Wilhelmy-balance and
demonstrated that these extracts had, in contrast to
detergents or plasma, genuine hysteresis properties (Fig. 2).

Phospholipids
Edited by I. Hanin and G. Pepeu
Plenum Press, New York, 1990

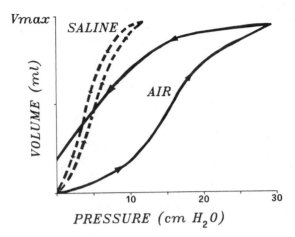

Fig. 1. Saline (dotted line) and air (solid line) volme
pressure diagrams of mammalian lung. Vmax = maximal volume.

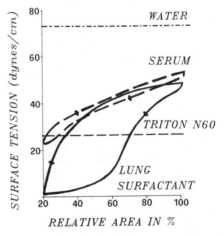

Fig. 2. Surface tension area diagrams from lung surfactant,
serum, a detergent Triton 60, and water. Whereas the
surface tension is constant with water and detergent during
compression and dilatation, there is hysteresis and large
alterations in surface tension with lung surfactant.

Avery and Mead (1) were the first, in 1959, to call attention to the possibility that lungs of infants with respiratory distress syndrome may have abnormal surface tension properties, due to a deficiency in lung surfactant.

BIOCHEMICAL COMPOSITION

Biochemical characterization has shown that surface-active phospholipids, proteins and mucopolysaccharides are the main constituents of the surfactant system of the lung. Generally, between 80% and 90% of surfactant lipid is phospholipid. The major representative (70% to 80%) of surfactant phospholipids is phosphatidylcholine (PC) and about 60% of the PC molecules contain two saturated acyl chains. This disaturated PC (DSPC) is largely dipalmitoylphosphatidylcholine (DPPC). Monoenoic PC molecules comprise most of the unsaturated PC in surfactant. In most adult mammalian species, phosphatidylglycerol (PG) is the second most abundant phospholipid class, accounting for up to 10% of total surfactant lipid. Surfactant contains, in addition, small proportions of phosphatidylinositol (PI), phosphatidylserine, phosphatidylethanolamine and sphingomyelin (for review see reference 13).

Phospholipids are synthesized in the type II alveolar cells. They are stored in these cells as so-called lamellar bodies, before being released to the surface of the alveoli. The surfactant lipids spread as a mono-layer or multi-laminar layer at the air-liquid boundary. They reduce the net contractile force of the alveolar surface, thus preventing the airspaces from collapsing at low lung volumes.

To date, the functional importance of the other components of the surfactant system, including proteins and muco-poly-saccharides, has not yet been fully explained. But, recently, it could be demonstrated that for optimal function of the lipids the presence of a few small proteins is a prerequisite for optimal in vivo function of artificial surfactant and surfactant extracted from mammalian lungs (13).

FUNCTION OF THE SURFACTANT SYSTEM

Mechanical stabilization of the lung

The integrity of the surfactant system of the lung is a prerequisite for normal breathing with the least possible effort. The surfactant system acts by decreasing surface tension of the interface between alveoli and air. This provides an explanation as to why we have to generate a "negative" pressure of only 5-10 cm H_2O during each inspiration; in the absence of surfactant the surface tension at the air-liquid interface would be that of plasma and the pressure needed to maintain lung aeration would be 25-30 cm H_2O (depending on the radius of the alveoli). This is a well-known problem in patients with respiratory failure (Fig. 3).

In alveoli with different radii an equal lowering of surface tension would not, however, produce stabilization of the alveolar system. It would instead, according to the law of Laplace, lead to the collapse of the smaller bubbles or alveoli, and to their emptying into larger bubbles (Fig. 4).

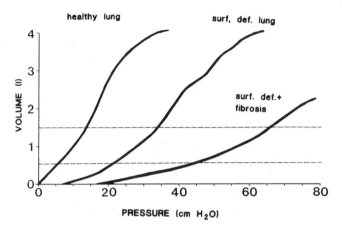

Fig. 3. Pressure-volume diagrams from a patient with
healthy lungs, a patient with ARDS (surfactant deficient
lung) and from a patient with ARDS 18 days after artificial
ventilation (surfactant deficient with fibrosis) who died
on day 22 after artificial ventilation due to hypoxemia.
The lower dashed line represents the pressure needed to
keep the lung at the same functional residual capacity; the
upper dashed line represents the peak airway which occurs
when these lungs are artificially ventilated with one
liter.

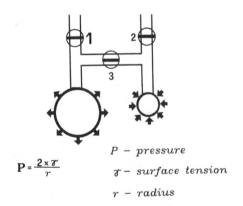

$$P = \frac{2 \times \tau}{r}$$

P – pressure

τ – surface tension

r – radius

Fig. 4. Dynamics from two soap bubbles of different size.
The bubbles are produced by open stopcocks (1, 2) and
closed stopcock (3). After closing 1, 2 and opening 3, the
small bubble will empty into the larger one – representing
the Laplace law.

Since alveoli in vivo do not exhibit such behaviour, one can conclude that the second remarkable quality of the alveolar lining layer is that it can change the surface tension in a manner related to the size of the alveoli. Meanwhile, there are also other explanations for this behaviour, but also these are not yet finally proved.

SURFACTANT AS ANTI-EDEMA FACTOR

Another function of the pulmonary surfactant system is stabilization of the fluid balance in the lung, and protection against lung edema. The normal plasma oncotic pressure of 37 cm H_2O is opposed by the oncotic pressure of interstitial fluid proteins of 18 cm H_2O, the capillary hydrostatic pressure of 15 cm H_2O and by the surface tension conditioned suction of 4 cm H_2O (Fig. 5).

In general, alveolar flooding will not occur so long as the suction force in the pulmonary interstitium exceeds the pressure gradient generated by surface tension in the alveolar air-liquid interface. Since this pressure gradient is inversely related to the radius of the alveolar curvature there is, for each combination of interstitial resorptive force and average surface tension, a critical value for surface tension and for alveolar radius, below which alveolar flooding occurs (3). From these facts it can also be concluded that established statements in the literature concerning the fluid exchange according to the law of Starling and the development of lung edema have to be reconsidered, at least for those cases where the lung edema appears to be due to a surfactant disturbance, because in this concept the surface tension is not included in the formula.

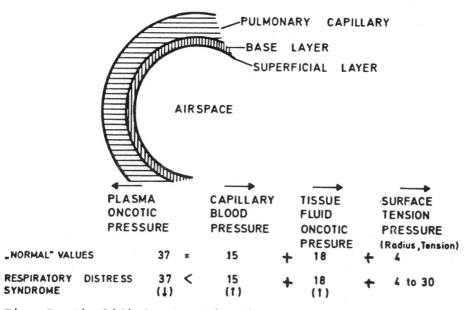

	PLASMA ONCOTIC PRESSURE	CAPILLARY BLOOD PRESSURE	TISSUE FLUID ONCOTIC PRESURE	SURFACE TENSION PRESSURE (Radius,Tension)
"NORMAL" VALUES	37 =	15 +	18 +	4
RESPIRATORY DISTRESS SYNDROME	37 < (↓)	15 (↑)	+ 18 (↑)	+ 4 to 30

Fig. 5. Simplified schematic diagram representing factors influencing fluid balance in the lung.

SURFACTANT AND LOCAL DEFENCE MECHANISMS

Observations in patients have shown that, following a decrease in lung compliance (thus, surfactant deficiency), pneumonia will often develop, despite the application of high doses of antibodies. Therefore, it is possible that the surfactant system is also involved in local defence mechanisms of the lung. It has been demonstrated that alveolar phagocytic macrophages ingest bacteria (or destroy them intracellularly) only in the presence of sufficient surface active material (5, 6). In this context surfactant seems to reduce the surface forces of bacterial membranes and it is also an energy-rich substrate which supports the macrophages' high rate of metabolism. Recently we have demonstrated that the pulmonary surfactant system may also be involved in protecting the lung against its own mediators, (e.g angiotensin II) and in protecting the cardiocirculatory system against mediators produced by the lung (Fig. 6) (4).

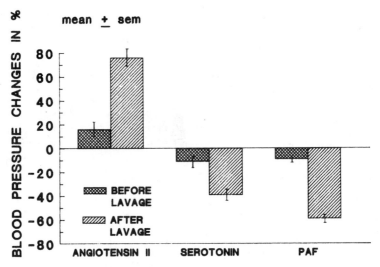

Fig. 6. Blood pressure changes following intratracheal application of angiotensin II (70 μg/kg), serotonin (700 μg/kg) and platelet activating factor (PAF) (10 μg/kg) before and after removing surfactant by lung lavage.

SURFACTANT AND AIRWAYS STABILIZATION

As early as 1970, Macklem et al. called attention to the significance of bronchial surfactant for the stabilization of the peripheral airways and hinted that lack of stabilization may cause airway obstruction or collapse of the small bronchi with air trapping (10). Just recently this could be proved in our laboratory in an animal model where bronchial surfactant was almost selectively destroyed (7). It was demonstrated that the pressure needed to open up the collapsed bronchi is about 20 cm H_2O (Fig. 7).

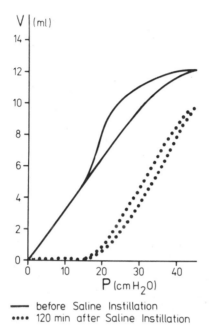

— before Saline Instillation
•••• 120 min after Saline Instillation

Fig. 7. Pressure-volume diagram from guinea pig before and after damaging bronchial surfactant.

Besides its role in mechanical stabilization, bronchial surfactant also has a transporting function for mucus and inhaled particles. This has been proven, in vitro, in a study showing that particles on a surface film move in one direction only if the surface film is compressed and dilated, comparable to the compression and expansion during expiration and inspiration (7). Furthermore, bronchial surfactant also acts as an antiglue factor preventing the development of large adhesive forces between mucus particles, as well as between mucus and the bronchial wall (Fig. 8) (12).

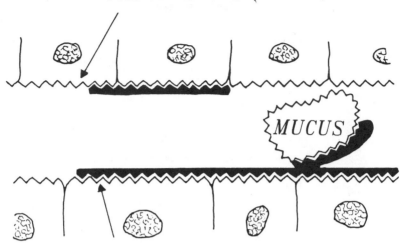

FREE CELL SURFACE (ADHESIVE)

MUCUS

CELL SURFACE COVERED BY SURFACTANT
(NON ADHESIVE)

Fig. 8. Schematic diagram demonstrating the interaction between mucus and airway. Note that there will be non-adhesive forces only if the mucus particle and the airways are covered with surfactant (thick solid line).

A further possible function of bronchial surfactant, which has scarcely been discussed, is its masking of receptors on smooth muscle with respect to substances which induce contraction and could lead to airway obstruction. We have recently shown that lining the airway with surfactant in ovalbumin-sensitized guinea pigs prevented significant bronchial obstruction during antigen challenge (Fig. 9). This means that bronchial surfactant could also be involved in asthma (8). This is further supported by the fact that the most effective bronchodilatory drugs (corticoids and betamimetics) lead to a release of surfactant.

FUNCTIONAL CHANGES DUE TO A "DISTURBED" SURFACTANT SYSTEM

When considering the main physiologic functions of the alveo-bronchial surfactant system it can easily be understood that alteration in its functional integrity will lead to: decreased lung distensibility and thus to increased work of breathing and increased oxygen demand by the respiratory muscles; atelectasis; transudation of plasma into the interstitium and into the alveoli with decreased diffusion for oxygen and CO_2; inactivation of the surfactant by plasma and specific surfactant inhibitors; hypoxemia; metabolic acidosis secondary to increased production of organic acids under

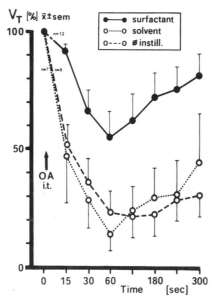

Fig. 9. Time course of changes in tidal volume (V_T) in ovalbumin sensitized guinea pigs during unchanged ventilation settings after challenge with 1 mg/kg ovalbumin intratracheally (OA) (percent of initial V_T). The difference between the surfactant pretreated, and solvent pretreated, respectively, and untreated animals was statistically significant at each time tested ($p < 0.05$).

anaerobic conditions; enlargement of functional right-to-left shunt due to perfusion of non-ventilated alveoli (the v Euler-Liljestrand reflex does not "work" in surfactant deficient alveoli); decreased production of surfactant as a result of hypoxemia, acidosis and hypoperfusion. This will finally lead to a vicious circle and the lung will fail as a gas exchange organ (Fig. 10) (9).

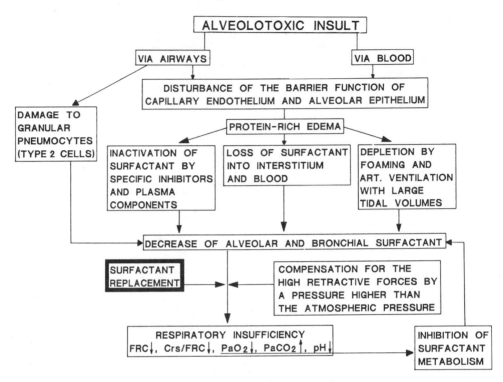

Fig. 10. Pathogenesis of ARDS with special reference to the surfactant system, including suggestions to compensate for a damaged surfactant system.

ACKNOWLEDGEMENT

This study was financially supported by a grant from the Dutch Foundation for Medical Research (SFMO).

REFERENCES

1. Avery, M.E. and Mead, J., 1959, Surface properties in relation to atelectasis and hyaline membrane disease, Am. J. Dis. Child., 97: 517-523.
2. Clements, J.A., 1957, Surface tension of lung extracts, Proc. Soc. Exp. Biol. Med., 95: 170-172.
3. Guyton, A.C., Moffatt, D.S., and Adair, T.A., 1984, Role of alveolar surface tension in transepithelial movement of fluids, in: Pulmonary Surfactant", B. Robertson, L.M.G. van Golde, and J.J. Batenburg, eds., Elsevier, Amsterdam, pp 171-185.
4. Hein, T., Lachmann, B., Armbruster, S., Smit, J.M., Voelkel, N., and Erdmann, W., 1987, Pulmonary surfactant inhibits the cardiovascular effects of platelet activating factor (PAF), 5-hydroxy-tryptamine (5-HT) and angiotensin II. Am. Rev. Resp. Dis., 135: A506.

5.   Huber, G., Mullane, J., and LaForce, F.M., 1976, The role of alveolar lining material in antibacterial defenses of the lung. Bull. Europ. Physiopath. Resp., 12: 178-179.
6.   Jarstrand, C., 1984, Role of surfactant in the pulmonary defence system, in: "Pulmonary Surfactant," B. Robertson, L.M.G. van Golde, J.J. Batenburg, eds., Elsevier, Amsterdam, pp 187-201.
7.   Lachmann, B, 1985, Possible function of bronchial surfactant, Eur. J. Respir. Dis., 67: 49-61.
8.   Lachmann, B., and Becher, G., 1986, Protective effect of lung surfactant on allergic bronchial constriction in guinea pigs, Am. Rev. Respir. Dis., 133: A118.
9.   Lachmann, B., 1988, Surfactant replacement in acute respiratory failure: animal studies and first clinical trials, in: "Surfactant Replacement Therapy in Neonatal and Adult Respiratory Distress Syndrome", B. Lachmann, ed., Springer Verlag, Berlin-Heidelberg, pp 212-223.
10. Macklem, P.T., Proctor, D.F., and Hogg, J.C., 1970, The stability of peripheral airways. Resp. Physiol., 8: 191-203.
11. Pattle, R.E., 1955, Properties, function, and origin of the alveolar lining layer, Nature, 175: 1125-1126.
12. Reifenrath, R., 1983, Surfactant action in bronchial mucus, in: "Pulmonary Surfactant System", E.V. Cosmi, E.M. Scarpelli, eds., Elsevier, Amsterdam, pp 339-347.
13. Van Golde, L.M.G., Batenburg, J.J., and Robertson, B, 1988, The pulmonary surfactant system: Biochemical aspects and functional significance, Physiol. Rev., 68: 374-455.
14. Von Neergaard, K., 1929, Neue Auffassungen über einen Grund-begriff der Atemmechanik, Z. Ges. Exp. Med. 66: 373-394.

NUTRITIONAL ASPECTS OF PHOSPHOLIPIDS AND THEIR INFLUENCE ON CHOLESTEROL

Walter Feldheim

Department of Human Nutrition and Food Science
University of Kiel
Düsternbrooker Weg 17-19, D-2300 Kiel 1, FR Germany

INTRODUCTION

Among prosperous societies, cardiovascular disease is the most common single cause of death. Important factors increasing the risk of the disease are mainly of nutritional origin, like over-nutrition, high fat and cholesterol intake, and too much salt or alcohol. There is no doubt that this is a disease of affluence, and increasing affluence makes it difficult to moderate the diet, and thus to decrease the death rate of this disease. The development of atherosclerotic lesions begins with a thickening of the intima and deposition of plaques, containing a lipid mass, particulary cholesterol and cholesterol-esters. In advanced stages, calcification may occur. The "lipid hypothesis" refers to such a central role of cholesterol, but there are many controversial viewpoints, reflecting the complexity in the initiation of the disease. 9,11,17

Atherosclerotic tissues contain lipid peroxidation products: ceroid and lipid peroxide are detectable. During peroxidation of polyunsaturated fatty acids, lipoxy radicals are formed, which may degrade to malondialdehyde. This reacts with phosphatidylethanolamine or phosphatidylcholine to form a Schiff-base, and as a result the membrane structure is altered and damaged. The increased membrane permeability to calcium leads to an elevated cytosolic calcium concentration with further injury and death of the cell. Here, lipid peroxidation is seen as the first step in the development of disease. 8,10, 19,21,25

While approaches to decrease plasma cholesterol concentrations were the central point of numerous studies over the last 40 years, only a few studies have investigated the effects of lecithins on cholesterol metabolism.

Phospholipids
Edited by I. Hanin and G. Pepeu
Plenum Press, New York, 1990

THE STATE OF THE ART: LECITHIN VERSUS CHOLESTEROL

In an early study of Kesten and Silbowitz[12] (1942) with adult chinchilla rabbits, first evidence was given about the possible influence of orally applied lecithin on cholesterol metabolism. In this study, all animals received a basic diet containing 150 mg of cholesterol daily. The animal were divided into three groups and rabbits in two of these three groups received daily supplements of 5 g and 1 g, respectively, of crude soya lecithin containing about 20% of pure lecithin. All animals were killed after 4 months of feeding. In the controls, after this time of feeding, a cholesterol-induced atherosclerosis was seen: seven of the 8 animals developed lesions of the aorta of moderate degree, but only 2 animals each in both of the lecithin supplemented experimental groups of 7 or 8 rabbits showed atherosclerotic changes (of minimal degree). The livers of all animals were normal histologically.

The levels of blood cholesterol were, on the average: 340 mg, 210 mg and 300 mg/100 ml, respectively. By supplementing the diet with 195 mg choline instead of the 5 g of crude soya lecithin as equivalent, the incidence of lesions was less when compared with the controls, but the hypercholesterolemia was more severe, when compared with the lecithin-group.

From this pioneer study, and up to the numerous experiments with different animal species[18,1,13] and humans[3] as well, which have been performed to date, there still is no clear information about the advantage of using dietary polyunsaturated phospholipids as prophylactic and/or therapeutic agents in atherosclerosis. Sometimes favorable effects of lecithin on cholesterol metabolism were reported, but in other studies no influence of lecithin was observed.

The main source of the discrepancy of these findings concerning the influence of phospholipids, given orally, on plasma lipid cholesterol-concentrations, and the incidence of atherosclerosis, is the variety in dietary conditions of the different studies, and the changing amount and composition of the lecithins used.

In earlier studies, the crude soya lecithins were less defined. Kesten and Silbowitz[12] used a product with approximately 20% lecithin, equivalent to about 40 mg of choline per g of the product. Newer products of purified lecithins contain much more phosphatidylcholine per g, but there are substancial differences in the composition of phospholipids and fatty acids as well of the products. An example of the components of the Lucas Meyer product is shown in Table 1.

The extent to which this protective effect of lecithin on cholesterol metabolism is due to the effect of the component choline itself has been discussed earlier in the literature.[26] We have to consider the efficiency of the whole lecithin molecule and that of the component choline as well. In such studies with lecithin and choline it is, however, difficult to compare

TABLE 1. PHOSPHOLIPID PRODUCTS* FROM SOYA

Composition of phospholipids (80-85%)

phosphatidyl ethanolamine   16-23
phosphatidyl choline   21-30
phosphatidyl inositol   12-18
others   15

Composition of fatty acids (57-69%)

saturated fatty acids   23-26
oleic acid   8-16
polyunsaturated fatty acids   62-68
P/S ratios 1.9, 2.75, 2.8

* e.g. Lucas Meyer
       Central Soya Corporation

the effective dose. Choline may be metabolized in the gut by bacteria and absorbed to a small extent.

On the other hand, it is still controversial to what extent applied phosphatidylcholine is absorbed from the gut. With a labelled lecithin, Lekim[14] was able to demonstrate that a high proportion of phosphatidylcholine is only hydrolyzed in the 2-position, but again reacylated in the mucosa, preferentially with an unsaturated fatty acid. In conclusion, it was calculated, that about 50% of the phosphatidylcholine is absorbed.

From other investigations it is known that, on the basis of the same amount calculated of choline, the application of lecithin is more effective than choline in increasing the serum choline concentration.

Looking at the composition of the lecithin preparations used, the amount of phosphatidylcholine was between 30 and 20% of total phospholipids with remarkable amounts of phosphatidylethanolamine and -inositol (Table 1). It is still uncertain, however, whether the plasma lipid changes are due to the content of phosphatidylcholine only. Studies with a higher content on purified true lecithin are required.

Another important question is, whether any possible effect of phosphatidylcholine on cholesterol metabolism is due to the unsaturated fatty acid moiety of the molecule or the phosphatidylcholine part of lecithin itself. In a study by Greten and coworkers,[6] in which 10 g soyabean oil was replaced by a corresponding amount of a lecithin preparation (18 g), no additional effect was seen, when compared with the isocaloric-fatmodified, low cholesterol diet in a 10 days experiment with 6 patients and 2 normal persons. The plasma cholesterol-level was not influenced.

But one surprising result of this study was a higher loss
of mostly neutral sterols, in the feces, due to the oral treatment with lecithin. It is difficult to interpret this result
of a negative sterol balance with unchanged plasma cholesterol concentration.The effect could be the consequence of undesirable increased cholesterol synthesis, a loss of choleste-
rols from other pools than the plasma, or a decreasing cholesterol absorption during phospholipid administration. In
addition, the experimental period was very short. Again, it
is not possible to draw any conclusions from these studies,
and further research is required.

In a study on normolipidemic and familiar hypercholesterolemic subjects carried out by Childs and coworkers,4 the
authors concluded hat, since the response of the subjects to
2 supplements - corn oil or lecithin (36 g), in a cross-over
design with an equivalent amount of polyunsaturated fatty
acids - was different, lecithin does not exert its lipoprotein
modification because of its polyunsaturated fatty acid content.

TABLE 2. EFFECT OF TREATMENT WITH SOYA LECITHIN ON CHOLESTEROL-
         CONCENTRATIONS IN PATIENTS AND HEALTHY SUBJECTS
         (1958-1988)

Number of patients	healthy subjects	daily amount of soya lecithin	duration of the treatment	decrease of total cholesterol concentration	(ref./ year)
15	-	36 g	2 months	+ (12 of 15)	(15/1958)
12	-	1.2g	4 months	-	(22/1974)
7	3	10-30g	2-11 months	+ (3 of 7) + (1 of 3)	(20/1977)
-	10	22.5g	1 month	-	(5/1980)
-	5	48 g	6 months	+	(24/1980)
6	2	18 g	10 days	-	(6/1980)
12	6	36 g	3 weeks 4 months	HDL LDL (+)	(4/1981)
-	6	6 g	2 weeks	+	(7/1981)
_	8	6 g	4 weeks	+	(7/1981)
18	-	12 g	4 months	+	(3/1982)
20	-	18 g	6 weeks	+	(16/1985)
-	50	10 g	6 weeks	+	(23/1988)

Cairella and coinvestigators[3] have prepared an excellent review of the available literature, which is summarized and enlarged in Table 2.

Among the studies mentioned in the table, the last two were carried out with placebos, as double blind, cross-over experiments.

In a clinical controlled study carried out by Ovesen and coworkers[16] in Denmark 20 out-patients with hyperlipidaemia (total cholesterol at least 250 mg/dl, mean 300 mg/dl), soya bean phospholipid or placebo, 18 g daily for 6 weeks were given orally. During the cross-over trial with a wash out phase of 4 weeks a standard therapeutic diet was consumed, with a cholesterol intake of less than 200 mg/day and a fat content of less than 30% of the energy intake. After 6 weeks the mean cholesterol concentration in serum decreased significantly by 22 mg/100 ml in the phospholipid group as compared with the placebo-treated patients. No effect on triglyceride and high-density lipoprotein cholesterol concentration was seen in this study.

A similar double-blind cross-over study was carried out by Tolonen and coworkers[23] in Finland, in 50 healthy men with blood cholesterol levels of at least 250 mg/100 ml, mean 280 mg/dl.

Compared with the study of Ovesen et al.,[16] the mean age of the probands was about 10 years lower (47 to 56 years), but the blood cholesterol level was similar. The dietary habits of the volunteers were not altered during the experimental period, and the wash out period was extended to 6 weeks. Three table-spoons (approximately 10 g) of the same phospholipid used in the study of Ovesen et al.[16] or placebo were given orally. Blood samples were taken at the beginning and after 6, 12 and 18 weeks under standardized conditions.

Total cholesterol, HDL-cholesterol and triglycerides were determined. During the first six seeks, serum cholesterol levels decreased in the lecithin group by an everage of 20 mg/100 ml, the difference being statistically significant. A slight, but not significant decrease was also observed in the placebo group. At the end of the wash out period, the mean cholesterol level was still below the baseline of the beginning in the phospholipid group. During the second part of the experiment after the cross-over, a slight decrease of the cholesterol levels was observed in the lecithin group. No change was observed on HDL-cholesterol nor triglyceride levels. As a side effect, all subjects who had eaten lecithin showed an increase of plasma linoleic acid levels.

In both studies purified soya lecithin was given. The placebo preparation was made from barley grits or defatted soya protein (52% protein). The results of both studies show similarities, despite the fact that the conditions were different: out-patients versus "normal" people, different doses of lecithins, types of placebos, and controlled diet versus free consumption of foods (Table 3).

TABLE 3. VARIABLE FACTORS OF THE STUDIES PERFORMED
WITH SOYA LECITHIN WITH REGARD TO
CHOLESTEROL-CONCENTRATIONS

---

probands: healthy people, patients with
hypercholesterolaemia/
hyperlipoproteinaemia

phospholipids: content of "true" lecithin, changing
amounts of other phospholipids

doses (applied daily) 1-50 g

length of the treatment: from 10 days up to
11 months

type of diet: free choice, 40 Cal% fat or fat-
restricted; type of placebo[26]

CONCLUSIONS

In conclusion, the question of effectiveness of orally
applied lecithin on blood levels of total cholesterol or HDL-
and LDL-concentrations is still unresolved. Checking the design
and the results of the previous studies in this field, we have
to pay more attention to keeping the experimental conditions
more precise and comparable. We also need more double blind
controlled studies.

From all the observations to date the cholesterol
lowering effect of lecithin may be in the range of a 10% decrease from the initial cholesterol value. In most cases,
this magnitude of effect will be not sufficient, although it
may be helpful toward approaching a safe level of cholesterol
concentration in blood. In a group of steps to improve the
health of the consumer, the consumption of lecithin may this
be an important part of the overal therapeutic strategy.

REFERENCES

1. C.W.M. Adams, R.W.R. Baker, and R.S. Morgan, 1969, The
   effect of oral polyunsaturated lecithin on the development of atheroma and fatty liver in the cholesterol-fed
   rabbit, J.Pathol. 97:35-41.
2. A.C. Beynen and C.E. West, 1987, Cholesterol metabolism in
   swine fed diets containing either casein or soyabean
   protein, JAOCS 64:1178-1182.
3. M. Cairella, V. del Balzo, R. Godi, R. Scatena, and L.D.
   Treves, 1982, Soya lecithin in therapy: clinical findings,
   in: "Soya Lecithin Dietetic Applications", J.N. Hawthorne
   and D. Lekim, eds., Semmelweis-Verlag, Hoya pp.93-106.
4. M.T. Childs, J.A. Bowlin, J.T. Ogilvie, W.R. Hazzard, and
   J.J. Albers, 1981, The contrasting effects of a dietary
   soya lecithin product and corn oil on lipoprotein lipids
   in normolipidemic and familiar hypercholesterolemic subjects, Atheroscler. 38:217-228.

5. M. Cobb, P. Turkki, W. Linscheer, and K. Raheja, 1980, Lecithin supplementation in healthy volunteers, Nutr. Metab. 24:228-237.
6. H. Greten, H. Raetzer, A. Stiehl, and G. Schettler, 1980, The effect of polyunsaturated phosphatidylcholine on plasma lipids and fecal sterol excretion, Atheroscler. 36:81-88.
7. J.N. Hawthorne, M. Hoccom, and J.E. O'Mullane, 1981, Soya lecithin and lipid metabolism, in: "Soya Lecithin.Nutritional and Clinical Aspects", M. Cairella and D. Lekim, eds., Società Editrice Universo, Roma.
8. H. Heinle and H. Liebich, 1980, The influence of diet-induced hypercholesterolemia on the degree of oxidation of glutathione in rabbit aorta, Atheroscler. 37:637-640.
9. H. Imai, N.T. Werthessen, V. Subramanyam, P.W. LeQuesne, A.H. Soloway, and M. Kanisawa, 1980, Angiotoxicity of oxygenated sterols and possible precursors, Science 207: 651-653.
10. M. Iwakami, Peroxides as a factor of atherosclerosis, 1965, Nagoya J.Med.Sci. 28:50-66.
11. R.L. Jackson and A.M. Gotto, 1976, Hypothesis concerning membrane structure, cholesterol and atherogenesis, Atheroscler.Rev. 1:1-
12. H.D. Kesten and R. Silbowitz, 1942, Experimental athero-sclerosis and soya lecithin, Proc.Soc.Exp.Biol.Med. 49: 71-73.
13. D. Kritchevsky, S.A. Tepper, and D.M. Klurfeld, 1979, Experimental atherosclerosis in rabbits fed cholesterol-free diets. 8. Effect of lecithin, Pharm.Res.Commun. 11: 643-647.
14. D. Lekim, 1976, On the pharmacokinetics of orally applied essential phospholipids, in: "Phosphatidylcholine - Biochemical and Clinical Aspects of Essential Phospholipids", H. Peeters, ed., Springer, Berlin, Heidelberg, New York.
15. L.M. Morrison, 1958, Serum cholesterol reduction with lecithin, Geriatr. 13:12-19.
16. L. Ovesen, K. Ebbesen, and E.S. Olesen, 1985, The effect of oral soyabean phospholipid on serum total cholesterol, plasma triglyceride, and serum high-density lipoprotein cholesterol concentrations in hyperlipidemia, J.Parent. Enter.Nutr. 9:716-719.
17. D. Papahadjopoulos, Cholesterol and cell membrane function: a hypothesis concerning the etiology of atherosclerosis, 1974, J.Theor.Biol. 43:329-337.
18. L. Samochowiec, D. Kadlubowska, and L. Rozewicka, 1976, The effects of phosphatidylcholine in experimental athero-sclerosis in white rats, Atheroscler. 23:305-317.
19. J.M. Shaw and T.E. Thompson, 1982, Effect of phospholipid oxidation products on transbilayer movement of phospho-lipids in single lamellar vesicles, Biochem. 21:920-927.
20. L.A. Simons, J.B. Hickie, and J. Ruys, 1977, Treatment of hypercholesterolemia with oral lecithin, Aust.New Zeal. J.Med. 7:262-266.
21. T.L. Smith and F.A. Kummerow, 1988, The role of oxidized lipids in heart disease and aging, in: "Nutrition and Heart Disease", Vol. I, R.R. Watson, ed., CRC-Press, Boca Raton.
22. H.F. ter Welle, C.M. van Gent, W. Decker, and A.F. Wille-brands, 1974, The effect of soya lecithin on serum lipid values in type II hyperlipoproteinemia, Acta Med.Scand. 195:267-271.

23. M. Tolonen, S. Sarna, and V. Knuutinen, 1988, Lecithin
    sønker høg kolesterolhalt i klodet hos møn. En rando-
    misrerad dubbelblind cross-over studie, Swed.J.Biol.Med.
    3:17-20.
24. R.K. Tompkins,and L.G. Parkin, 1980, Effect of longterm
    ingestion of soya phospholipids in serum lipids in
    humans, Am.J.Surg. 140:360-364.
25. G. van Duijn, A.J. Verkleij, and B. de Kruijff, 1984,
    Influence of phospholipid peroxidation on the phase
    behavior of phosphatidylcholine and phosphatidylethanol-
    amine in aqueous dispersions, Biochem. 23:4969-4977.
26. R.J. Wurtman, M.J. Hirsch, and J.H. Growdon, 1977, Lecithin
    consumption elevates serum free choline levels, Lancet 2:
    68-69.

BIOLOGICAL MARKERS OF ALZHEIMER'S DISEASE: A VIEW FROM THE

PERSPECTIVE OF PHOSPHOLIPIDS IN MEMBRANE FUNCTION

George S. Zubenko

Western Psychiatric Institute and Clinic
University of Pittsburgh School of Medicine
3811 O'Hara Street, Room E-1230, Pittsburgh, PA 15213

INTRODUCTION

Considerable variability exists in the clinical expression of Alzheimer's disease, yet the biological underpinnings of this variability are not well understood. Significant differences in clinical presentation include both global characteristics such as age at onset and rate of progression of dementia (39,40,60), as well as the manifestation of selective cognitive deficits (4,6,11,63), neurological symptoms (21,31), and behavioral syndromes (27,33,38,66). Of these, age at onset of dementia may be the clinical characteristic that has received the most attention. While Alzheimer's disease appears to have a typical onset between 75 and 85 years (26,34,64), a subpopulation of patients who meet clinical consensus and neuropathological criteria for this disorder develop dementia several decades earlier. Age at onset of dementia appears to be a familial characteristic that is influenced by at least two genes, FAD (47) and PMF (7). However, twin studies indicate that this variable is also affected by stochastic processes or unidentified environmental factors (61). Furthermore, an early onset of Alzheimer's disease is associated with more global distributions of brain morphologic lesions and neurochemical deficits than are the more typical later-onset cases (39,40,67).

Over the past 5 years, a growing body of evidence has addressed the observation of increased platelet membrane fluidity in Alzheimer's disease (59), which appears to identify a subgroup of patients with distinct clinical features including an early age of onset. In this chapter, we review the evidence bearing on the clinical and biological significance of this cell membrane alteration and present a model that relates this observation to the biology of Alzheimer's disease.

MEASUREMENT OF PLATELET MEMBRANE FLUIDITY

"Membrane fluidity" is a term that has been used to refer to a variety of biophysical measurements that provide inferential information on the dynamic state of biological membranes. In the studies described in this chapter, the steady-state fluorescence anisotropy (at 37°C) of 1,6-diphenyl-1,3,5-hexatriene (DPH) dissolved at low concentration in platelet membranes was used as an index of membrane fluidity (45). Briefly, platelets were quantitatively isolated from fasting blood samples, platelet membranes were isolated from platelet homogenates by differential centrifugation, labeled with DPH, and examined in a temperature-controlled, computer-assisted spectrofluorometer (60,65). Subjects with medical disorders or medication histories that confound this measurement were excluded (60,65). The fluorescence anisotropy of DPH-labeled platelet membrane corresponding to the 10th percentile for neurologically-healthy elderly controls was used to stratify patients into subgroups with increased (<0.1920) or normal (≥ 0.1920) membrane fluidity (70).

CLINICAL FEATURES OF PATIENTS WITH INCREASED PLATELET MEMBRANE FLUIDITY

As a group, patients with increased platelet membrane fluidity suffer from an earlier symptomatic onset, a more rapidly progressive decline, and are more likely to have a family history of dementia, especially early-onset dementia (60,64). Furthermore, the prevalence of focal EEG slowing and risk factors for stroke is lower in the subgroup of patients with increased platelet membrane fluidity (58). In light of these findings, the subgroup with increased membrane fluidity may represent a more homogeneous group of patients whose dementias are more likely to result from a degenerative etiology, rather than ischemia or a mixture of degeneration and ischemia. Evaluation of the cognitive impairments of patients with Alzheimer's disease has not revealed any specific cognitive deficits, or spectrum of cognitive deficits, associated with increased platelet membrane fluidity (63).

SPECIFICITY AND STABILITY OF INCREASED PLATELET MEMBRANE FLUIDITY

Platelet membrane fluidity has been examined in a study population (n=300) including 123 patients with probable Alzheimer's disease, 38 patients with affective illness, 25 patients with Parkinson's disease, and 114 elderly controls (57,60,70). Of these disease groups, only the mean DPH anisotropy value of the group with Alzheimer's disease differed significantly from the elderly control group. Moreover, 45% of the group with Alzheimer's disease exhibited increased platelet membrane fluidity as determined by the 10th percentile for the controls (0.1920). Hicks and colleagues have replicated our findings of increased platelet membrane fluidity in Alzheimer's disease and also found no significant change in the fluorescence anisotropy of DPH-labeled platelet membranes from patients with multi-infarct dementia (19). These results indicate that the increase in platelet membrane fluidity associated with a subgroup of patients with Alzheimer's disease is not shared by patients with affective disorders, cognitively-intact patients with Parkinson's disease, or patients with multi-infarct dementia. However, the comparison groups of demented patients with Parkinson's disease or multiple infarcts showed a trend toward increased platelet membrane fluidity. This may have resulted from the coexistence of Alzheimer's disease, which has been diagnosed by neuropathological examination in a significant fraction of such patients at autopsy (1,5,35). As expected, if increased platelet membrane fluidity were relatively specific for Alzheimer's disease, this membrane characteristic predicted a poorer antidepressant response in elderly patients who presented with mixed symptoms of depression and dementia, for whom the primary diagnosis was unclear (68). Furthermore, the dichotomized platelet membrane fluidity phenotype was stable in 10 controls and 14 of 15 patients with Alzheimer's disease during a 1-year follow-up study (69).

FAMILY AND GENETIC STUDIES OF INCREASED PLATELET MEMBRANE FLUIDITY IN ALZHEIMER'S DISEASE

Family and epidemiologic studies have uniformly found first-degree relatives of patients with Alzheimer's disease to be at increased risk for developing primary dementia (2,6,16,18,26,30,34,64). Estimates of increased risk from these studies have ranged from 3.6 to 10.5 times those for the general population. Since platelet membrane fluidity appeared to be a stable trait with diagnostic specificity, we conducted a family study to determine whether increased platelet membrane fluidity aggregated in the families of patients with Alzheimer's disease, and if so, whether this platelet abnormality selectively ran in the families of probands who exhibited the abnormality (70). The prevalence of increased platelet membrane fluidity was found to be 3.2 to 11.5 times higher in asymptomatic, first-degree relatives (n=75) of clinically-diagnosed (n=23) or autopsy-confirmed (n=15) cases of Alzheimer's disease than in the control population. This result excluded nonspecific concomitants of chronic illness or neurodegeneration, as well as medication exposure, as potential causes of increased platelet membrane fluidity in Alzheimer's disease. Moreover, 87% of the relatives with increased platelet membrane fluidity were related to probands with this phenotype. In a subsequent study employing 421 first-degree relatives of patients with Alzheimer's disease and healthy controls,

Alzheimer's disease developed earlier among the relatives of demented patients with increased platelet membrane fluidity than among those of patients with normal platelet membrane fluidity (64).

In view of the evidence indicating that increased platelet membrane fluidity is a stable, familial trait that is vertically transmitted in families of patients with Alzheimer's disease, we conducted a complex segregation analysis of this phenotype in 26 nuclear families from 14 pedigrees identified by probands with Alzheimer's disease (7). The segregation analysis suggested that platelet membrane fluidity, as reflected by a continuum of DPH anisotropy values, is controlled by the inheritance of a single locus, PMF, with two alleles whose effects are additive. Moreover, the estimated frequency of the allele causing increased fluidity was 0.56, suggesting that genes affecting the expression of Alzheimer's disease may not be rare in the population. We have also studied a 50 year old patient with Down's syndrome who was admitted to our hospital for behavioral decline (62). Her karyotype revealed an unbalanced (14q,21q) translocation and the DPH anisotropy value of her labeled platelet membranes was 3.7 standard deviations below the mean for controls and 2.4 standard deviations below the mean for our cohort with Alzheimer's disease. Although other explanations are possible, a parsimonious explanation for this extreme value is that the patient carried three PMF alleles for increased fluidity. If this interpretation is correct, it would localize the PMF locus to 21q.

CELLULAR BASIS FOR INCREASED PLATELET MEMBRANE FLUIDITY IN ALZHEIMER'S DISEASE

At the cellular level, several lines of evidence suggest that the increase in platelet membrane fluidity associated with Alzheimer's disease results from an accumulation of internal membranes rather than a generalized abnormality of cell membranes. Ultrastructural studies have revealed an excess of atypical cells containing an overabundant system of trabeculated cisternae bounded by smooth membrane in platelet preparations from patients with Alzheimer's disease (65). Menashi and coworkers have reported that internal membranes exhibit higher membrane fluidity than external platelet membranes, as reflected by DPH anisotropy (32). Therefore, a relative increase in internal membranes may account for the increase in platelet membrane fluidity associated with Alzheimer's disease. In support of this hypothesis, platelets that exhibit increased membrane fluidity manifest a reduction in the cholesterol:phospholipid ratio that could be accounted for by an approximate doubling of the usual mass of internal membranes per cell (9). In addition, when intact platelets from patients with Alzheimer's disease are labeled with DPH, a process that preferentially labels external membrane, they fail to exhibit an alteration in fluorescence anisotropy (65). Moreover, erythrocyte ghost preparations, which lack internal membranes, also fail to exhibit an alteration in membrane fluidity as reflected by fluorescence (65) or electron spin resonance spectroscopy (29). The weight of this evidence suggests that the increase in platelet membrane fluidity associated with Alzheimer's disease may result from a dysregulation of platelet membrane biogenesis or turnover.

More recently, we have found a selective reduction in the specific activity of antimycin A-insensitive NADH cytochrome c reductase, an enzyme marker for smooth endoplasmic reticulum, in homogenates of platelets with increased membrane fluidity (56). In contrast, the fluidity change was not associated with significant alterations in the specific activities of NADH-dehydrogenase or leucine aminopeptidase, two enzymes that mark multiple internal membrane compartments, or bis(p-nitrophenyl) phosphate phosphodiesterase, a plasma membrane marker. These results further suggest that the internal membrane that accumulates in platelets with increased membrane fluidity may be abnormal endoplasmic reticulum.

DISCUSSION

Independent lines of evidence from clinical, epidemiologic, family, genetic, ultrastructural, and biochemical studies described in this section suggest that increased platelet membrane fluidity is a reliable biological index that defines a valid subtype of Alzheimer's disease. Alterations in cell membrane composition and structure have also been found in brain tissue from patients who died with Alzheimer's disease (3,8,37,55). Furthermore, abnormalities of endoplasmic reticulum have been observed in ultrastructural (46,48) and biochemical (10,53) studies of neurons within the neocortices of patients with Alzheimer's disease. Since the endoplasmic reticulum plays a central role in both the maturation and localization of cellular proteins (20,36), an inherited functional defect in this organelle as the result of the genotype at the PMF locus would likely lead to the synthesis of mismodified proteins along with the improper localization of proteins within or outside of neurons or their supporting cells. Indeed, mutations that give rise to defects in endoplasmic reticulum or Golgi function in other eucaryotic cells accumulate internal membranous structures and exhibit these expected pleiotropic effects (41,43,51). This model provides a potential mechanism whereby a single loss of function could give rise to many of the cellular changes that have been reported in the brains of patients with Alzheimer's disease.

The neurofibrillary tangle and senile plaque, two of the histopathologic hallmarks of Alzheimer's disease (23), are interesting to consider from this perspective. The neurofibrillary tangle is an intracellular aggregate of cytoskeletal proteins composed of paired helical filaments (24,49,52), and abnormal cytoskeletal elements associated with endoplasmic reticulum have been observed in autopsy brain tissue from patients with Alzheimer's disease (13,46). The extraneuronal senile plaque consists of a core of ß-amyloid (14,44,54) whose precursor is highly glycosylated (42) and contains two possible sites for N-glycosylation (22), a process that occurs in the endoplasmic reticulum along with other glycoconjugate processing steps (12,15,17,25). Moreover, morphometric studies have found that the density of both tangles and plaques is greatest in patients with early-onset dementia (28,50,67), a characteristic shared with the increased platelet membrane fluidity phenotype. Based on these similarities and the previous arguments, it is tempting to speculate that the expression of the PMF locus may play a role in the formation of these morphologic lesions in the brains of patients with Alzheimer's disease. Furthermore, since the biological activity of many essential proteins (enzymes, structural proteins, receptors, lectins, etc.) is dependent upon an effective system for marshalling the traffic of proteins through intracellular membrane systems, the PMF genotype may also contribute directly to the neuronal loss that occurs in Alzheimer's disease.

ACKNOWLEDGEMENT

The studies described in this review were supported by research grant MH43261, program grant AG03705, and center grants AG05133 and MH30915. Dr. Zubenko was the recipient of NIMH Research Scientist Development Award MH00540.

REFERENCES

1.   Alafuzoff, I., Iqbal, K., Friden, H., Adolfsson, R., and Winblad, B., 1987, Histopathological criteria for progressive dementia disorders: Clinical-pathological correlates and classification by multivariate data analysis. Acta. Neuropathol. 74:209-225.
2.   Amaducci, L. A., Fratiglioni, L., Rocca, W. A., Fieschi, C., Livrea, P., Pedone, D., Bracco, L., Lippi, A., Gandolto, C., Bino, G., Prencipe, M., Bonatti, M. L., Girotti, F., Carella, F., Tavolato, B., Ferla, S., Lenzi, G. L., Carolei, A., Gambi, A., Grigoletto, F., and Schoenberg, B. S., 1986, Risk factors for clinically diagnosed Alzheimer's disease: A case-control study of an Italian population. Neurology 36:922-931.
3.   Barany, M., Chang, Y., Arus, C., Rustan, T., and Frey, W. H., 1985, Increased glycerol-3-phosphorylcholine in post-mortem Alzheimer's brain (letter). Lancet 1:517.
4.   Becker, J. T., Huff, F. J., Nebes, R. D., Holland, A., and Boller, F., 1988, Neuropsychological function in Alzheimer's disease. Arch. Neurol. 45:263-268.
5.   Boller, F., Lopez, O. L., and Moossy, M., 1989, Diagnosis of dementia: Clinicopathologic correlates. Neurology 38:76-79.
6.   Breitner, J. C. S., and Folstein, M. F., 1984, Familial Alzheimer dementia: A prevalent disorder with specific clinical features. Psychol. Med. 14:63-80.

7.   Chakravarti, A., Slaugenhaupt, S., and Zubenko, G. S., 1989, Inheritance pattern of platelet membrane fluidity in Alzheimer's disease. Am. J. Hum. Genet. 44:799-805.

8.   Chia, L. S., Thompson, J. E., and Moscarello, M.A., 1984, X-ray diffraction evidence for myelin disorder in brain from humans with Alzheimer's disease. Biochim. Biophys. Acta. 775:308-312.

9.   Cohen, B. M., Zubenko, G. S., and Babb, S., 1987, Abnormal platelet membrane composition in Alzheimer's disease. Life Sci., 40:2445-2451.

10.   Cross, A. J., Crow, T. J., Dawson, J. M., Ferrier, I.N., Johnson, J. A., Peters, T. J., and Reynolds, G. P., 1986, Subcellular pathology of human neurodegenerative disorders. J. Neurochem. 47:882-889.

11.   Crystal, H. A., Horoupian, D. S., Katzman, R., and Jotkowitz, S., 1981, Biopsy-proved Alzheimer disease presenting as a right parietal lobe syndrome. Ann. Neurol. 12:186-188.

12.   Czichi, U., Lennarz, W. J., 1977, Localization of the enzyme system for glycosylation of proteins via the lipid-linked pathway in rough endoplasmic reticulum. J. Biol. Chem. 252:7901-7904.

13.   Ellisman, M., Ranganathan, R., Deerinck, T., Young, S., Terry, R., Mirra, S., 1987, Neuronal fibrillary cytoskeleton and endomembrane system organization in Alzheimer's disease, in "Alterations in the Neuronal Cytoskeleton in Alzheimer's Disease", G. Perry, ed., Plenum Publishing Corporation.

14.   Glenner, G. G., and Wong, C. W., 1984, Alzheimer's disease: Initial report of the purification and characterization of a novel cerebrovascular amyloid protein. Biochem. Biophys. Res. Commun. 120:885-890.

15.   Grinna, L. S., Robbins, P. W., 1979, Glycoprotein biosynthesis. J. Biol. Chem. 254:8814-8818.

16.   Heston, L. L., Mastri, A. R., Anderson, V. E., and White, J. L., 1981, Dementia of the Alzheimer type. Arch. Gen. Psychiatry 38:1085-1090.

17.   Hettkamp, H., Gegler, G., Bause, E., 1984, Purification by affinity chromatography of glucosidase I, an endoplasmic reticulum hydrolase involved in the processing of asparagine-linked oligosaccharides. Eur. J. Biochem. 142:85-90.

18.   Heyman, A., Wilkinson, W. E., Hurwitz, B. J., Heyman, A., Wilkinson, W. E., Hurwitz, B. J., Schmechel, D., Sigmon, A. H., Weinberg, T., Helms, M. J., and Swift, M., 1983, Alzheimer's disease: Genetic aspects and associated clinical disorders. Ann. Neurol. 14:507-515.

19.   Hicks, N., Brammer, M. J., Hymas, N., and Levy, R., 1987, Platelet membrane properties in Alzheimer and multi-infarct dementas. J. Alzheimer Dis. Assoc. Disord. 14:507-515.

20.   Hubbard, S. C., and Ivatt, R. J., 1981, Synthesis and processing of asparagine-linked oligosaccharides. Ann. Rev. Biochem. 50:555-583.

21.   Huff, F. J., Growdon, J. H., Corkin, S., and Rosen, T. J., 1987, Age at onset and rate of progression of Alzheimer's disease. JAGS 35:27-30.

22.   Kang, J., Lemaire, H.-G., Unterbeck, A., Salbaum, J. M., Masters, C. L., Grzeschik, K.-H., Multhaup, G., Beyreuther, K., and Muller-Hill, B., 1987, The precursor of Alzheimer's disease amyloid A4 protein resembles a cell-surface receptor. Nature 325:733-736.

23.   Khachaturian, Z., 1985, Diagnosis of Alzheimer's disease. Arch. Neurol. 42:1097-1105.

24.   Kidd, M., 1963, Paired helical filaments in electron microscopy of Alzheimer's disease. Nature, 197:192-193.

25.   Kiely, M. C., McKnight, G. S., Schimke, R. T., 1976, Studies on the attachment of carbohydrate to ovalbumin nascent chains in hen oviduct. J. Biol. Chem. 251:5490-5495.

26.   Larsson, T., Sjogren, T., and Jacobson, G., 1963, Senile dementia: A clinical, socio-medical and genetic study. Acta. Psychiat. Scand. 39(167):1-259.

27.   Lazarus, L. W., Newton, N., Cohler, B., Lesser, J., and Schweon, C., 1987, Frequency and presentation of depressive symptoms in patinets with primary degenerative dementia. Am. J. Psychiatry 144:41-45.

28.   Mann, D. M. A., Yates, P. O., Marcyniuk, B., 1985, Some morphometric observations on the cerebral cortex and hippocampus in presenile Alzheimer's

disease, senile dementia of the Alzheimer-type, and Down's syndrome in middle age. J. Neurol. Sci. 69:139-159.

29. Markesbery, W. R., Leung, P. K., and Butterfield, D. A., Spin label and biochemical studies of erythrocyte membranes in Alzheimer's disease. J. Neurol. Sci. 45:323-330.

30. Martin, R. L., Gerteis, G., and Gabrielli, W. F., 1988, A family genetic study of dementia of the Alzheimer type. Arch. Gen. Psychiatry 45:894-900.

31. Mayeux, R., Stern, Y., and Spanton, S., 1985, Heterogeneity in dementia of the Alzheimer type: Evidence of subgroups. Neurology, 35:453-461.

32. Menashi, S., Weintroub, H., and Crawford, N., 1981, Characterization of human platelet surface and intracellular membranes isolated by free flow electrophoresis. J. Biol. Chem. 256:4095-4101.

33. Merriam, A. E., Aronson, M. K., Gaston, P., 1988, The psychiatric symptoms of Alzheimer's disease. J. Am. Geriatr. Soc. 36:7-12.

34. Mohs, R. C., Breitner, J. C. S., Silverman, J. M., and Davis, K. L., 1987, Alzheimer's disease: Morbid risk among first-degree relatives approximates 50% by 90 years of age. Arch. Gen. Psychiatry 44:405-408.

35. Morris, J. C., McKeel, D. W., Fulling, K., Torack, R. M., and Berg, L., 1988, Validation of clinical diagnostic criteria for Azheimer's disease. Ann. Neurol. 24:17-22.

36. Palade, G., 1975, Intracellular aspects of the process of protein secretion. Science 189:347-358.

37. Pettegrew, J. W., Kopp, S. J., Minshew, N. J., Glonek, T., Feliksik, J. M., Tow, J. P., and Cohen, M. M., 1987, 31P nuclear magnetic resonance of phosphoglyceride metabolism in developing and degenerating brain: Preliminary observation. J. Neuropathol. Exp. Neurol. 46:419-430.

38. Reding, M., Haycox, J., Blass, J., 1985, Depression in patients referred to a dementia clinic: A three-year prospective study. Arch. Neurol. 42:894-896.

39. Rossor, M. N., Iverson, L. L., Reynolds, G. P., Mountjoy, C. Q., Roth, M., 1984, Neurochemical characteristics of early- and late-onset types of Alzheimer's disease. Br. Med. J. 288:961-964.

40. Roth, M., 1986, The association of clinical and neurological findings and its bearing on the classification and aetiology of Alzheimer's disease. Br. Med. Bull. 42:42-50.

41. Schekman, R., and Novick, P., 1982, The secretory process and yeast cell-surface assembly, in: "Molecular Biology of the Yeast Saccharomyces, Volume 2," J. N. Strathern, E. W. Jones, J. R. Broach, eds., Cold Springs Harbor, New York, pp. 361-398.

42. Schubert, D., Schroeder, R., LaCorbiere, M., Saitoh, T., Cole, G., 1988, Amyloid ß protein precursor is possibly a heparan sulfate proteoglycan core protein. Science 241:223-226.

43. Seger, N., Mulholland, J., Botstein, D., 1988, The yeast GTP-binding YPT1 protein and a mammalian couterpart are associated with the secretion machinery. Cell 52:915-924.

44. Selkoe, D. J., Abraham, C. R., Podlisny, M. B., Duffy, L. K., 1986, Isolation of low-molecular-weight proteins from amyloid plaques fibers in Alzheimer's disease. J. Neurochem. 46:1820-1834.

45. Shinitzky, M., and Barenholz, Y., 1978, Fluidity parameters of lipid regions determined by fluorescence polarization. Biochim. Biophys. Acta 515:367-394.

46. Sloper, J. J., Powell, T. P. S., Barnard, R. O., and Eglin, R. P., 1986, Ultrastructure abnormality in Alzheimer neurocortex. Lancet 1:511-512.

47. St. George-Hyslop, P. H., Tanzi, R. E., Polinsky, R. J., Haines, J. L., Nee, L., Watkins, P. C., Myers, R. H., Feldman, R. G., Pollen, D., Drachman, D., Growdon, J., Bruni, A., Foncin, J. F., Salmon, D., Frommelt, P., Amaducci, L., Sorbi, S., Placentini, S., Stewart, G. D., Hobbs, W. J., Conneally, P. M., and Gusella, J. F., 1987, The genetic defect causing familial Alzheimer's disease maps on chromosome 21. Science 235:885-890.

48. Sumpter, P. Q., Mann, D. M., Davies, C. A., Yates, P. O., Snowden, J. S., and Neary, D., 1986, A quantitative study of the ultrastructure of pyramidal neurons of the central cortex in Alzheimer's disease in relationship to the degree of dementia. Neuropathol. Appl. Neurobiol. 12:321-329.

49. Terry, R. D., 1963, The fine structure of neurofibrillary tangles in Alzheimer's disease. J .Neuropathol. Exp. Neurol. 22:629-642.
50. Terry, R. D., Hansen, L. A., DeTeresa, R., Davies, P., Tobias, H., Katzman, R., 1987, Senile dementia of the Alzheimer type without neocortical neurofibrillary tangles. J. Neuropathol. Exp. Neurol. 46:262-268.
51. Verner, K., and Schatz, G., 1988, Protein translocation across membranes. Science 241:1307-1313.
52. Wisniewski, H. M., Narang, H. K., and Terry, R. D, 1976, Neurofibrillary tangles of paired helical filaments. J. Neurol. Sci. 27:173-181.
53. Wolfe, L. S., Ng Ying Kin, N. M., Palo, J., and Haltia, M., 1982, Raised levels of cerebral cortex dolichols in Alzheimer's disease. Lancet 1:99.
54. Wong, C., W., Quaranta, V., Glenner, G., 1985, Neuritic plaques and cerebrovascular amyloid in Alzheimer's disease are antigenically related. Proc. Natl. Acad. Sci. USA 83:8729-8732.
55. Zubenko, G. S., 1986, Hippocampal membrane alteration in Alzheimer's disease. Brain Research, 385:115-121.
56. Zubenko, G. S., 1989, Endoplasmic reticulum abnormality in Alzheimer's disease: Selective alteration in platelet NADH-cytochrome c reductase activity. J. Geriatric Psychiatry Neurol. 2:3-10.
57. Zubenko, G. S., in press, Increased platelet membrane fluidity in Alzheimer's disease: An initial assessment of specificity. Neuropsychopharmacology.
58. Zubenko, G. S., Brenner, R. P., and Teply, I., 1988, Electroencepholographic correlates of increased platelet membrane fluidity in Alzheimer's disease. Arch. Neurol., 45:1009-1013.
59. Zubenko, G. S., Cohen, B. M., Growden, J., and Corkin, S., 1984, Cell membrane abnormality in Alzheimer's disease. Lancet, 2:235.
60. Zubenko, G. S., Cohen, B. M., Reynolds, C. F., Boller, F., Malinakova, I., and Keefe, N., 1987, Platelet membrane fluidity in Alzheimer's disease and major depression. Am. J. Psychiatry, 144:860-868.
61. Zubenko, G. S., and Ferrell, R. E., 1988, Monozygotic twins concordant for probable Alzheimer's disease and abnormal platelet membrane fluidity. Am. J. Med. Genet., 29:431-436.
62. Zubenko, G. S., and Howland, R., 1988, Markedly increased platelet membrane fluidity in Down's syndrome with a (14q,21q) translocation. J. Geriatric Psychiatry Neurol., 1:218-219.
63. Zubenko, G. S., Huff, R. J., Becker, J., Beyer, J., and Teply, I., 1988, Cognitive function and platelet membrane fluidity in Alzheimer's disease. Biol. Psychiatry, 24:925-936.
64. Zubenko, G. S., Huff, F. J., Beyer, J., Auerbach, J., and Teply, I., 1988, Familial risk of dementia associated with a biologic subtype of Alzheimer's disease. Arch. Gen. Psychiatry, 45:889-893.
65. Zubenko, G. S., Malinakova, I., and Chojnacki, B., 1987, Proliferation of internal membranes in platelets from patients with Alzheimer's disease. J. Neuropath. and Exp. Neurol., 46:407-418.
66. Zubenko, G. S. and Moossy, J., 1988, Major depression in primary dementia: Clinical and neuropathologic correlates. Arch. Neurol., 45:1182-1186.
67. Zubenko, G. S., Moossy, J., Martinez, A. J., Rao, G. R., Kopp, U., and Hanin, I., 1989, A brain regional analysis of morphologic and cholinergic abnormalities in Alzheimer's disease. Arch. Neurol. 46:634-638.
68. Zubenko, G. S., Reynolds, C. F., Perel, J., Decker, C., and Teply, I., 1988, Platelet membrane fluidity and treatment response in cognitively-impaired, depressed elderly. Psychopharmacology, 94:347-349.
69. Zubenko, G. S. and Teply, I., 1988, Longitudinal study of platelet membrane fluidity in Alzheimer's disease. Biol. Psychiatry, 24:918-924.
70. Zubenko, G. S., Wusylko, M., Boller, F., Cohen, B. M., and Teply, I., 1987, Family study of platelet membrane fluidity in Alzheimer's disease. Science, 238:539-542.

THERAPEUTIC PROPERTIES OF PHOSPHATIDYLSERINE IN THE AGING BRAIN

Maria Grazia Nunzi, Fabrizio Milan, Diego Guidolin,
Adriano Zanotti and Gino Toffano

Fidia Research Laboratories, Abano Terme, Italy

INTRODUCTION

The structural and regulatory function of phospholipids in biological membranes has stimulated investigation on the pharmacological
properties of these compounds, particularly at cerebral level. In the
aging brain, changes in lipid composition or content have been related
to alterations of cerebral membranes, such as reduction of membrane
fluidity and enzymatic activities, loss of receptors and decreased
efficiency of signal transduction mechanisms (21). This has led to the
proposal that administration of endogenously occurring phospholipids
may preserve the structural and functional integrity of central nervous
system membranes, and prevent or reverse neuronal dysfunctions that
occur in the course of aging, and age-associated neurodegenerative
disorders. In addition, administered phospholipids may participate in
phospholipid metabolism, yielding biologically active intermediates in
response to physiopathological phenomena (22).

Phosphatidylserine constitutes the major acidic phospholipid in
the brain. Besides its role in (Na^+-K^+)-ATPase activity (18), this
phospholipid is the membrane component responsible for activation of
protein kinase C (6). Its role in fusogenic processes and on cell
activation (16) has also been reported.

Exogenous phosphatidylserine, extracted and purified from bovine
brain (BC-PS), has been shown to affect neurotransmission and behavioral
performance in experimental animals. BC-PS administration stimulates
brain catecholaminergic turnover (23) and normalizes the impaired release of acetylcholine in the cerebral cortex of aged rats (19).
Treatments with BC-PS increase learning and memory functions in aged
rodents (8,11) and prevent the age-related decay in active avoidance
behavior (25).

Normal and pathological aging has been characterized neurochemi-
cally by a decrease, among others, of cholinergic and monoaminergic
neurotransmission. Furthermore, it is well known that in neurodegenerative conditions which affect the elderly, selective populations of neurons are at risk, e.g. the cholinergic neurons in the basal forebrain.
In this study we have concentrated on documenting, in aged rats,
structural deterioration of the cholinergic neurons of the basal

forebrain, and the counteracting effect of BC-PS administration. We demonstrate that morphological data may be related to the capability of BC-PS treatment to reverse deficits in spatial behavior in aged rats.

Pharmacological treatment and methods

The animals used in the study were male Sprague-Dawley rats (Charles River, Italy). Animals were maintained on a 12hr light: 12hr dark cycle, with access to drinking water and food (Standard Diet No. 4RF18, Italiana Mangimi, Milano) _ad libitum_. An aqueous suspension of BC-PS was given to the rats in place of drinking water. The concentration of the phospholipid was adjusted throughout the course of the treatment, in order to ensure an average daily intake of 50 mg/kg of BC-PS per rat (17). For the morphological analysis BC-PS was administered from the age of 15 months to sacrifice. In the behavioral study, BC-PS administration started 1 week after screening, and lasted until the end of behavioral testing.

Brains of young-adult (4 months), aged (27 months) and aged matched BC-PS-treated rats, were processed for choline acetyltransferase (ChAT)-immunocytochemistry. After perfusion with a mixture of aldehydes (4% paraformaldehyde and 0.05% glutaraldehyde in 0.1 M sodium phosphate buffer, pH 7.4), the brains were cut on a Vibratome. Sections through the septal complex (medial septum and diagonal band) were incubated with a monoclonal antibody to ChAT (Boehringer, Type I). They were then processed for peroxidase-antiperoxidase immunocytochemistry. The sections were then mounted on glass slides coated with chrome alum gelatin, reacted with 2% O_sO_4, and processed by standard procedures for light microscopic observation.

For morphometric analysis of ChAT-immunoreactive neurons, four parameters were considered: number of ChAT-positive cells, area covered by all immunoreactive profiles (somata and neuronal processes), somatic cross-sectional area and maximum diameter. Morphometric analysis was carried out by means of a computerized image analysis system (IBAS, Kontron, Zeiss). The analyzed sections contained the greatest extension of the medial septum and diagonal band. For each animal the mean of the sections was considered. Differences across the animal groups were tested by means of the Student's t test.

In the behavioral study, young-adult (5 months) and old (21-24 months) rats were tested in the Morris' water maze (adapted from Morris, 15) for assessing spatial memory. The swim path and latency to the hidden platform were automatically recorded. Old impaired rats were selected for the study were those whose mean escape latencies were above the 99% confidence limits of the young-adult group. The remaining rats constituted the old non-impaired group. Rats were trained in 2 blocks of 4 trials each, every day of each test week (7th and 12th week). After the last trial of test week 7, the platform was removed and rats were submitted to a single "spatial probe" trial for evaluation of searching behavior.

RESULTS

The morphometric immunohystochemical study showed regressive changes in the cholinergic neuronal population of forebrain nuclei of aged rats. The cholinergic cell number was markedly decreased in aged animals with respect to young-adult subjects (-24.5%, $p<.01$), suggesting a high vulnerability of cholinergic cells to age-related compromising processes. Similarly, the area covered in the sections by all immunoreactive structures decreased significantly in old-untreated rats as compared to younger controls (-31.7%, p<0.05). This is consistent with the reduction in soma size and maximum diameter of ChAT-positive neurons observed in the aged group of animals with respect to young-adult rats (-14.8%, p<.05 and -6.7%, p<.01, respectively). No statistically significant differences in the above reported parameters were however, found between aged BC-PS-treated rats and young animals (Fig. 1).

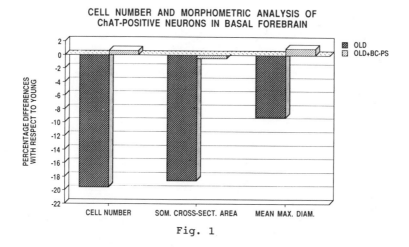

Fig. 1

As to the behavioral performance in the Morris water maze, mean escape latencies during the screening test indicated that a subpopulation of aged rats was impaired in the acquisition of the spatial task.

Mean escape latencies of young-adult, old nonimpaired, old impaired and old impaired BC-PS-treated rats are shown in Fig.2.

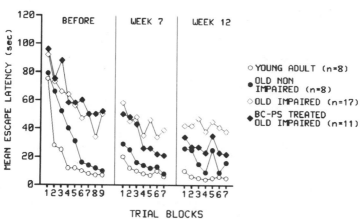

Fig.2

The performance of old impaired control rats did not change over the last four trial blocks (test week 7 and test week 12) as compared to the screening test, and continued to be significantly different from that of both young-adult and old nonimpaired rats over the 2 test weeks. On the contrary, treatment with BC-PS improved the performance of old impaired rats at both retesting weeks, as shown by the signifi-cant decrease in escape latencies when compared to screening (p<.05; Dunnett's test).

The searching behavior of all animal groups (i.e. the ability of rats to use spatial cues to locate the platform in the pool) was evaluated in a single "spatial probe" trial. Old impaired control rats did not show any spatial bias, suggesting that higher escape latencies were due to impaired ability to use spatial cues in locating the hidden platform. On the contrary, the searching behavior of old impaired BC-PS-treated rats was focused on the previous platform location, similarly to young-adult and old non-impaired animals.

DISCUSSION

Atrophy and cell loss in the cholinergic neuronal population of the aged rat basal forebrain, may be hypothesized to be related to age-induced retrograde changes in cortical cholinergic afferents. In this regard, spatial memory deficit in aged rodents have been recently correlated with degeneration of fibers in the hippocampus (14). Fur-thermore, reduced synaptic efficacy (9) and loss of synaptic contact (1) are known to occur in the hippocampal formation of aged rats. Decreased production or sensitivity to neuronotrophic factors in target areas and/or slowing of synaptic renewal, may be phenomena involved in the deterioration of the cholinergic basal forebrain neurons. Since spatial memory deficit in aged animals has been related to decreased cholinergic function (12,13), the age-associated impairment in spatial behavior reported in this study, may be a consequence of the age-in-duced morpho-functional derangement of the septo-hippocampal pathway.

Long-term, oral BC-PS administration improves spatial behavior in aged impaired rats and prevents atrophy of cholinergic neurons. These effects may be, at least in part, related to the capability of BC-PS treatment to counteract loss of axo-spinous connections in the hippo-campus of aged rodents (17). Given the relevance of synaptic renewal to cognitive functions, the beneficial effects of BC-PS on age-mediated decay in memory function may reflect improved interneuronal connec-tivity. As to the mechanism of action, administration of BC-PS to aged rats has been shown to normalize the cholesterol to phospholipid ratio and to enhance Na^+-K^+-ATPase activity in synaptosomal membranes of aged rats (4). More recently, we found that BC-PS treatment is capable of restoring the balance between the cytosolic and particulate forms of protein kinase C in the cerebral cortex of aged rodents (5). These effects may be factors relevant in old age for renewal and maintenance of synaptic connections since synaptic function has a major role in stabilization of synaptic sites (7). Maintenance of trophic degree of cholinergic neurons in old age is also consistent with the finding that serine phospholipids show synergism with nerve growth factor (2). We are considering, therefore, that small amounts of serine phospholipids crossing the blood-brain barrier can stimulate membrane-bound events relevant to neuronal function. This is supported by the observation that micromolar concentrations of lysophosphatidylserine stimulate the incorporation of polyunsaturated fatty acids into membrane phospho-lipids (20). Since several membrane activities (e.g. channel function and receptor mobility) depend on the proper degree of acyl chain unsaturation, this effect *per se* might explain the beneficial effects of BC-PS administration on age-dependent changes in neuronal structure and function.

In addition, with regard to the effect of serine phospholipids on cells of the immune system (2,3) we are further suggesting a role for serine phospholipids in mediating cell defense reactions and repair mechanisms.

In conclusion, the experimental outcome obtained with BC-PS may justify the use of this endogenous phospholipid as a therapeutic agent for the treatment of pathological brain aging (10,24).

REFERENCES

1. Bertoni-Freddari, C., Giuli, C., Pieri, C., and Paci, D., 1986, Quantitative investigation of the morphological plasticity of synaptic junctions in rat dentate gyrus during aging, Brain Res. 366:187-192

2. Bruni, A., 1988, Autacoids from membrane phospholipids, Pharmacol. Res. Commun., 20:529-544

3. Bruni, A., Bigon, E., Boarato, E., Mietto, L., Leon, A., and Toffano G., 1982, Interaction between nerve growth factor and lysophosphatidylserine on rat peritoneal mast cells, Febs. Lett. 138:190-192

4. Calderini, G., Aporti, F., Bellini F., Bonetti, A.C., Rubini, R., teolato, S., Xu, C., Zanotti, A. and Toffano, G., 1985, Phospholipids as pharmacological tools in the aging brain. in: "Phospholipids in the Nervous System" Vol. 2, Physiological Roles. L.A. Horroksn J.N. Kanfer and G. Porcellati, eds., Raven Press, New York, pp. 11-19

5. Calderini, G., Bellini, F., Bonetti A.C., Galbiati, E., Guidolin, D., Milan, F., Nunzi, M.G., Rubini, R., Zanotti, A., and Toffano, G, 1987, Pharmacological properties of phosphatidylserine in the aged brain, Clin. Trials 24:9-17

6. Castagna, M., Takai, Y., Kaibuchi, K., Sano, K., Kikkawa, V., and Nishizuka, Y., 1982, Direct activation of calcium-activated, phospholipid-dependent protein kinase by tumor-promoting phorbol esters, J. Biol. Chem. 257:7847-7851

7. Changeux, J., and Danchin, P., 1976, Selective stabilization of developing synapses as a mechanism for the specification of neuronal networks. Nature 264:705-712

8. Corwin, J., Dean III, R.L., Bartus, R.T., Rotrosen, J., and Watkins, D.L., 1985, Behavioral effects of phosphatidylserine in the aged Fischer 344 rat: amelioration of passive avoidance deficits without changes in psychomotor task performance, Neurobiol. Aging 6:11-15

9. De Toledo-Morrel, L., and Morrel, F., 1985, Electrophysiological markers of aging and memory loss in rats, Ann. N.Y. Acad. Sci. 444:296-311

10. Delwaide, P.J., Hurlet, A., Hambourg, A.M., and Klieff, M., 1986, Double-blind randomized study of phosphatidylserine in senile demented patients, Acta Neurol. Scand. 73:136-140

11. Drago, F., Canonico, P.L., and Scapagnini, U., 1981, Behavioral effects of phosphatidylserine in aged rats, Neurobiol. Aging, 2:209-213

12. Gage, F.H., and Bjorklund, A., 1986, Cholinergic septal grafts into the ippocampal formation improve spatial learning and memory in aged rats by an atropine-sensitive mechanism, J. Neurosci. 6:2837-2847

13. Gallagher, M., and Pelleymounter, M.A., 1988, An age-related spatial learning deficit: choline uptake distinguishes "impaired" and "unimpaired" rats, Neurobiol. Aging 9:363-369

14. Greene, E., and Narajo, J.N., 1987, Degeneration of hippocampal fibers and spatial memory deficit in the aged rat, Neurobiol. Aging 8:35-43

15. Morris, R., 1984, Developments of a water-maze procedure for studying spatial learning in the rats, J. Neurosci. Methods 11:47-60

16. Nishizuka, Y., 1984, Turnover of inositol phospholipids and signal transduction. Science 225:1365-1370

17. Nunzi, M.G., Milan, F., Guidolin, D., and Toffano, G., 1987, Dendritic spine loss in hippocampus of aged rats. Effect of brain phosphatidylserine administration, Neurobiol. Aging 8:501-510

18. Palatini, P., Dabboni-Sala, F., and Bruni, A., 1972, Reactivation of a phospholipid-depleted sodium potassium-stimulated ATP-ase, Biophys. Acta 288:413-422

19. Pedata, F., Giovannelli, L., Spignoli, G., Giovannini, M.G., and Pepeu,G., 1985, Phosphatidylserine increases acetylcholine release from cortical slices in aged rats, Neurobiol. Aging 6:337-339

20. Sbaschnig-Agler, M., and Pullarkat, R.K., 1985, Lysophosphatidylserine-dependent incorporation of Acyl-CoA into phospholipids in rat brain microsomes, Neurochem. Int. 7:295-300

21. Schroeder, F., 1984, Role of membrane lipid asymmetry in aging, Neurobiol. Aging 5:323-333

22. Toffano, G., and Bruni, A., 1980, Pharmacological properties of phospholipid liposmes. Pharmacol. Res. Commun. 12:408-417

23. Toffano, G., Leon, A., Mazzari, S., Teolato, S., and Orlando, P., 1978, Modification of noradrenergic hypothalamic system in rats injected with phosphatidylserine liposomes, Life Sci. 23:1093-1102

24. Villardita, C., Grioli, S., Salmeri, G., Nicoletti, F., and Pennisi, G., 1987, Multicentre chemical trial of brain phosphatidylserine in elderly patients with intellectual deterioration, Clin. Trials J. 24:84-93

25. Zanotti, A., Rubini, R., Calderini, G., and Toffano, G., 1987, Pharmacological properties of phosphatidylserine: effects on memory function, in: "Nutrients and Brain Function," W.B. Essman, ed., Karger, Basel, pp. 95-102

PHOSPHOLIPIDS AND CHOLINE DEFICIENCY.

Steven H. Zeisel

Nutrient Metabolism Laboratory
Departments of Pathology and Pediatrics
Boston University School of Medicine
85 East Newton Street, Room M1002
Boston, Massachusetts, U.S.A.

INTRODUCTION

Lecithin (phosphatidylcholine; PtdCho) is the major phospholipid constituent of most membranes, and it is vital to the normal function of every cell and organ. PtdCho is synthesized from choline in all organs[29,30]. Much of this choline comes from the diet. PtdCho, in foods such as liver, eggs, soybeans and peanuts, is the most important source of choline in the human diet[29]. This dietary choline interacts with methionine and folate metabolism so that changes in the availability of choline alter biological methylation reactions. Choline is also used by tissues to make acetylcholine. This chapter will focus on PtdCho as a source of choline in the diet, and upon the biological consequences of diminished availability of choline.

CHOLINE DEFICIENCY

There are several lines of evidence that indicate that choline may be an essential nutrient in man: 1) many species of animal become ill if fed a choline-deficient diet[30], 2) human cells in culture require choline[10], 3) choline stores are depleted in blood of malnourished humans[24], and 4) humans fed by vein with solutions containing little or no choline develop liver dysfunction that is similar to that seen in choline deficient animals[24].

Choline deficiency is associated with fatty infiltration of the liver in the rat, dog, hamster, pig, baboon and chicken[1,4,17]. Fatty acids can be transported out of the liver only in the form of triglycerides within lipoproteins (VLDL, HDL). An essential constituent of these lipoproteins is PtdCho. When adequate supplies of PtdCho are not available, the liver is unable to export triglyceride and becomes infiltrated with fat[28]. This is an absolute dependence for PtdCho, no other phospholipid will suffice. Renal function may also be impaired due to membrane PtdCho deficiency; choline-deficient animals have abnormal concentrating ability, free water reabsorption, glomerular filtration rate, renal plasma flow and gross renal hemorrhaging[20]. Choline deficiency in animals has also been associated with infertility, growth retardation, bony abnormalities, spontaneous hepatocarcinogenesis and increased sensitivity to carcinogens[30], (see discussion below). The expression of choline deficiency varies, and depends upon dietary calorie source and amino acid content, and upon the rate of growth of the animal. The rat may have greater requirements than humans do for methionine, as cystine is needed for hair growth. Methyl-donors (such as methionine) can spare some of the choline requirement, as choline can be formed *de novo* by the methylation of phosphatidylethanolamine within liver and other organs, (see discussion below).

Humans often develop fatty liver during total parenteral nutrition (TPN) therapy[24]. The amino acid solutions used during TPN contain no choline, and, therefore, do not rehabilitate their low choline stores[24]. It would be simplistic to claim that all hepatic damage can be related to a single etiology. However, the hepatic dysfunction associated with choline deficiency appears to be similar to that seen during therapy with TPN. In rats, oral or intravenous supplements of choline reversed hepatic lipid accumulation associated with TPN[16]. Other investigators, however, did not observe such an effect of choline[15]. Choline deficiency is not the sole explanation for TPN-associated hepatic dysfunction, but rather it is possible that choline deficiency may contribute to the problem. The extent of this contribution will depend upon methionine-status, calorie load, choline demands related to growth, and many other factors. The symptoms of choline deficiency will be expressed when the demands for choline molecules exceed the body's capacity for choline biosynthesis. This would explain why undernourished individuals, who are not increasing tissue mass, and who are not synthesizing much triglyceride, do not develop fatty liver despite having choline-deficiency. Their demands for choline have been diminished due to starvation.

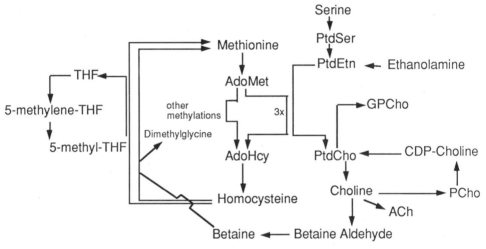

Figure 1. Metabolic Pathways for PtdCho and Choline
Abbreviations used: THF = tetrahydrofolate, AdoMet = S-adenosylmethionine, AdoHcy = S-adenosylhomocysteine, PtdSer = phosphatidylserine, PtdEtn = phosphatidylethanolamine, GPCho = glycerophosphocholine, DAG = diacylglycerol, PCho = phosphocholine, TMA = trimethylamine.

SPECIAL REQUIREMENTS IN THE NEONATE

A major portion of the choline needed by mammals is used to sustain tissue growth. PtdCho is the major phospholipid in all membranes, amounting to 0.1 to 1% of dry body weight. Neonates grow very quickly, and therefore must have increased demands for choline. In addition, availability of choline for acetylcholine synthesis is critical as brain develops. Rats exposed perinatally to supplemental choline have long-term enhancement of spatial memory capacity when, as adults, they are tested on radial arm maze tasks[18]. These behavioral changes are associated with characteristic neuroanatomical changes. The activity of the pathway which makes choline molecules from phosphatidylethanolamine and AdoMet (methylation pathway catalyzed by phosphatidylethanolamine methyltransferase, PEMT; figure 1) in newborn mammals is very low, while choline requirements for use in growth related membrane formation are very high[32]. For this reason, the sequelae of choline deficiency are most easily elicited in young, growing animals.

Milk is the first, and often the sole food for the mammalian neonate. It contains large amounts of free choline, PtdCho and sphingomyelin (a choline-containing phospholipid)[31] (see figure 2). Artificial formulas can have a choline content that is very different from that of mother's milk[31]. There are characteristic changes in the choline content of human milk that occur during lactation (see figure 2). We have observed that mammary is capable of concentrative uptake of choline from maternal blood[8], and of *de novo* synthesis of choline molecules via the transmethylation of phosphatidylethanolamine[27].

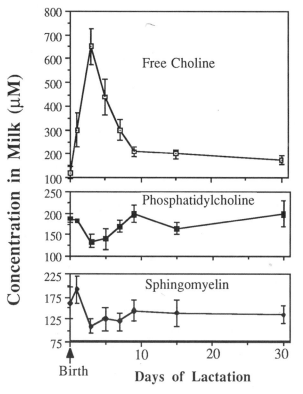

Figure 2. Changes in the concentrations of choline-containing compounds in human milk during stages of lactation.

Mothers collected 1 ml samples of milk by manual expression after the midday feeding of their infant (hind-milk). Birth day was designated as day 1. Data are expressed as mean concentration in nmoles/ml ± SEM. (n= 5-6 per point).

From Zeisel *et al.*[31].

CHOLINE, METHIONINE AND FOLATE METABOLISM

In rats, methionine or folate deficiency exacerbates the hepatic effects of choline deficiency[30]. The regulation of methionine metabolism is complex, and we do not have room to discuss it completely, rather we will focus on the potential interactions between choline, folate and methionine metabolism (figures 1 & 3). Several mechanisms exist for the introduction of methyl-groups into mammalian biochemical pathways. S-adenosylmethionine

(AdoMet) is required for most biological methylation reactions, and it is made from methionine. The key reactions in the regeneration of methionine from homocysteine lie at the intersection of choline and 1-carbon metabolic pathways. Betaine, a metabolite of choline, serves as the methyl donor in a reaction converting homocysteine to methionine (catalyzed by betaine:homocysteine methyltransferase). The only alternative methyl-donor for regeneration of methionine is 5-methyltetrahydrofolate, via a reaction catalyzed by 5-methyltetrahydro-folate:homocysteine methyltransferase (a vitamin B_{12} dependent reaction).

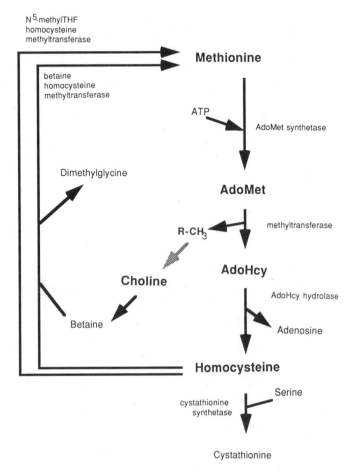

Figure 3. Methionine Metabolism
Abbreviations used: THF = tetrahydrofolate, AdoMet = S-adenosylmethionine, AdoHcy = S-adenosylhomocysteine, R-CH3 = methylated product.

There are several reasons to believe that a disturbance in 1-carbon/methionine metabolism would result in changes in choline metabolism and vice versa. Hepatic AdoMet concentrations are decreased in animals ingesting diets deficient in choline[25,33]. In part, this is because the 5-methyltetrahydrofolate:homocysteine methyltransferase reaction alone cannot fulfill the total requirement for methionine when the betaine dependent remethylation of homocysteine is limited by the availability of betaine[2,11,12]. Betaine concentrations in livers of choline deficient rats are markedly diminished[33]. In addition, we observed that AdoMet concentrations in choline-deficient liver fell before, and to a greater extent than did methionine concentrations[33]. We believe that an additional reason that AdoMet levels decrease during choline deficiency is that utilization of AdoMet for the synthesis of PtdCho increases in order to maintain membrane integrity. The key question - does the use of AdoMet to make PtdCho amount to a significant sink for AdoMet, or is it only a tiny portion of the methylation activity within the cell?

CHOLINE DEFICIENCY AND PtdCho BIOSYNTHESIS

PtdCho is the major phospholipid in mammalian cell membranes. There are two distinct pathways for the biosynthesis of PtdCho (see figure 1): the CDP-pathway using preformed choline, and the methylation pathway using phosphatidylethanolamine and AdoMet (PEMT-pathway)[30]. Both are present with high activity in the liver, though normally the CDP-pathway makes the greatest contribution. In the CDP-pathway, choline is first phosphorylated (choline kinase) and then CDP-choline is formed from CTP and phosphocholine (CTP phosphocholine cytidylyltransferase). This step is usually rate limiting (see discussion below)[22]. Finally CDP-choline and diacylglycerol (DAG) are used to form PtdCho (CDP-choline:DAG phosphotransferase).

PtdCho concentrations in membranes are determined by the activity of these two pathways. If mechanisms exist to maintain PtdCho, then a decrease in the contribution of one pathway would be made up for by an increase in the contribution of the other. The key to control of the CDP-pathway seems to be the regulation of cytidylyltransferase activity. This enzyme exists in a relatively inactive form in the cytosol, and the process of activation involves translocation to the membrane where it can associate with anionic phospholipids and lysophosphatidylethanolamine which are required for maximal activity. It has been suggested that decreased membrane PtdCho causes increased membrane binding (and hence activation) of cytidylyltransferase[26]. Free fatty acids, monoacylglycerol, or DAG can associate with the cytidylyltransferase, giving it an hydrophobic domain, thereby making it easier for the enzyme to bind to membranes, though high levels of DAG or of free fatty acids are not an absolute requirement for translocation[22,26]. PtdCho concentrations within liver decreased in choline deficient rats[9,14]. Treatment of cells with phospholipase C depletes membranes transiently of PtdCho, generates DAG and free fatty acids, and activates cytidylyltransferase[26]. Total CDP-choline:diacylglycerol-phosphocholine transferase activity was normal, while cytosolic cytidylyltransferase activity decreased during choline deficiency[23]. Since the level of total activatable cytidylyltransferase remained constant during choline deficiency, decreased cytosolic enzyme activity must mean that membrane bound activity increased.

Choline deficiency, by decreasing the absolute contribution of the CDP-pathway (despite activation of cytidylyltransferase, the pathway must be limited by choline availability), increases the requirement for phosphatidylcholine synthesis via the sequential methylation of phosphatidylethanolamine. Choline deficiency does increase the utilization of AdoMet for phosphatidylcholine synthesis[23]. Despite such compensatory mechanisms, rats deficient in choline have reduced hepatic PtdCho concentrations, probably because diminished AdoMet pools limit PEMT activity.

CHOLINE DEFICIENCY, DIACYLGLYCEROL AND CANCER

Many studies have demonstrated that feeding a choline-devoid diet to experimental animals is an effective promoter of carcinogenesis[21]. In the mouse, choline deficiency enhances the development of hepatocarcinomas in the absence of addition of any known carcinogen[21]. In the rat, 51% of animals fed a choline-methionine deficient diet for from 13-24 months developed hepatocarcinoma, despite being exposed to no known carcinogen; a diet containing 0.8% added choline completely prevented the development of cancer in these animals[13]. Choline deficiency also makes it much more likely that administration of a known carcinogen, such as aflatoxin, will result in cancer formation, suggesting that choline-deficiency acts as a promoter of carcinogenesis[21].

There are several mechanisms which have been suggested for the cancer-promoting effect of a choline-deficient diet. These include increased cell proliferation, decreased methylation of DNA, and formation of lipid peroxides[21]. None of the above hypotheses is entirely satisfactory, and given our new awareness of the potential role of diacylglycerol in carcinogenesis, we propose a new mechanism.

In choline deficient animals extremely large amounts of lipid (mainly triglycerides, but also diacylglycerol) can accumulate in liver, eventually filling the entire hepatocyte[6]. This occurs because triglyceride must be packaged as very low density lipoprotein (VLDL) to be exported from liver. PtdCho is an obligatory component of VLDL, and choline is required for PtdCho biosynthesis[28]. Thus, choline deficient liver cannot export VLDL. We have measured DAG concentrations with a stereoisomer-specific technique, and observed that large amounts of 1,2-sn-DAG accumulate in choline deficient livers[6].

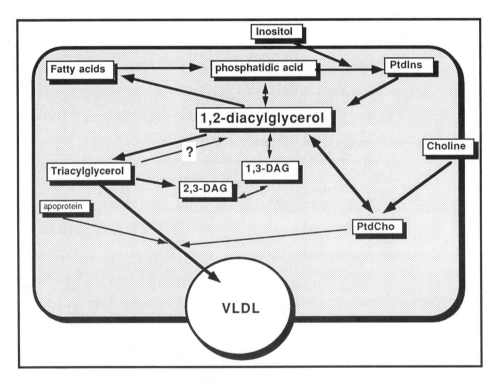

Figure 4. Synthesis of Diacylglycerol in Liver
Diacylglycerol is formed from phosphatidic acid, triacylglycerol and membrane phospholipids within liver.
Abbreviations used: DAG = diacylglycerol; PtdCho = phosphatidylcholine; PtdIns = phosphatidylinositol bisphosphate

We have discussed how DAG is an important intermediate for the biosynthesis of membranes. 1,2-*sn*-DAG is also an important second messenger, generated when membrane PtdCho and phosphatidylinositides (phosphatidylinositol bisphosphate; PtdIns) are metabolized. Signal transduction via generation of DAG and metabolites of PtdIns has been extensively reviewed elsewhere[3]. Briefly, PtdCho and PtdIns breakdown are triggered after a membrane receptor is excited (figure 4). Messengers that induce phospholipid turnover include: acetylcholine (M1), norepinephrine (α1), epinephrine (α1), dopamine, histamine (H1), serotonin, vasopressin (V1), angiotensin II, cholecystekinin, gastrin, pancreozymin, substance P, bradykinin, thromboxane, thrombin, collagen, platelet activating factor, secretagogues, growth factors, mitogens. Products generated after such receptor activation include inositol-1,4,5-trisphosphate (Ins-1,4,5-P_3) and 1,2-*sn*-DAG. Ins-1,4,5-P_3 is a water soluble product, which acts to release calcium from stores in the endoplasmic reticulum. This increase in cytosolic calcium makes more calcium available for activation of protein kinase C (PKC). The neutral diacylglycerol molecule remains within the membrane after hydrolysis of PtdIns and increases the affinity of PKC for calcium, thereby making it easier to activate PKC. The appearance of diacylglycerol in membranes is usually transient, and therefore PKC is activated only for a short time after a receptor has been stimulated.

There are several other points at which the lecithin and PtdIns metabolic pathways intersect. Agonists which activate PKC activate a lecithin-specific phospholipase C[5]. Thus, phorbol esters, which are analogs of diacylglycerol and activate PKC without triggering PtdIns breakdown, stimulate the release of phosphocholine, diacylglycerol and arachidonate from lecithin[5]. In addition, there appear to be receptors which, through a mechanism which is independent of PKC, stimulate hydrolysis of lecithin to form diacylglycerol and phosphocholine [5]. Thus, lecithin breakdown can act to sustain a message which was initially transmitted via inositide breakdown, and lecithin breakdown can generate second messengers independent of inositide breakdown. This is important, as diacylglycerol can activate PKC in the absence of an increase in intracellular calcium[5]. The fatty acid species in lecithin are different from those in PtdIns, therefore the diacylglycerols generated from each will differ. PKC is present in several forms within cells. It is possible that diacylglycerol generated from lecithin is recognized by a special form of PKC, which acts in a different domain than does inositide-stimulated PKC, and that this may provide a mechanism for maintenance of signal specificity. Other products of lecithin hydrolysis, such as phospho-choline, could also be second messengers [5].

Some of the most potent mitogens and tumor promoters, the phorbol esters, are analogs of 1,2-DAG which have higher affinity than 1,2-DAG for the same site on PKC; they cause PKC translocation to membranes and long lasting activation. Prolonged activation of PKC by these compounds leads to down regulation of the enzyme (i.e. proteolysis to a form which is not bound to the membrane). It is believed that the carcinogenic effects of phorbol esters may be explained by their interactions with PKC (though it is not clear whether activation or down-regulation is more important in this regard). The activation of PKC by mitogens can be very impressive. Buckley *et al.*[7] found that prolactin, a mitogen for liver, stimulated PKC activity several hundred-fold in rat liver nuclear membrane; probably by a phospholipid-1,2-DAG mediated pathway. Choline deficiency is also mitogenic in the liver. Transfection of cells with a mutant PKC that is always activated results in the transformation of the cell[19]. Gene expression abnormalities that are often associated with tumors, can also be associated with alterations in 1,2-DAG mediated pathways. For example, NIH 3T3 cells transformed with Ha-*ras* or Ki-*ras,* v-*src,* and v-*fms* oncogenes have elevated 1,2-DAG levels as well as tonic activation and partial down regulation of PKC[6]. Activated PKC, in turn, may participate in mechanisms leading to the induction of expression of the *c-myc* oncogene[6]. Thus, we propose that the accumulation of

1,2-DAG in choline deficient liver may cause prolonged activation of PKC, acting as an endogenous tumor promoter.

CHOLINE DEFICIENCY IN HUMANS

At the current time we are characterizing the effects of making normal humans choline deficient. For a week at the beginning and end of the study, subjects eat a diet that contains choline, while for 3 weeks in the middle of the study the subjects may, or may not, eat choline (see figure 5). The choline content of the deficient diet is 13 mg/70 kg body weight/day, the choline-containing diet has 713 mg/70 kg bw/day. Both diets deliver 40 Kcal/kg bw (10% protein, 35% fat, 55% carbohydrate) in the form of liquid shakes. The diet meets the RDA for all amino acids, vitamins and minerals; of special interest - folate (300 μg//70 kg bw/day) and vitamin B12 (9 μg//70 kg bw/day). The protein source is a-soy protein (STA-PRO 3200, Central Soya) which contains adequate amounts of methionine (921 mg/70 kg bw/day). We report preliminary data from 7 subjects {Control (n =3; CONTROL); Choline deficient (n= 4; DEFICIENT)}.

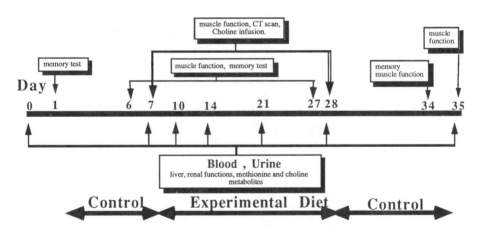

Figure 5. Design of Human Choline Deficiency Study

Plasma concentrations of choline dropped 30% in the DEFICIENT group (decreased in all subjects) between day 7 and day 28 (period on deficient diet); there were no changes observed in the CHOLINE group (figure 6). Plasma phosphatidylcholine concentrations decreased 15% in the DEFICIENT group and increased slightly in the CHOLINE group. We conclude that humans fed a choline deficient diet for 3 weeks have decreased circulating levels of phosphatidylcholine and choline.

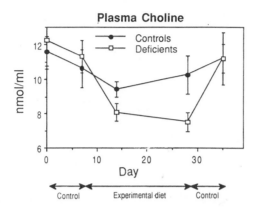

Figure 6. Plasma choline concentrations in humans fed a control or choline deficient diet.

Humans were treated as described in figure legend 5. Plasma choline was measured using a gas chromatography/mass spectrometric method.

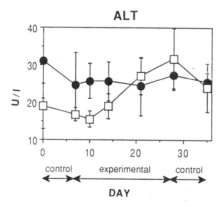

Figure 7. Plasma alanine aminotransferase (ALT) activity in humans fed a control or choline deficient diet.

Humans were treated as described in figure legend 5. Alanine aminotransferase activity (an indicator of hepatocellular damage) was measured in plasma using a spectrophotometric method. □ = deficient subjects; ● = control subjects.

Enzymes which leak from liver cells into blood are sensitive indicators of hepatotoxicity. Alanine aminotransferase (ALT) and aspartate aminotransferase activities in plasma increased 90% and 65% respectively in the DEFICIENT group from day 7 (end of control diet) to day 28 (end of experimental diet), and did not increase in the CHOLINE group (figure 7). We can detect these changes because each subject serves as his own control, however, at no time did any of these values exceed the upper range of normal for the general population. We conclude that we are detecting changes consistent with modest hepatic dysfunction.

As discussed earlier, choline deficiency impairs the excretion of cholesterol by the liver. We found that choline deficient humans had significantly lower plasma low density lipoprotein (LDL) cholesterol than did control subjects (figure 8).

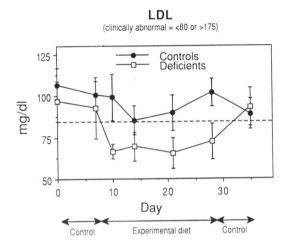

Figure 8. Plasma low density lipoprotein cholesterol (LDL) in humans fed a control or choline deficient diet.
Humans were treated as described in figure legend 5. Total cholesterol concentrations decreased in a similar manner in the DEFICIENT group.

We examined the effects of choline deficiency upon cholinergic neurotransmission at the neuromuscular junction. Muscle function testing (anterior tibialis) using surface electromyography was performed. Using surface electromyography (EMG), the extraction of individual motor unit information from data collected on a population of concurrently active motor units requires decomposition of the composite ME signal into its component motor unit action potentials (MUAPs). This requires the recognition and differentiation between shapes and firing patterns of individual MUAPs, and is accomplished using a computer assisted decomposition algorithm because the data are very complex. For the purposes of initial evaluation we have analyzed several parameters. The m-wave describes the summation of individual MUAPs. Its configuration and area is influenced by the number of motor units firing and/or by changes in the number of quanta of acetylcholine released at each myoneural junction. The median frequency is a Fourier transformation of the frequencies of the MUAPs, and is influenced by variability in the firing rate of motor units as well as by the conduction velocity of muscle. The muscle fiber conduction velocity is the velocity of propagation of action potentials along muscle fibers. The force exerted by the muscle is measured as torque applied to a pressure transducer. In myasthenia gravis (deficient cholinergic transmission at the myoneural junction) there is increased variability of MUAP frequency, resulting in a decrease in median frequency, and a decrease in m-wave area. We hypothesized that choline deficiency might have similar effects. All of these measured parameters decay during sustained muscle contractions, and the changes are characteristic of muscle fatigue. The rate of change in a parameter (slope) provides a measure of muscle fatigue. We observed a significant increase (2-3 fold) in the rate of decay (slope) of median frequency and conduction velocity during voluntary contractions in the DEFICIENT group (see table 1). This suggests an increased rate of fatigue in the DEFICIENT group as compared to the CHOLINE group. Creatinine phosphokinase activity in plasma was not different among the groups.

Table 1. Muscle Function Testing in Choline Deficient Humans

Ratio of test values (day 28/day 7)

stimulated (30 Hz)

group	area of m-wave	median frequency
CHOLINE	0.976±0.074	1.06±0.020
DEFICIENT	0.839±0.079	0.82±0.160

maximum voluntary contraction

	median freq.	slope median frequency
CHOLINE	1.113±0.113	1.17±0.340
DEFICIENT	1.184±0.120	**1.77±0.360**

	conduction velocity	slope conduction velocity
CHOLINE	1.046±0.124	0.80±0.190
DEFICIENT	1.005±0.041	**2.94±0.850**

	force exerted	slope of force
CHOLINE	1.094±0.109	1.16±0.190
DEFICIENT	0.923±0.018	1.30±0.330

There were no significant differences in serum creatinine, BUN, or serum calcium concentrations between the DEFICIENT and CHOLINE groups. Urine creatinine concentration dropped 30% in the DEFICIENT group (dropped in every subject but stayed within normal range for the general population). We suggest that this is consistent with decreased availability of methyl-groups. Serum phosphorus and uric acid concentrations rose slightly (15%; stayed within normal range for general population) on the soy diets, probably related to the composition of the diet. Urine specific gravity, urinalysis, urine sediments, and specific gravity of first morning voids were not changed by choline deficiency. We have not seen significant changes in renal function associated with choline deficiency.

IN SUMMARY: We found that humans who had been fed a choline deficient diet for 3 weeks developed changes suggestive of modest hepatic dysfunction and developed subtle abnormalities in muscle function. We have discussed other consequences of choline deficiency which have been identified in studies using experimental animals.

Acknowledgements

I would like to thank L. Schurman, K. daCosta, N. Xia, N. Sheard, P. Gorman, and V. Siouris for their help with the human choline deficiency study. The work described in this review was supported by grants from the National Institutes of Health (HD16727, RR00533), the United States Department of Agriculture (CRCR 1-2464), the American Institute for Cancer Research, the Central Soya Corporation, and the Lucas Meyer Corporation.

References

1. Atsushi, I., Hellerstein, E. E., Hegsted, D. M. and . , 1963, Composition of dietary fat and the accumulation of liver lipid in the choline-deficient rat. J. Nutr. 79: 488-92.
2. Barak, A. J. and Tuma, D. J. , 1983, Betaine, metabolic by-product or vital methylating agent? [Review]. Life Sci. 32: 771-4.
3. Berridge, M. J. and Taylor, C. W. , 1988, Inositol trisphosphate and calcium signaling. Cold Spring Harbor Symposia On Quantitative Biology. 2: 927-33.
4. Best, C. H. and Huntsman, M. E. , 1932, The effects of the components of lecithin upon the deposition of fat in the liver. J. Physiol. 75: 405-12.
5. Besterman, J. M., Duronio, V. and Cuatrecasas, P. , 1986, Rapid formation of diacylglycerol from phosphatidylcholine: a pathway for generation of a second messenger. Proc. Nat. Acad. Sci. U. S. A. 83: 6785-9.
6. Blusztajn, J. K. and Zeisel, S. H. , 1989, 1,2-sn-diacylglycerol accumulates in choline-deficient liver. A possible mechanism of hepatic carcinogenesis via alteration in protein kinase C activity? FEBS Lett. 243: 267-70.
7. Buckley, A., Crowe, P. and Russell, D. , 1988, Rapid activation of protein kinase C in isolated rat liver nuclei by prolactin, a known hepatic mitogen. Proc. Natl. Acad. Sci. USA. 85: 8649-53.
8. Chao, C. K., Pomfret, E. A. and Zeisel, S. H. , 1988, Uptake of choline by rat mammary-gland epithelial cells. Biochem. J. 254: 33-8.
9. Chen, S. H., Estes, L. W. and Lombardi, B. , 1972, Lecithin depletion in hepatic microsomal membranes of rats fed on a choline-deficient diet. Exp. Mol. Pathol. 17: 176-86.
10. Eagle, H. , 1955, The minimum vitamin requirements of the L and Hela cells in tissue culture, the production of specific vitamin deficiencies, and their cure. J. Exptl. Med. 102: 595-600.
11. Finkelstein, J. D., Martin, J. J., Harris, B. J. and Kyle, W. E. , 1982, Regulation of the betaine content of rat liver. Arch. Biochem. Biophys. 218: 169-73.
12. Finkelstein, J. D., Martin, J. J., Harris, B. J. and Kyle, W. E. , 1983, Regulation of hepatic betaine-homocysteine methyltransferase by dietary betaine. Journal of Nutrition. 113: 519-21.
13. Ghoshal, A. K. and Farber, E. , 1984, The induction of liver cancer by dietary deficiency of choline and methionine without added carcinogens. Carcinogenesis. 5: 1367-1370.
14. Haines, D. S. and Rose, C. I. , 1970, Impaired labelling of liver phosphatidylethanolamine from ethanolamine-14C in choline deficiency. Can. J. Biochem. 48: 885-92.
15. Hall, R. I., Ross, L. H., Bozovic, M. G. and Grant, J. P. , 1985, The effect of choline supplementation on hepatic steatosis in the parenterally fed rat. J. Parent. Ent. Nutr. 9: 597-9.

16. Kaminski, D. L., Adams, A. and Jellinek, M. , 1980, The effect of hyperalimentation on hepatic lipid content and lipogenic enzyme activity in rats and man. Surgery. 88: 93-100.

17. Lombardi, B., Pani, P. and Schlunk, F. F. , 1968, Choline-deficiency fatty liver: impaired release of hepatic triglycerides. J. Lipid Res. 9: 437-46.

18. Meck, W. H., Smith, R. A. and Williams, C. L. , 1988, Pre- and postnatal choline supplementation produces long-term facilitation of spatial memory. Dev. Psychobiol. 21: 339-53.

19. Megidish, T. and Mazurek, N. , 1989, A mutant protein kinase C that can transform fibroblasts. Nature. 342: 807-811.

20. Michael, U. F., Cookson, S. L., Chavez, R. and Pardo, V. , 1975, Renal function in the choline deficient rat. Proc. Soc. Exp. Biol. Med. 150: 672-76.

21. Newberne, P. M. and Rogers, A. E. , 1986, Labile methyl groups and the promotion of cancer. [Review]. Ann. Rev. Nutr. 6: 407-32.

22. Pelech, S. L. and Vance, D. E. , 1984, Regulation of phosphatidylcholine biosynthesis. [Review]. Biochim. Biophys. Acta. 779: 217-51.

23. Schneider, W. J. and Vance, D. E. , 1978, Effect of choline deficiency on the enzymes that synthesize phosphatidylcholine and phosphatidylethanolamine in rat liver. Eur. J. Biochem. 85: 181-187.

24. Sheard, N. F., Tayek, J. A., Bistrian, B. R., Blackburn, G. L. and Zeisel, S. H. , 1986, Plasma choline concentration in humans fed parenterally. Am. J. Clin. Nutr. 43: 219-24.

25. Shivapurkar, N. and Poirier, L. A. , 1983, Tissue levels of S-adenosylmethionine and S-adenosylhomocysteine in rats fed methyl-deficient, amino acid-defined diets for one to five weeks. Carcinogenesis. 4: 1051-1057.

26. Sleight, R. and Kent, C. , 1983, Regulation of phosphatidylcholine biosynthesis in mammalian cells. I. Effects of phospholipase C treatment on phosphatidylcholine metabolism in Chinese hamster ovary cells and L.M. mouse fibroblasts. J. Biol. Chem. 258: 824-830.

27. Yang, E. K., Blusztajn, J. K., Pomfret, E. A. and Zeisel, S. H. , 1988, Rat and human mammary tissue can synthesize choline moiety via the methylation of phosphatidylethanolamine. Biochem. J. 256: 821-8.

28. Yao, Z. M. and Vance, D. E. , 1988, The active synthesis of phosphatidylcholine is required for very low density lipoprotein secretion from rat hepatocytes. J. Biol. Chem. 263: 2998-3004.

29. Zeisel, S. H. , 1981, Dietary choline: biochemistry, physiology, and pharmacology. [Review]. Ann. Rev. Nutr. 1: 95-121.

30. Zeisel, S. H. , 1988, "Vitamin-like" molecules. 440-452.

31. Zeisel, S. H., Char, D. and Sheard, N. F. , 1986, Choline, phosphatidylcholine and sphingomyelin in human and bovine milk and infant formulas. J. Nutr. 116: 50-8.

32. Zeisel, S. H. and Wurtman, R. J. , 1981, Developmental changes in rat blood choline concentration. Biochem. J. 198: 565-70.

33. Zeisel, S. H., Zola, T., daCosta, K. and Pomfret, E. A. , 1989, Effect of choline deficiency on S-adenosylmethionine and methionine concentrations in rat liver. Biochem. J. 259: 725-729.

PHOSPHATIDYLINOSITOL DERIVATIVES AS CELL SIGNALLING MOLECULES

John N. Hawthorne

Department of Biochemistry
Nottingham University Medical School
Queen's Medical Centre
Nottingham NG7 2UH, U.K.

INTRODUCTION

Interest in phosphatidylinositol metabolism began with the observations of Hokin and Hokin (19) which linked turnover of this lipid with the activation of receptors. Folch (14) had shown that brain contained a more complex phosphoinositide and this product proved later to be a mixture of three such lipids, phosphatidylinositol itself (PtdIns), phosphatidylinositol 4-phosphate (PtdIns 4P) and phosphatidylinositol 4,5-bisphosphate (PtdIns(4,5)P_2). Dawson (9) showed that ^{32}P was rapidly incorporated into brain PtdIns 4P and PtdIns(4,5)P_2, suggesting that these so-called polyphosphoinositides might also have an important function.

PHOSPHOINOSITIDE METABOLISM

Agranoff et al. (2) and Paulus and Kennedy (35) showed that CDP-diacylglycerol, a new type of activated lipid, was the key intermediate in the biosynthesis of PtdIns. It was formed from phosphatidic acid and CTP and reacted with free inositol, not an inositol phosphate, to form PtdIns and CMP. The polyphosphoinositides were formed by successive phosphorylations of PtdIns with ATP (8,25). Their association with myelin suggested that the polyphosphoinositides were plasma membrane lipids and this was confirmed by the localization of PtdIns kinase in that membrane (33).

Receptor activation seemed likely to involve hydrolysis of phosphoinositides. Labelling studies suggested that inositol and phosphate were removed together by an enzyme of the phospholipase C type and the isolation of inositol phosphate from liver supported this idea (21). The enzyme was first detected in pancreas (10) and liver (26,27). It requires Ca^{2+} for activity and hydrolyses the polyphosphoinositides at cytosolic concentrations of this cation (100nM). Hydrolysis of PtdIns requires Ca^{2+} concentrations in the millimolar range. Several

Phospholipids
Edited by I. Hanin and G. Pepeu
Plenum Press, New York, 1990

laboratories have now purified this enzyme to homogeneity and there seem to be several forms of it differing in amino acid sequence and molecular weight. Our own preparation from rat brain soluble fraction (7) gave two enzyme species of Mr 151,000 and 147,000. Antibodies to the similar enzymes from bovine brain (36) cross reacted, but the membrane-bound phospholipase C of rat brain did not. Two different enzymes have been isolated from rat brain by Homma et al. (20) having Mr 85,000. As yet, we do not know whether they are immunologically distinct from our rat brain enzymes.

PHOSPHOINOSITIDES AND CELL CALCIUM

Changes in cytosolic Ca^{2+} concentration can have important effects on cells, the best known being muscle contraction. Michell (32) suggested that receptors linked to phosphoinositide metabolism all acted by raising the intracellular concentration of Ca^{2+}. He proposed a calcium-gating theory in which the hydrolysis of PtdIns in the plasma membrane led to the opening of channels which allowed calcium to enter the cell. At that time PtdIns was the only phosphoinositide which could reliably be shown to respond to receptor activation but the polyphosphoinositides appeared more suitable candidates for receptor-linked events since they were specifically associated with the plasma membrane while PtdIns was not. However, Abdel-Latif et al. (1) showed that acetylcholine caused the breakdown of PtdIns$(4,5)P_2$ in iris muscle. This was at first thought to be a consequence of increased cytosolic Ca^{2+} concentration, but later work with hepatocytes (28) showed that this was not the case and suggested that PtdIns$(4,5)P_2$ hydrolysis might cause the calcium increase.

Attention now shifted to the concept that calcium was released from intracellular stores, inositol 1,4,5-trisphosphate acting as a second messenger (5). Working with Irvine, who prepared a number of different inositol phosphates, Berridge showed that calcium release was most effective with D-inositol 1,4,5-trisphosphate (6). The current view then is that activation of the relevant cell surface receptors leads to the stimulation of phospholipase C which hydrolyses PtdIns$(4,5)P_2$ to release the 1,4,5-trisphosphate and diacylglycerol. The enzyme and the receptor appear to be linked by a GTP-binding protein, but this G-protein differs from those involved in cyclic AMP function.

As would be expected, the messenger action is terminated by hydrolysis of inositol 1,4,5-trisphosphate by a 5-phosphatase, producing inositol 1,4-bisphosphate which does not mobilize Ca^{2+} (11). This bisphospate is released from PtdIns 4P by phospholipase C and studies in vitro indicate that at low concentrations of Ca^{2+} this lipid is as good a substrate as PtdIns$(4,5)P_2$. Receptor experiments with intact cells show loss of both lipids, but it is not clear whether PtdIns 4P is lost by hydrolysis or by phosphorylation to PtdIns$(4,5)P_2$.

Some years ago Griffin and Hawthorne (18) showed that influx of Ca^{2+} caused hydrolysis of the polyphosphoinositides in isolated brain synaptosomes. There is now evidence that such hydrolysis may take place in adrenal medullary chromaffin cells after nicotinic Ca^{2+} influx (12) and in other nerve cells (13). In view of the earlier arguments, it is interesting that as well as receptor-mediated polyphosphoinositide hydrolysis leading to release of internal Ca^{2+}, there is also hydrolysis as a consequence of Ca^{2+} entry into cells. The physiological significance of this latter effect is as yet unknown.

DIACYLGLYCEROL AS A SECOND MESSENGER

In addition to the inositol phosphates discussed above, phospholipase C releases diacylglycerol from the phosphoinositides. This can also function as a second messenger by activating protein kinase C (34), which in turn phosphorylates various cellular proteins. Kinase C is also activated by phorbol esters and the inappropriate signals given through kinase C probably account for the tumour-promoting characteristics of these esters.

How the two messengers of the phosphoinositide signalling system, inositol 1,4,5-trisphosphate and diacylglycerol, work together is not well understood. In some systems, kinase C may provide feedback inhibition. Work with single rat hepatocytes has shown that vasopressin causes repetitive free calcium transients rather than a sustained rise in cytosolic calcium. Phorbol esters reduce the frequency of the transients and abolish these altogether at higher concentrations (43). Since vasopressin mobilizes Ca^{2+} by the phosphoinositide mechanism this work suggests that activation of kinase C by the phorbol esters is producing feedback inhibition. In other cells the phosphoinositide response appears to mobilize little or no Ca^{2+} and in these instances the activation of kinase C by diacylglycerol may be the significant effect. Muscarinic receptors in heart and bovine chromaffin cells come into this category. The muscarinic response in the heart is inhibitory and associated with reduced cyclic AMP concentration. The phospho-inositide effect requires effect requires higher concentrations of acetylcholine and is unlikely to involve Ca^{2+} release since the overall muscarinic response is to reduce the rate and force of the heart-beat. The bovine chromaffin cell is equally enigmatic. Nicotinic receptors allow entry of Ca^{2+} and release of catecholamines. Muscarinic receptors cause release of inositol trisphosphate, but the resulting mobilization of Ca^{2+} is insufficient for catecholamine release. In these tissues it is possible that release of diacylglycerol for activation of kinase C is more important than the inositol trisphosphate release.

MORE COMPLEX INOSITOL PHOSPHATES

Activation of muscarinic receptors in rat parotid glands produces an inositol trisphosphate which differs from the 1,4,5-compound (24). Chemical studies showed that it was inositol 1,3,4-trisphosphate and subsequent work indicates its production in many tissues. Its origin was elucidated by Batty et al., (4), who discovered a 1,3,4,5-tetrakis-phosphate which is attached by a specific 5-phosphatase to give the 1,3,4-compound. The tetrakisphospate is formed by a 3-kinase acting on inositol 1,4,5-trisphosphate (23).

Irvine and his colleagues have shown that inositol 1,3,4,5-tetrakisphosphate may have a second-messenger function in making extracellular Ca^{2+} available to the endoplasmic reticulum store which is sensitive to the 1,4,5-trisphosphate. The evidence came from studies of the sea urchin egg and mouse lacrimal cells and the field has recently been reviewed briefly by Irvine (22).

For many years avian erythrocytes have been known to contain appreciable amounts of an inositol pentakisphosphate. Small quantities of inositol 1,3,4,5,6-pentakisphosphate have now been detected in mammalian cells (39). It is produced from inositol 3,4,5,6-tetrakisphosphate by a 1-hydroxy kinase which occurs in rat brain, liver, heart and parotid gland, as well as mouse bone macrophages. The origin of the 3,4,5,6-tetrakisphosphate is unknown. It should be noted that all optically active inositol phosphates are numbered for the D-series. This D-3,4,5,6-tetrakisphosphate can also be referred to as the L-1,4,5,6-compound. It seems unlikely to arise from $PtdIns(4,5)P_2$ but it should be remembered that in the biosynthesis of myo-inositol glucose 6-phosphate is converted to D-inositol 3-phosphate. It is also of interest that a phosphatidylinositol 3-kinase has been discovered in fibroblasts (42) and that this kinase can utilize $PtdIns(4,5)P_2$ as a substrate to make $PtdIns(3,4,5)P_3$, a phospholipid which appears briefly in stimulated neutrophils (40). Hydrolysis of such a lipid by phospho-inositidase C would produce inositol 1,3,4,5-tetrakisphosphate, a compound already discussed as a possible second messenger.

A possible physiological function for the 1,3,4,5,6-pentakisphosphate and the hexakisphosphate is suggested by the work of Vallejo et al. (41). Both compounds, though not the tetrakisphosphates, produced dose-dependent changes in heart rate and blood pressure when injected into a specific brain stem nucleus. The action appeared to be extracellular.

GLYCOSYL-PHOSPHATIDYLINOSITOL

Low (30) has reviewed this new field. His work opened it up by showing that a bacterial phospholipase specific for PtdIns could release various plasma membrane enzymes such as acetylcholinesterase. These enzymes and other proteins in mammalian tissues, protozoal parasites and Torpedo electric tissue are linked covalently to the inositol of PtdIns by a glycan bridge containing mannose and glucosamine, but varying according to the source. This is the only example yet recorded of a membrane phospholipid being covalently bonded to a protein. An early paper by Klenk and Hendricks (36) indicates that such compounds occur in brain.

The action of phosphoinositidase C on the PtdIns glycan releases a phospho-oligosaccharide which appears to modulate some of the actions of insulin. Saltiel and Cuatrecasas (37) obtained such a compound when insulin interacted with hepatic plasma membranes and showed that it activated cyclic AMP phosphodiesterase. Similarly, the phospho-oligosaccharide mimics the action of insulin on adipocytes to change the phosphorylation of target enzymes such as ATP citrate lyase (3).

DIABECTIC NEUROPATHY

Inositol has been classed as a vitamin but it is not clear that there is a dietary requirement except in germ-free animals. Nevertheless, synthesis from glucose 6-phosphate in mammalian tissues seems inadequate for the body's requirements. Thus phosphatidylinositol is an important dietary constituent as a source of inositol. This final section will also indicate that dietary inositol could be specifically important in preventing the nerve damage seen in long-standing diabetes mellitus.

Diabetic neuropathy is seen particularly in peripheral nerves and leads to lack of sensation and ulceration in the feet and legs. The autonomic nervous system can also be affected. Before morphological changes such as segmental demyelination of nerves are seen, reduced conduction velocity can be detected in experimental diabetes. In both human and experimental diabetes, a reduced concentration of free inositoland of inositol lipids can be seen in the sciatic nerve (17,31). The sciatic nerve of diabetics also contains considerable amounts of sorbitol, formed from glucose by the enzyme aldose reductase. Several pharmaceutical companies have produced compounds which inhibit this enzyme, in the hope that they may be of therapeutic value in treating diabetic neuropathy. One such compound, sorbinil (Pfizer Ltd.), restored nerve conduction velocity to normal in streptozotocin-diabetic rats, and as well as reducing nerve sorbitol to the low level seen in normal rats, raised myo-inositol to control levels (15). Dietary inositol also raised the nerve level to normal and at the same time corrected the reduced conduction velocity without the use of aldose reductase inhibitors. It seems then, that inositol rather than sorbitol has some function in connection with nerve conduction.

The relatively high concentration of inositol in peripheral nerve is maintained by sodium-dependent active transport. In diabetes however, the sodium pump, as measured by Na^+/K^+-ATPase (16) or by the transport of $^{86}Rb^+$ (38), has reduced activity. This activity, like the reduced conduction velocity, is restored to normal by aldose reductase inhibitors or dietary inositol. Since inositol transport is linked to the sodium pump, reduced activity of the pump will reinforce the inositol deficit, which in turn will slow the pump. This self-reinforcing cycle is considered to initiate the nerve damage seen in diabetes.

Exactly how reduced levels of inositol affect the sodium pump is unknown as yet. It seems likely that the phosphoinositides are involved and possibly protein kinase C. Like the work on receptor-linked signalling, these studies emphasize the importance of phosphatidyl-inositol in mammalian biochemistry. The biosynthesis of this lipid requires free myo-inositol and since the body cannot produce sufficient quantities of this compound a dietary source is required. Soya lecithin is one such source, since it contains phosphatidylinositol.

REFERENCES

1. Abdel-Latif, A.A., Akhtar, R.A. and Hawthorne, J.N., 1977, Acetylcholine increases the breakdown of trisphosphoinositide of rabbit iris muscle prelabelled with [^{32}P]phosphate, Biochem. J., 162: 61-73.
2. Agranoff, B.W, Bradley, R.M. and Brady, R.O., 1958, The enzymatic synthesis of inositol phosphatide, J. Biol. Chem. 233:1077-1083.
3. Alemany, S., Mato, J.M., and Stralfors, P., 1987, Phospho-dephospho-control by insulin is mimicked by a phospho-oligosaccharide in adipocytes, Nature (London), 330:77-79.
4. Batty, I.R., Nahorski, S.R., and Irvine, R.F., 1985, Rapid formation of inositol 1,3,4,5-tetrakisphosphate following muscarinic receptor stimulation of rat cerebral cortical slices, Biochem. J., 232:211-215.

5. Berridge, M.J., 1983, Rapid accumulation of inositol trisphosphate reveals that agonists hydrolyze polyphosphoinositides instead of phosphatidylinositol, Biochem. J., 212:849-858.
6. Berridge, M.J., amd Irvine, R.F., 1984, Inositol trisphosphate, a novel second messenger in cellular signal transduction, Nature (London), 312:315-321.
7. Blank, J.L., and Hawthorne, J.N., 1989, Purification of two phosphoinositidase C species from rat brain, Biochem. Soc. Trans., 17:96-97.
8. Colodzin, M., and Kennedy, E.P., 1965, Biosynthesis of diphosphoinositide in brain, J. Biol. Chem. 240:3771-3780.
9. Dawson, R.M.C., 1954, Measurement of ^{32}P labelling of individual kephalins and lecithin in a small sample of tissue, Biochim. Biophys. Acta, 14:374-379.
10. Dawson, R.M.C., 1959, Studies on the enzymic hydrolysis of monophospho-inositide by phospholipase preparations from P. notatum and ox pancreas, Biochim. Biophys. Acta, 33:68-77.
11. Downes, C.P., Mussat, M.D., and Michell, R.H., 1982, The inositol trisphosphate phosphomonoesterase of the human erythrocyte membrane, Biochem. J., 203:169-177.
12. Eberhard, D.A., and Holzl, R.W., 1987, Cholinergic stimulation of inositol phosphate formation in bovine adrenal chromaffin cells: distinct nicotinic and muscarinic mechanisms, J. Neurochem, 49:1634-1643.
13. Eberhard, D.A., and Holzl, R.W., 1988, Intracellular Ca^{2+} activates phospholipase C. Trends Neurosci., 11:517-520.
14. Folch, J., 1949, Complete fractionation of brain cephalin: isolation from it of phosphatidylserine, phosphatidylethanolamine and diphosphoinositide, J. Biol. Chem., 177:497-504.
15. Gillon, K.R.W., Hawthorne, J.N., and Tomlinson, D.R., 1983, myo-inositol and sorbitol metabolism in relation to peripheral nerve function in experimental diabetes in the rat: effect of aldose reductase inhibition, Diabetologia, 25:365-371.
16. Greene, D.A., and Lattimer, S.A., 1983, Impaired rat sciatic nerve Na^+/K^+-ATPase in acute streptozotocin diabetes and its correction by dietary myo-inositol supplementation, J. Clin. Invest., 72:1058-1063.
17. Greene, D.A., de Jesus, P.V., and Winegrad, A.J., 1975, Effects of insulin and dietary myo-inositol on impaired peripheral nerve conduction velocities in acute streptozotocin diabetes, J. Clin. Invest., 55:1326-1336.
18. Griffin, H.D., and Hawthorne, J.N., 1978, Calcium-activated hydrolysis of PtdIns 4P and PtdIns(4,5)P_2 in guinea-pig synaptosomes, Biochem. J., 176:541-552.
19. Hokin, M.R., and Hokin, L.E., 1953, Enzyme secretion and the incorporation of ^{32}P into phospholipids of pancreas slices, J. Biol. Chem. 203:967-977.
20. Homma, Y., Imaki, J., Nakanishi, O., and Takenawa, T., 1988, Isolation and characterization of two different forms of inositol phospholipid-specific phospholipase C from rat brain, J. Biol. Chem., 263:6592-6598.
21. Hubscher, G., and Hawthorne, J.N., 1957, The isolation of inositol mono-phosphate from liver, Biochem. J., 67:523-527.
22. Irvine, R.F., 1989, How do inositol 1,4,5-trisphosphate and inositol 1,3,4,5-tetrakisphosphate regulate intracellular Ca^{2+}? Biochem. Soc. Trans., 17:6-9.

23. Irvine, R.F., Letcher, A.J., Heslop, J.P., and Berridge, M.J., 1986, The inositol tris/tetrakisphosphate pathway - demonstration of inositol 1,4,5-trisphosphate 3-kinase activity in animal tissues, Nature (London), 320:631-634.
24. Irvine, R.F., Letcher, A.J., Lander, D.J., and Downes, C.P., 1984, Inositol trisphosphates in carbachol-stimulated rat parotid glands, Biochem. J., 223:237-243.
25. Kai, M., Salway, J.G., and Hawthorne, J.N., 1968, The diphosphoinositide kinase of rat brain, Biochem. J., 106:791-801.
26. Kemp, P., Hubscher, G., and Hawthorne, J.N., 1959, A liver phospholipase hydrolysing phosphoinositides, Biochim. Biophys. Acta, 31:585-586.
27. Kemp, P., Hubscher, G., and Hawthorne, J.N., 1961, Enzymic hydrolysis of inositol-containing phospholipids, Biochem. J., 79:193-200.
28. Kirk, C.J., Creba, J.A., Downes, C.P., and Michell, R.H., 1981, Hormone-stimulated metabolism of inositol lipids and its relationship to hepatic receptor function, Biochem. Soc. Trans., 9:377-379.
29. Klenk, E., and Hendricks, U.W., 1961, An inositol phosphatide containing carbohydrate isolated from human brain, Biochim. Biophys. Acta, 50:602-603.
30. Low, M.G., 1987, Biochemistry of the glycosyl-phosphatidylinositol membrane protein anchors, Biochem. J., 244:1-13.
31. Mayhew, J.A., Gillon, K.R.W., and Hawthorne, J.N., 1983, Free and lipid inositol, sorbitol and sugars in sciatic nerve obtained post-mortem from diabetic patients and control subjects, Diabetologia, 24:13-15.
32. Michell, R.H., 1975, Inositol phospholipids and cell surface receptor function, Biochim. Biophys. Acta, 415:81-147.
33. Michell, R.H., and Hawthorne, J.N., 1965, The site of diphospho-inositide synthesis in rat liver, Biochem. Biophys. Res. Commun., 21:333-338.
34. Nishizuka, Y., 1988, The molecular heterogeneity of protein kinase C and its implications for cellular regulation, Nature (London), 334:661-665.
35. Paulus, H., and Kennedy, E.P., 1960, The enzymatic synthesis of inositol monophosphatide, J. Biol. Chem., 235:1303-1311.
36. Ryu, S.H., Cho, K.S., Lee, K.-Y., Suh, P.-G., and Rhee, S.G., 1987, Purification and characterization of two immunologically distinct phosphoinositide-specific phospholipases C from bovine brain, J. Biol. Chem., 262:12511-12518.
37. Saltiel, A.R., and Cuatrecasas, P., 1986, Insulin stimulates the generation from hepatic plasma membranes of modulators derived from an inositol glycolipid, Proc. Natl. Acad. Sci. USA, 83:5793-5797.
38. Simpson, C.M.F., and Hawthorne, J.N., 1988, Reduced Na$^+$/K$^+$-ATPase activity in peripheral nerve of streptozotocin-diabetic rats: a role for protein kinase C? Diabetologia, 31:297-303.
39. Stephens, L.R., Hawkins, P.T., Morris, A.J., and Downes, P.C., 1988, L-myo-Inositol 1,4,5,6-tetrakisphosphate (3-hydroxy) kinase, Biochem. J., 249:283-292.
40. Traynor-Kaplan, A.E., Harris, A.L., Thompson, B.L., Taylor, P. and Sklar, L.A., 1988, An inositol tetrakisphosphate-containing phospholipid in activated neutrophils, Nature (London), 334:353-356.
41. Vallejo, M., Jackson, T., Lightman, S., and Hanley, M.R., 1987, Occurrence and extracellular actions of inositol pentakis-and hexakisphosphate in mammalian brain, Nature (London), 330:656-658.

42. Whitman, M., Downes, P.C., Keeler, M., Keller, T., and Cantley, L.,
    1988, Type I phosphatidylinositol kinase makes a novel inositol
    phospholipid phosphatidylinositol 3-phosphate, Nature (London),
    332:644-646.
43. Woods, N.M., Cuthbertson, K.S.R., and Cobbold, P.H., 1987, Phorbol-
    ester-induced alterations of free calcium ion transients in
    single rat hepatocytes, Biochem. J., 246:619-623.

DIETARY EGG YOLK-DERIVED PHOSPHOLIPIDS: RATIONALE FOR THEIR

BENEFITS IN SYNDROMES OF SENESCENCE, DRUG WITHDRAWAL, AND AIDS

Wolfgang Huber and Parris M Kidd

HK Biomedical, Incorporated
1200 Tevlin Street
Berkeley, CA 94706, USA

INTRODUCTION

Phospholipids are coming of age as dietary supplements, having diverse potential applications and an enviably good ratio of benefit to risk. Phospholipids are indispensable components of cellular membrane systems, whatever their level of complexity, and the phospholipids of cellular membranes can be significantly manipulated in situ by manipulating dietary lipid intake. Thus, through provision of the appropriate types and amounts of lipids in the daily diet, the potential exists to profoundly affect cellular membrane performance, and with it cellular function overall (53).

Several decades have passed since lipids were first recognized as making a structural contribution to cellular membranes (20). More recently, it has become clear that phospholipids are important not only for membrane structural integrity, but for functional modulation of the membrane and with it that of the cell as a whole (Table 1). Membrane phospholipid makeup influences the behavior of the non-lipid membrane macromolecules, as well as *signal transduction* from the environment into the cell and the *release of cytokine messenger molecules* from the cell into the environment, with both processes utilizing phospholipids as starting substrates either wholly or in part.

Major membrane macromolecules embedded partially or wholly in the lipid bilayer, including receptors, antigens, ion-substrate transport enzymes, electron transport enzymes, and lipid-metabolizing enzymes, are subject to influence by the lipid composition of this bilayer. Also, each membrane macromolecule tends to be associated with a particular spectrum of membrane lipids in its immediate environment. Such short-range lipid-protein interactions strongly influence the spectrum of properties for that macromolecule. As a final consequence, the shorter-range and longer-range lipid-protein interactions altogether provide for the homeostatic maintenance of optimal membrane functionality.

Apart from their primarily structural contributions to the membrane, phospholipids within the membrane can act as starting substrates for the production of second-messenger substances that help mediate transmembrane signal transduction through influencing the behavior of the cell in which they are produced. As summarized in Table 1, a list of second messengers produced from membrane phospholipids would include arachidonic acid (AA), diacylglycerol (DAG), phosphatidic acid (PA), and phosphocholine (PhC) (22); inositol phosphates (IP) and glycans (IG); and sphingosine and lysosphingolipids (SPH/Lyso) (23).

TABLE 1. THE KNOWN MEMBRANE FUNCTIONS OF PHOSPHOLIPIDS

MEMBRANE STRUCTURAL INTEGRITY
Outer leaflet of bilayer: PC dominant
Inner leaflet of bilayer: PE, PI dominant

TRANSMEMBRANE SIGNAL TRANSDUCTION
Intracellular: AA, DAG, PA, IP, IG, SPH/LYSO, other
Extracellular: PG, LT, Tx, PAF, other

FUNCTIONAL MODULATION OF MEMBRANE PROTEINS
Receptors, antigens, enzymes - transport, metabolic

In addition to their permissive and regulatory roles in the transduction of signals into the cell, membrane phospholipids also are involved in transduction from the interior of the cell to the outside, by acting as starting substrates for cytokine production. Prostaglandins, leukotrienes, thromboxanes, and lipoxins are membrane lipid metabolites with cytokine action. Some cytokines, rather than being lipid-derived, can be nonlipid in character yet lipid-associated and releasable following enzymatic processing of the lipid moiety. For example, decay-accelerating factor (DAF) is an integral membrane protein apparently covalently bonded to membrane phosphatidylinositol (PI), from which it can be released by the action of phospholipase C (10).

The degree of ordering of the lipids of the living membrane can be summed as its "fluidity." Although the validity of fluidity as a descriptor for the state of the lipids in the membrane as a whole has been rendered somewhat questionable by the findings that lipids tend to segregate into "microdomains," homeostatic regulation of membrane fluidity appears to be relatively strict.

Fluidity can be difficult to measure, due to the danger of artifact (57). Properly measured, the fluidity of a cell's plasma membrane reflects, to a large degree, the relative content of lipids which "fluidize" the membrane, mainly phosphatidylcholine (PC), relative to the lipids which "rigidify" the membrane, mainly cholesterol (Chol). Thus, changes in "fluidity" as measured by DPH polarization can be considered somewhat indicative of change in the lipid makeup of the membrane (58). Often the converse will also apply: changes in membrane lipid makeup are reflected in fluidity.

In the course of our explorations of the structures and functions of phospholipids and Chol in membranes, we have developed a multifaceted rationale for certain of the observed as well as the potential beneficial health care effects of dietary egg yolk phospholipids, which generally have been administered as egg yolk lipid extracts (ELE).

ROLE OF EXOGENOUS PHOSPHOLIPIDS IN IMMUNE CELL ACTIVATION

It has been established for some time that the maintenance of immune cells in culture requires exogenous fatty acids (57, for a review). Alterations in phospholipid and fatty acid metabolism are early responses to the "turning on" of cells following appropriate stimulation (15, 31, 37), and phospholipids and fatty acids facilitate DNA synthesis during this process (12). Following mitogenic stimulation of lymphocytes, changes in phospholipid metabolism are some of the earliest changes observed (14). The evidence that is currently available clearly supports the interpretation that exogenous phospholipids can promote the activation of lymphocytes. It points to exogenous phospholipids, properly presented to the body, as having a significant role in the boosting of antigen-induced immune response in mammals.

242

During the course of lymphocyte activation, in what may be an amplification process to facilitate a spectrum of changes in the plasma membrane, polyunsaturated fatty acids (PUFA) can be taken up preferentially from the medium (19). Intracellular triacylglycerols can serve as a reserve source of PUFA for redistribution to phospholipids, with PE and PI becoming preferentially enriched in arachidonic acid. In the macrophage, another type of immune cell, dietary manipulation of fatty acids becomes reflected in the acyl chain profiles of the phospholipids, except for PI in which the AA content remains stable (11).

Thus it appears that membrane AA is conserved preferentially in those phospholipids that are most involved in signal transduction functions, as AA is a precursor of second messengers in addition to being a second messenger itself (Fig.1). Arachidonic acid (C20:4 w-6) is thus one of the most prominent, along with inositol phosphates and diacylglycerol, of the "wheels within wheels" that according to Julius Axelrod (59) make up the multilevel cellular information systems.

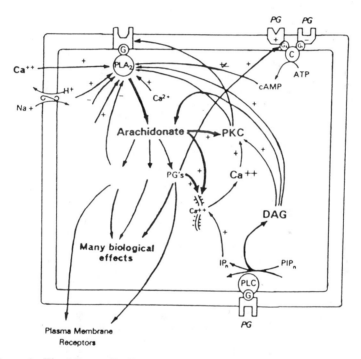

Figure 1. The "Crosstalk" Between Different Second-Messenger Systems. From Axelrod, in Vaughan (59).

In the sections that follow, having in mind a general mechanistic framework with which to interpret what has been observed to date, we discuss available results on the applications of exogenous phospholipids in specific human syndromes.

PHOSPHOLIPIDS CAN HELP RESTORE MEMBRANE FUNCTION IN AGING

The aging syndrome in humans and other mammals is often associated with dysfunctionality of body tissues, including those of the immune system. Aging of the

243

immune system is associated with increasing dysfunctionality of B lymphocytes, T lymphocytes, and macrophages (13). With aging comes increased autoantibody complex formation; decreased B-cell responsiveness to stimulation by T-lymphocytes; decreased proliferative capacity of T lymphocytes; and decreased phagocytic ability of macrophages. The topic of age-related decline in human immune function is particularly deserving of attention because infection is a major cause of death in the aged.

The Lipid Research Clinics Prevalence Study established that in American males, aging was associated with increased plasma Chol content (60, for a review). Increased Chol content in the plasma membrane has been documented for cells belonging to the immune system and to other tissues in several mammalian species, including the human (4, 26, 47, 48). These findings (summarized in Table 2) add credence to the claims made from studies on murine and human circulating lymphocytes, namely that the Chol content of the plasma membrane of such cells reflects that of the plasma in which the cells are bathed (44).

TABLE 2. PLASMA MEMBRANE CHOL INCREASES WITH AGING

MOUSE	LYMPHOCYTES, BRAIN SYNAPSES
RAT	ADIPOCYTES, BRAIN SYNAPSES AND MYELIN
RABBIT	BRAIN AND SPINAL CORD MYELIN
HUMAN	LYMPHOCYTES, ERYTHROCYTES, NERVE TISSUES

Plasma membrane-based dysfunctionalities reminiscent of some of those seen in aging can be imposed on immune cells experimentally, by increasing the Chol content of this membrane (1). Whether documented chemically (as a decrease in the Ch/PL mole ratio), or physically (as an increase in membrane fluidity), restoration of membrane Chol to the normal range in such cells appears to be correlated with restoration of functionality to a significant degree (45, 46).

It may be possible, using dietary phospholipids, to partially reverse age-related declines in the functionality of cells of the immune system. In a preliminary human trial carried out in Israel, the administration of a formulation of hen egg yolk lipids to 10 aged human subjects (average > 83 years) resulted in partial restoration of immune cell functionality in a majority (6) of them (44). Lymphocytes from these aged humans had abnormally high membrane Chol and abnormally low proliferative ability. Dietary administration of this lipid mixture (10g/day) resulted in lowering of the membrane Chol, and proliferative response to mitogen was partially restored. The effect faded after dosing was ceased.

According to Rabinowich and his collaborators, the formulation used to achieve this effect was "a potent membrane fluidizing lipid mixture comprised of 70% neutral lipids, 20% phosphatidylcholine, and 10% phosphatidyl-ethanolamine, all derived from hen egg yolk." Rabinowich and his collaborators referred to Lyte and Shinitzky (35) for a description of the formulation. These authors (35) defined their formulation as a mixture for which the neutral lipids (7 parts) were prepared from hen egg yolk by isopropanol-chloroform extraction (0.9:1.0), and the phospholipids were >99% pure PC (2 parts) and >99% pure PE (1 part), prepared from hen egg yolk by Lipid Products, Nutfield, England.

In related <u>ex vivo</u> experiments, macrophages were drawn from the peripheral blood of aged men, then were incubated in culture for two hours with this lipid mixture prior to challenge with a bacterial inoculation. Macrophages so treated subsequently displayed partial enhancement of their phagocytic/killing activity, in some cases to levels comparable with those of young men (U.S. Pat. No. 4,474,773 - Shinitzky, Heron & Samuel, 1984). Interestingly, control macrophages drawn from young, healthy donors did not show enhanced phagocytic responses; the action of this lipid mixture must have been restorative rather than stimulatory.

CELL MEMBRANE CHOL MODULATION BY PHOSPHOLIPIDS IN VIVO

Excessive Chol content in cellular plasma membranes has been linked to a variety of human diseases and syndromes. Dietary phospholipids can be used to influence membrane lipid content, through influencing the concentrations of lipids in proximity to the membrane of the target cell. Under experimental conditions, liposomal or micellar preparations of phospholipids can be used for this purpose. Liposomes have been shown to fuse directly with cellular plasma membranes in vivo. Under physiological conditions, however, the major vehicles for the transport of lipids to their sites of action are the plasma lipoprotein particles.

The plasma lipoprotein system both delivers needed lipids to the tissues and influences cellular membrane lipid content (39). Plasma lipoprotein particles consist essentially of lipids— phospholipids, Chol, Chol esters— associated with one or more specific proteins which are characteristic for each class of particle. The basic layout of a plasma lipoprotein particle is illustrated in Fig. 2.

Figure 2.   Molecular Organization of a Lipoprotein Particle. Left: surface view. Right: cross-sectional view. Ellipses represent Chol (open, no tail) and cholesteryl esters (closed, with tail). Surface shell primarily monolayer of phospholipids (two tails); core primarily cholesteryl esters, triglycerides (not shown). Extended apoprotein molecule(s) insert primarily into the surface shell. Left, from Stryer (56). Right, from Mims and Morrisett, in Yeagle (66).

Circulating lipoprotein assemblies are responsible for the transport of the various classes of lipids to and from the gut, the liver, and the other tissues by way of the lymphatic and blood circulations. Two major functional cycles are involved in lipid transport. The *exogenous* cycle involves the absorption of lipids after ingestion, their packaging into chylomicrons and, after delivery of their triglycerides to solid tissues, into low-density lipoproteins. The *endogenous* cycle involves continuous turnover of plasma lipoproteins, in the process removing Chol from the tissues and recycling the apoprotein constituents of the lipoprotein particles (Fig. 3).

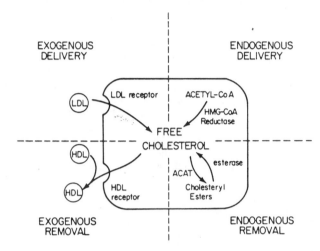

Figure 3. Summary Scheme of Chol Pathways in Extrahepatic Cells.
From Bierman and Oram (6).

After dietary ingestion, phospholipids diffuse across the enteric brush border and enter the enterocyte cells which line the intestinal epithelium. Phospholipids with polar head groups, such as PC, phosphatidylethanolamine (PE), phosphatidylserine, and sphingomyelin, may not necessarily require enzymatic hydrolysis in order to cross the enteric brush border (5, 8, 27; but see 63). Triglycerides, in contrast, must undergo enzymatic hydrolysis to render them sufficiently polar to cross the brush border. Chol, due to its amphipathic nature, is absorbed unchanged; cholesteryl esters need to be hydrolyzed for absorption across the brush border.

Once within the enterocyte, the absorbed lipids become transported to the endoplasmic reticulum of the cell. There, those phospholipids which have been hydrolyzed become re-esterified, largely with fatty acids identical to those originally present prior to hydrolysis. The triglyceride breakdown products are also re-esterified. Mammals have solved the transport problem for both ingested and liver-produced lipids by packaging them into plasma lipoprotein particles. These include chylomicrons produced by the enterocytes, and *very low-density lipoproteins* (VLDL) produced by the hepatocytes. In both, triglycerides and cholesteryl esters form a hydrophobic, densely-packed core, surrounded by a surface monolayer of phospholipids and apolipoprotein to achieve solubilization. Unesterified Chol acts as a subsurface stabilizer.

The chylomicron particles are secreted into the lymph, and are subsequently delivered into the blood circulation, via the thoracic duct into the clavicular venous system. The VLDL are delivered directly into the circulation from the liver.

The initial function of the chylomicrons and the VLDL is to transport triglycerides to the tissues via receptors located in the vascular linings of these tissues. After becoming depleted of their triglycerides, the chylomicron and VLDL remnants travel to the liver. There they become transformed into *low-density lipoprotein particles* (LDL) which are, relatively speaking, much higher in phospholipid and Chol content.

LDL interact with the circulating cells of the body, primarily via receptors, and become internalized as small vesicles (endosomes). These migrate toward, and subsequently fuse with, the endoplasmic reticulum, from which their lipid contents become distributed to the rest of the cell. LDL are a reservoir of phospholipids, but also provide their target cells with a sizeable amount of cholesteryl esters and some free Chol.

In contrast to the chylomicrons and the LDL, which transport primarily lipids derived from the diet, the *high-density lipoproteins* (HDL) are continually produced via the endogenous cycle. The HDL are both high in phospholipids and low in Chol. HDL are the lipoproteins responsible for the uptake of Chol from extrahepatic tissues for subsequent excretion.

The HDL particle is synthesized by the liver as a "discoid-nascent" particle having, as its name implies, a discoid shape. Initially the HDL particle lacks lipids at its core, but cholesteryl esters come to make up the core as Chol is transferred from cell membranes to the HDL particle, being esterified en route (i.e., in the plasma) by the enzyme *lecithin-Chol acyltransferase* (LCAT).

The mode of interaction of the HDL particle with a target cell is markedly different from that of the chylomicron or the LDL particle. The HDL particle does not become internalized. Rather, through proximity-mediated, high-affinity binding to the target cell, it seemingly can "dock" with the plasma membrane (6). A temporary, "channel" type of union is established which permits a bi-directional exchange of phospholipids and Chol between the HDL particle and the plasma membrane of the cell. This exchange, which is concentration-dependent (66) leads to a flux of phospholipids into, and a flux of Chol out of, the membrane (7, 41).

As equilibrium is approached, the HDL particle undocks from the membrane and recaptures Chol that has left the membrane and been newly esterified by the LCAT enzyme. The cholesteryl ester-enriched HDL particle is then processed by the liver, which depletes it of its Chol and returns the Chol-depleted particle to the circulation. Chol extracted from HDL is subsequently excreted into the intestinal tract.

HDL particles appear, therefore, to be a major means for maintaining or reconstituting the lipid makeup of plama membranes in situ. An extensive literature on HDL-plasma membrane interactions in vivo and in vitro supports the possibility that plasma lipoprotein concentration gradients contribute directly to regulating cellular Chol content in vivo. The content of PC and Ch in the plasma membrane of cells in vivo appears to be the balance between their rate of migration out of the membrane and their rate of migration into the membrane, as effected by their interactions with lipoproteins. This perspective, embodied in the *Reverse Cholesterol Transport* hypothesis (63), is supported by several lines of evidence:

• *Phospholipids in the medium extract membrane Chol and insert PL.* Mixtures of phospholipids extracted from chicken egg yolks, and containing mostly PC and PE, have long been used for this purpose. The phospholipids used can be in the form of micelles, liposomes, or serum-type lipoproteins, provided they are lower in Chol than the membranes with which they are incubated (9, 55, 63).

• *Phospholipids given i.v. caused experimental atherosclerosis to regress.* In rabbits, both the lesions and the Chol deposits which they encompassed were affected (16). This finding was subsequently replicated in baboons (28), and quail (54).

• *Phospholipid liposomes given i.v. lowered tissue Chol.* Liposomes prepared from egg PL and injected into dogs or rabbits tended to acquire Chol while in the circulation. They also acquired apoproteins which targeted them for recycling through the liver, thus favoring the subsequent excretion from the body of their acquired Chol (62, 65). In one recent study, the authors concluded "Our data support the view that injection of phospholipids has an antiatherogenic effect by enhancing the reverse cholesterol transport mechanism" (38).

A RATIONALE FOR THE ANTIVIRAL ACTION OF EGG PHOSPHOLIPIDS

A good deal of evidence now available also suggests that dietary egg yolk phospholipids can confer a degree of control over viral infectivity in vivo. The mechanisms by which such control may be effected very likely involve some combination of the following:

- Interference with viral envelope formation
- Protection or restoration of transmembrane signalling functions of infected cells
- Protection against metabolic derangements to the cell from the presence of virus

Interference with Virus Envelope Formation

Of the viruses pathogenic for humans, a great many are enveloped viruses. That is, the infectious virus particle is separated from the exterior by a continuous envelope. The viral envelope is organized along the bilayer scheme, much like the host cell membranes from which it is typically derived as the virus particle assembles in the host cell. The viral envelope has a lipid bilayer matrix made up of phospholipids and Chol, in which are embedded proteins. Many of the latter are glycoproteins which serve as conformationally - specific "docking" sites to enable the virus particle to attach to its target cell.

Among the enveloped viruses that infect humans (36) are the herpesviruses, including herpes simplex virus (HSV), Epstein-Barr Virus (EBV), and cytomegalovirus (CMV); the paramyxoviruses (measles viruses); and the orthomyxoviruses (influenza viruses). The retroviruses are also enveloped viruses. This virus class includes the human T-Cell lymphotropic viruses (HTLV-1 and -2); feline leukemia virus (FeLV), associated with cancers; and the human immunodeficiency viruses (HIV-1 and -2), associated with acquired immune system deficiencies. The envelope configuration of HIV-1 is illustrated in Fig. 4.

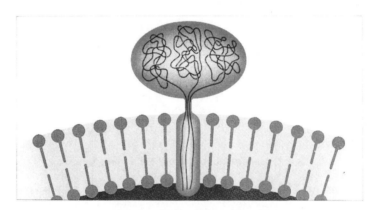

Figure 4. The Envelope of HIV-1 in Cross Section.
This is a bilayer membrane acquired from the plasma membrane of the host cell. Embedded in the bilayer is the gp41 protein, to which is attached the gp120 glycoprotein which helps bind virus particle to target cell. From Gallo and Montagnier (17).

Vesicular stomatitis virus (VSV) is a rhabdovirus, rod-shaped and containing single-stranded RNA. VSV usually infects livestock, but has infected humans exposed to it. VSV has been the subject of extensive investigations by R. R. Wagner's group at the University of Virginia's School of Medicine as a "model" for enveloped viruses.

As with all other enveloped viruses that have been studied, the envelope is critical for infectivity. Wagner's group has established beyond doubt that the content of Chol in the VSV envelope is critical for the infectivity of the virus particle. Stages in the membrane assembly and release of enveloped viruses from their host cells are illustrated in Fig. 5; Semliki, influenza, VSV, and HIV-1 all seem to follow the same general scheme.

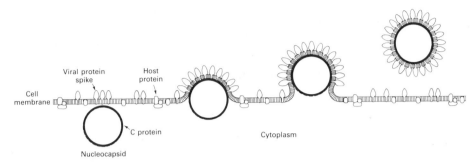

Figure 5. Scheme of Membrane Assembly and Release of an Enveloped Virus. From Simons et al (51), as modified by Stryer (56). The "nucleocapsid" core includes the genetic material of the virus.

The phospholipid profile of the VSV envelope closely parallels that of the host cell plasma membrane, and also responds readily to manipulation of lipids in the growth medium (43). However, as with other enveloped viruses including HIV-1 (2), the VSV envelope contains a much higher content of Chol and a significantly higher cholesterol/phospholipid (Chol/PL) mole ratio than the cell membrane from which it is derived. Growing the virus in Chol-depleted host cells results in the production of virus particles with significantly lower envelope Chol and markedly lowered infectivity (43).

Depletion of the Chol of the virus envelope by incubation with egg lecithin micelles (40), or with serum lipoproteins enriched with phospholipids (42) also results in disruption of the envelope and loss of viral infectivity. These results with VSV offer the prospect that the infectivity of HIV-1 and other enveloped viruses (cytomegalovirus, herpesviruses, influenza viruses, to mention just a few) could be modulated in vivo through modulating the Chol content of the viral envelope.

Protection Against Damage to Host Cell Metabolism by the Virus

Lymphoma cells in culture have been reported to develop derangements in their metabolism of triglycerides and fatty acids following infection with HIV-1 (61). These preliminary findings are supported by the findings of a broader in vitro study (34). A human T-cell line (ERIC) was infected with HIV-1, and triacylglycerol and phospholipid turnover was measured beginning 72-96 hours after infection, when viral antigen production was at its peak and cellular pathology was beginning to manifest. The study found synthesis of triacylglycerols was enhanced, as was the membrane Chol content. Membranes of peripheral leukocytes drawn from subjects with advanced AIDS (Walter - Reed Stage 5 or 6) are reported to have significantly higher Chol content than subjects at stages 0-4, in the absence of higher serum Chol or higher dietary Chol intake. The investigators (50) suggested that "the results may relate to the obligatory Chol content of this enveloped virus."

Protection or Restoration of Host Cell Transmembrane Signalling Functions

The above described results from Lynn et al (34) also bear on the issue of possible derangement of signal transduction mechanisms in cells infected with HIV-1. There was a sizeable decrease in the synthesis of all phospholipids measured (PC, PE, PI, and PA) after infection. Further, the levels of diacylglycerol (DAG), a phospholipid metabolite and "second messenger," were reduced by 92%. At this point the cells were beginning to die, as assessed by increasing permeability to Ca^{2+}. Addition of DAG to the culture medium slowed this process.

The CD4 protein present on the outer surface of the T4/helper cell has the capacity to alter multiple functions of the cell that normally come into play following antigen recognition. Binding of any of a number of histocompatibility molecules or "growth factors" to the CD4 receptor results in transmembrane signal transduction and subsequent activation of the cell. One molecule which successfully "turns on" signal transduction is gp120, the envelope glycoprotein "receptor" of HIV-1 (32). Unfortunately, AA turnover was not examined in this study.

Activation of signal transduction in the host cell plasma membrane as a result of infection also appears to be a feature of enveloped viruses other than HIV-1. An extensive study conducted on students who were slow to recover from infection with EBV (an enveloped virus of the herpes group), documented abnormalities in their serum profiles of fatty acids, including AA (64). The activities of desaturase enzymes also were found to be depressed, indicating there could be difficulty in elaborating AA from dietary precursor fatty acids. Herpes simplex, another enveloped virus closely related to EBV, has been reported to activate plasma membrane phospholipase A_2 during the course of infection of cells in culture (33); an effect of this kind would be expected to increase the cellular turnover of AA.

Preliminary Results with Egg Phospholipids Against HIV-1 Disease

An "AL 721" formulation described as "composed of neutral glycerides, phosphatidylcholine, and phosphatidylethanolamine in a 7:2:1 ratio," was reported to inhibit HIV-1 in an in vitro culture system (49). It is not clear whether such 7-2-1 mixtures were used for other published studies, because commercial ELE are prepared by solvent extraction (from chicken egg yolks) and do not approach a 7-2-1 ratio.

Small clinical trials to evaluate commercial ELE in HIV-associated disease have been conducted in the USA, Israel, and West Germany. Overall, these trials have suffered from the use of too few patients also too diverse in clinical status, and from aspects of inadequate design. Also, too often the ELE used was poorly characterized or contained constituents likely to compromise its efficacy. Nevertheless, we think the available data do indicate that daily ingestion of egg yolk phospholipids as part of a structured dietary regimen can benefit the quality of life for many people with AIDS or other symptomatologies related to HIV-1 infection.

Clinically, ELE have been reported to provide symptomatic improvement in a majority (60-80%) of the subjects studied (18, 52, 67), as listed below:

- Fever, night sweats, diarrhea, skin rashes, and opportunistic infections were favorably affected in some subjects. Others reported improved wellbeing.
- A single dose of 10g once per day after an overnight fast appears to be effective.
- Benefits became evident by the end of the first month, and faded within weeks after discontinuation of the dosing.
- Subjects questioned in two mail polls (29) rated egg yolk-derived lipid mixtures (PC+PE) more efficacious than those derived from soy (PC only). This subjective evaluation is consistent with the AA of the egg product having a central role in the benefits observed after taking the product.

Laboratory data on the administration of ELE to HIV-infected subjects indicate that:

- On the average, red cell, white cell, and platelet counts tend to improve (21, 52).
- Counts of T4/helper and T8/suppressor cells may increase marginally, and in any case did not decrease significantly during treatment.
- Lymphocyte mitogenic responses seem to improve (21). This finding is consistent with the effects of ELE seen in aged men, as discussed above (44).
- HIV-1 activity is reduced, sometimes markedly, as judged from measurements of the HIV-1 reverse transcriptase enzyme in vivo (21). Viral antigen titers also were reduced in seropositive subjects (67).

The phospholipids of chicken egg yolks, taken in the form of egg lipid extract, therefore emerge as an important component of a *combination therapy for containment* of HIV-1 disease syndromes. The ELE clearly is <u>not</u> a cure for HIV-1 infection or for HIV-1 disease. However, from the preponderance of the evidence we conclude that when properly prepared and administered the ELE can have virustatic and immune-restorative effects.

These evident clinical benefits of the egg phospholipids, combined with their enviable lack of toxicity, argue strongly for the inclusion of the ELE as a dietary component of a *multicomponent combination therapy for HIV-1 disease*. We suggest that clinical trials be initiated on this basis, with sufficient subjects included to permit valid statistical assessment. One avenue yet to be explored is a non-crossover trial design which compares subjects taking a virustatic drug (AZT or combination of it with other drugs) with subjects taking the drug <u>plus</u> the phospholipids at various dosages.

DIETARY PHOSPHOLIPIDS CAN ALLEVIATE DRUG WITHDRAWAL SYMPTOMS

In animal models of drug addiction, ethanol and morphine were found to have a hyperfluidizing effect on membranes, which is compensated for homeostatically (at least in part) by addition of Chol into the fluidized membranes (30). After test animals (rodents) were exposed chronically to these drugs, their brain cell membrane fluidity at first increased, probably as a consequence of the fluidizing effect of the drug, then progressively decreased concomitant with an increased Chol/PL mole ratio. During the period of daily dosing with the drug, membrane Chol/PL remained within the normal range. When exposure to the drug was terminated, however, the animals experienced withdrawal symptoms.

Withdrawal symptoms upon removal from daily dosing with the drug were linked to above-normal availability of receptors for the drug in the brain as a consequence of residual hyperviscosity in the plasma membranes. Seemingly, as administration of the drug ceased its fluidizing effect also ceased, the membranes became effectively hyperviscous, and the receptor functions became abnormal, resulting in the withdrawal symptoms.

Administration of an ELE to mice suffering drug withdrawal alleviated much of the symptomatology (24). This formulation was described as "active lipid," "a special lipid mixture from egg yolk where the active ingredients are phospholipids, mainly PC (AL Labs Inc., Ojai, California)". The results were rationalized in terms of membrane fluidity affecting the binding affinity of membrane receptors for these drugs.

In mice, the affinity of certain brain receptors for their natural agonists (dopamine, serotonin) was found to decline with aging. Concomitantly, membrane fluidity decreased, as did cyclic AMP - dependent protein phosphorylation which is linked to signal transduction across the membrane (25). The administration of the abovementioned lipid mixture was stated to restore the affinity of the receptors to a level close to that of young mice.

According to Antonian et al (3), in comparison with the egg lipid extract soy extracts were considerably less effective in experiments aimed at recovering the function of serotonin receptors in aged mice. Thus egg yolk phospholipid formulations could be explored as possible adjunctive therapies in human withdrawal syndromes related to addiction to membrane -fluidizing drugs.

CONCLUDING REMARKS

Based upon the existing literature, egg yolk phospholipids given as dietary supplements show promise for the prevention or alleviation of syndromes related to membrane lipid insufficiencies or abnormalities. In this respect, supplementation of the diet with ELE would seem to offer superior promise to phospholipid mixtures prepared from other sources, or to polyunsaturated fatty acid preparations. In our opinion, dietary phospholipids deserve further exploration in human subjects for:

- Modulation of cell membrane Chol in various tissues;
- Maintenance of essential "second-messenger" functions;
- Maintenance of receptor and transport functions of membrane proteins;
- Restoration of the decline of membrane functions with aging.

The *structural effects* of dietary phospholipids seemingly are accomplished both through removing Chol from the membrane system of the cell by way of the plasma membrane and through contributing PC and possibly other phospholipids to the outer membrane leaflet. Both these influences of dietary phospholipids have the end-result of maintaining membrane Chol and membrane fluidity within the physiological range.

The *functional effects* of dietary phospholipids appear to be linked mainly to providing animal-source PE and smaller amounts of other phospholipids - such as PI, PS, SPH - to the inner leaflet of the plasma membrane, mainly to maintain the signal transduction function of the membrane. The PE that is one of the major constituents of the competently-prepared ELE is an excellent dietary source of AA (especially since plants do not make AA), and also of other fatty acids which can be metabolized to eicosanoids (prostaglandins, thromboxanes, leukotrienes), cytokines (diacylglycerol, platelet-activating factor, phosphatidic acid, lipoxins), and other hormone-like factors.

From the experimental and clinical evidence available, dietary phospholipid mixtures have an enviably good risk-benefit ratio. However, the choice of solvent for their manufacture, the presence of solvent residues and their byproducts, the stability and content of the active constituents, the mode of dosing, and the dietary habits and serum lipid status of the recipient, all can adversely affect the effectiveness and safety of such preparations. As part of their overall regimen, subjects taking dietary phospholipids should be maintained on a low-fat diet, including overnight fasting before dosing, to avoid competitive inhibition of the dietary phospholipids by other lipids in the circulation.

In the course of our work, we have become enthralled by the utility inherent in the known structures and functions of phospholipids. We feel certain that phospholipid formulations will continue to gain importance as dietary supplements for therapeutic as well as preventive applications in humans.

REFERENCES

1. Alderson, J. C. E., and Green, C., 1975, Enrichment of lymphocytes with cholesterol and its effect on lymphocyte activation, FEBS Lett., 52:208-211.
2. Aloia, R. C., Jenson, F. C., Curtain, C. C., Mobley, P. W., and Gordon, L. M., 1988, Lipid composition and fluidity of the human immunodeficiency virus, Proc. Natl. Acad. Sci. USA, 85:900-904.
3. Antonian, L., Shinitzky, M., Samuel, D., Lippa, A.S., 1987, AL721, a novel membrane fluidizer, Neurosci. Behav. Revs., 11:399-413.
4. Araki, K., and Rifkind, J.M., 1980, Erythrocyte membrane cholesterol: an explanation of the aging effect on the rate of hemolysis, Life Sciences, 26:2223-2230.
5. Artom, C., and Swanson, M. A., 1948, On the absorption of phospholipides, J.Biol. Chem., 175:871-881.
6. Bierman, E. L., and Oram, J. F., 1987, The interaction of high-density lipoproteins with extrahepatic cells, Am. Ht. J., 113 (2, pt.2):549-560.
7. Biesbroeck, R., Oram, J. F., Albers, J. J., and Bierman, E. L., 1983, Specific high affinity binding of high density lipoproteins to cultured human skin fibroblasts and arterial smooth muscle cells, J. Clin. Invest., 71:525-530.
8. Boucrot, P., 1972, Is there an entero-hepatic circulation of the bile phospholipids?, Lipids, 7:282-288.
9. Bruckdorfer, K. R., Demel, R. A., De Gier, J., and van Deenen, L. L. M., 1969, The effect of partial replacements of membrane cholesterol by other steroids on the osmotic fragility and glycerol permeability of erythrocytes, Biochim. Biophys. Acta, 183:334-345.

10. Caras, I. W., and Weddell, G. N., 1989, Signal peptide for protein secretion directing glycophospholipid membrane anchor attachment, Science, 243:1196-1198.

11. Chapkin, R. S., Somers, S. D., and Erickson, K. L., 1988, Dietary manipulation of macrophage phospholipid classes: Selective increase of dihomogammalinolenic acid, Lipids, 23 (8):766-770.

12. Cuthbert, J. A., and Lipsky, P. E., 1986, Promotion of human T lymphocyte activation and proliferation by fatty acids in low density and high density lipoproteins, J. Biol. Chem., 261:3620-3627.

13. Dogget, D. L., Chang, M.-P., Makinodan, T., and Strehler, B. L., 1981, Cellular and molecular aspects of immune system aging, Mol. Cell. Biochem., 37:137-156.

14. Ferber, E., De Pasquale, G. G., and Resch, K., 1975, Phospholipid metabolism of stimulated lymphocytes: composition of phospholipid fatty acids, Biochim. Biophys. Acta, 398:364-368.

15. Fisher, D. B., and Mueller, G. C., 1968, An early alteration in the phospholipid metabolism of lymphocytes by phyohemagglutinin, Proc. Natl. Acad. Sci. USA, 60:1396-1402.

16. Friedman, M., Byers, S. O., and Rosenman, R. H., 1957, Resolution of aortic atherosclerotic infiltration in the rabbit by phosphatide infusion, Proc. Soc. Exp. Biol. Med., 95:586-588.

17. Gallo, R. C., and Montagnier, L., 1988, AIDS in 1988, Sci. Am, 259:41-48.

18. Goebel, F.-D., Bogner, J., Matuschke, A., Nerl, C., and Gurtler, L., 1988, Clinical findings after administration of lipids in AIDS - A pilot study, IVth Intl. Conf. AIDS, Stockholm, Abst. No. 3531.

19. Goppelt, M., Kohler, L., and Resch, K., 1985, Functional role of lipid metabolism in activated T-lymphocytes, Biochim. Biophys. Acta, 833:463-472.

20. Gorter, E., and Grendel, F., 1925, On bimolecular layers of lipoids on the chromocytes of the blood, J. Exp. Med., 41:439-443.

21. Grieco, M. H., Lange, M., Buimovici-Klein, M., Reddy, M., England, A., McKinley, G. F., Ong, K., and Metroka, C., 1988, Open study of AL 721 treatment of HIV-infected subjects with generalized lymphadenopathy syndrome: An eight week open trial and follow-up, Antiviral Res., 9:177-190.

22. Grillone, L. R., Clark, M. A., Godfrey, R. W., Stassen, F., and Crooke, S. T., 1988, Vasopressin induces V1 receptors to activate phosphatidylinositol-and phosphatidylcholine-specific phospholipase C and stimulates the release of arachidonic acid by at least two pathways in the smooth muscle cell line, A-10, J. Biol. Chem., 263:2658-2663.

23. Hannun, Y. A., and Bell, R. M., 1989, Functions of sphingolipids and sphingolipid breakdown products in cellular regulation, Science, 243:500-507.

24. Heron, D. S., Shinitzky, M., and Samuel, D., 1983, Alleviation of drug withdrawal symptoms by treatment with a potent mixture of natural lipids, Eur. J. Pharmacol., 83:253-261.

25. Hershkowitz, M., Heron, D. S., Samuel, D., and Shinitzky, M., 1983, Modulation of protein phosphorylation and receptor binding in synaptic membranes by changes in lipid fluidity: Implication for aging, in: "Progress in Brain Research," Gispen, W. H. and Routtenberg, A., ed, p:419-434, Elsevier Biomedical Press, Amsterdam.

26. Hitzemann, R.J., Harris, R.A., Loh, H.H., 1984, Synaptic membrane fluidity and function, in: "Physiology of Membrane Fluidity," Shinitzky, M., ed, p:110-119, CRC Press, Boca Raton, Florida.

27. Hoelzl, J., and Wagner, H., 1971, Über den Einbau von intraduodenal appliziertem ^{14}C/^{32}P-Polyen-Phosphatidylcholin in die Leber von Ratten und seine Ausscheidung durch die Galle, Naturforsch, 26B:1151-1158.

28. Howard, A. N., Patelski, J., Bowyer, D. E., and Gresham, G. A., 1971, Atherosclerosis induced in hypercholesterolaemic baboons by immunological injury; and the effects of intravenous polyunsaturated phosphatidyl choline, Atherosclerosis, 14:17-29.

29. James, J. S., 1987, Al 721 survey results: Preliminary report, AIDS Treatment News, No. 39:1-7.

30. Johnson, D. A., Lee, N. M., Cooke, R., and Loh, H. H., 1979, Ethanol-induced fluidization of brain lipid bilayers: required presence of cholesterol in membranes for the expression of tolerance, Mol. Pharmacol., 15:739-746.

31. Kay, J. E., 1968, Phytohaemagglutinin: An early effect on lymphocyte lipid metabolism, Nature, 219:172-173.

32. Kornfeld, H., Cruikshank, W. W., Pyle, S. W., Berman, J. S., and Center, D. M., 1988, Lymphocyte activation by HIV-1 envelope glycoprotein, Nature, 335:445-448.

33. Lehtinen, M., Koivisto, V., Lehtinen, P., Aaran, R.-K., and Leinikki, P., 1988, Phospholipase A2 activity is copurified together with herpes simplex virus-specified Fc receptor proteins, Intervirology, 29:50-56.

34. Lynn, W. S., Tweedale, A., and Cloyd, M. W., 1988, Human immunodeficiency virus (HIV-1) cytotoxicity: perturbation of the cell membrane and depression of phospholipid synthesis, Virology, 163:43-51.

35. Lyte, M., and Shinitzky, M., 1985, A special lipid mixture for membrane fluidization, Biochim. Biophys. Acta, 812:133-138.

36. Matthews, R. E. F., 1985, Viral taxonomy for the nonvirologist, Ann. Rev. Microbiol., 39:451-474.

37. Meade, C. J., and Mertin, J., 1978, Fatty Acids and Immunity, Adv. Lipid Res., 16:127-165.

38. Mendez, A. J., He, J. L., Huang, H. S., Wen, S. R., and Hsia, S. L., 1988, Interaction of rabbit lipoproteins and red blood cells with liposomes of egg yolk phospholipids, Lipids, 23:961-967.

39. Miller, K. W., and Small, D. M., 1987, Structure of triglyceride-rich lipoproteins: An analysis of core and surface phases, in: "Plasma Lipoproteins," Gotto, A. M., ed, p:1-75, Elsevier Science Publishers, New York.

40. Moore, N. F., Patzer, E. J., Shaw, J. M., Thompson, T. E., and Wagner, R. R., 1978, Interaction of vesicular stomatitis virus with lipid vesicles: Depletion of cholesterol and effect on virion membrane fluidity and infectivity, J. Virol, 27:320-329.

41. Oram, J. F., Brinton, E. A., and Bierman, E. L., 1983, Regulation of high-density-lipoprotein receptor activity in cultured human skin fibroblasts and human arterial smooth muscle cells, J. Clin. Invest., 72:1611-1615.

42. Pal, R., Barenholz, Y., and Wagner, R. R., 1981, Depletion and exchange of cholesterol from the membrane of vesicular stomatitis virus by interaction with serum lipoproteins or poly(vinylpyrrolidone) complexed with bovine serum albumin, Biochem., 20:530-539.

43. Pal, R., Petri, W. A., and Wagner, R. R., 1980, Alteration of the membrane lipid composition and infectivity of vesicular stomatitis virus by growth in a chinese hamster ovary cell sterol mutant and in lipid-supplemented baby hamster kidney clone 21 cells, J. Biol. Chem., 255:7688-7693.

44. Rabinowich, H., Lyte, M., Steiner, Z., Klajman, A., and Shinitzky, M., 1987, Augmentation of mitogen responsiveness in the aged by a special lipid diet AL 721, Mechs. Ageing Dev., 40:131-138.

45. Rivnay, B., Globerson, A., and Shinitzky, M., 1978, Perturbation of lymphocyte response to concanavalin A by exogenous cholesterol and lecithin, Eur. J. Immunol., 8:185-189.

46. Rivnay, B., Globerson, A., and Shinitzky, M., 1979, Viscosity of lymphocyte plasma membranes in aging mice and its possible relation to serum cholesterol, Mechs. Ageing Dev., 10:71-76.

47. Rivnay, B., Bergman, S., Shinitzky, M., Globerson, A., 1980, Correlations between membrane viscosity, serum cholesterol, lymphocyte activation, and aging in man, Mechs. Ageing Dev., 12:119-126.

48. Rouser, G., Kritchevsky, G., 1972, Lipids in the nervous system of different species as a function of age, Adv. Lipid Res., 10:261-360.

49. Sarin, P. S., Gallo, R. C., Scheer, D. I., Crews, F., and Lippa, A. S., 1985, Effects of a novel compound (AL 721) on HTLV-III infectivity in vitro, New Engl. J. Med., 313:1289-1290.

50. Shoemaker, J. D., Millard, M. C., and Johnson, P. B., 1987, Mixed leukocyte cholesterol in HTLV-III (HIV-1) infection, Fed. Proc., 46:1318 (Abstract).

51. Simons, K., Garoff, H., and Helenius, A., 1982, How an animal virus gets into and out of its host cell, Sci. Am., 246:58-63.

52. Skornick, Y., Yust, I., Zakuth, V., Asner, A., Vardinon, N., and Shinitzky, M., 1988, Treatment of AIDS patients with AL 721, IVth Intl. Conf. AIDS, Stockholm, Abst. No. 3529.

53. Spector, A. A., and Yorek, M. A., 1985, Membrane lipid composition and cellular function, J. Lipid Res., 26:1015-1035.

54. Stafford, W. W., and Day, C. E., 1975, Regression of atherosclerosis effected by intravenous phospholipid, Artery, 1:106-114.

55. Stein, O., and Stein, Y., 1973, The removal of cholesterol from Landschutz Ascites cells by high-density apolipoprotein, Biochim. Biophys. Acta, 326:232-244.

56. Stryer, L., 1988, "Biochemistry" (3rd Edit.), W. H. Freeman, New York.

57. Traill, K. N., Offner, F., Winter, U., Paltauf, F., and Wick, G., 1988, Lipid requirements of human T lymphocytes stimulated with mitogen in serum-free medium. Membrane fluidity changes are an artefact of lipid (AL 721) uptake by monocytes, Immunobiol., 176:450-464.

58. Traill, K. N., and Wick, G., 1984, Lipids and lymphocyte function, Immunol. Today, 5:70-76.

59. Vaughan, C., 1987, Second wind for second-messenger research, BioScience, 37:642-650.

60. Vega, G. L., and Grundy, S. M., 1987, Mechanisms of primary hypercholesterolemia in humans, Am. Heart J., 113 (2, pt. 2):493-502.

61. Willer, A., Buff, K., Goebel, F.-D., and Erfle, V., 1988, HIV-1 induced membrane alterations increase susceptibility to cytolysis by lipid formulations, IVth Intl. Conf. AIDS, Stockholm, Abst. No. 3535.

62. Williams, K. J., and Scanu, A. M., 1986, Uptake of endogenous cholesterol by a synthetic lipoprotein, Biochim. Biophys. Acta, 875:183-194.

63. Williams, K. J., Werth, V. P., and Wolff, J. A., 1984, Intravenously administered lecithin liposomes: A synthetic antiatherogenic lipid particle, Persp. Biol. Med., 27:417-431.

64. Williams, L. L., Doody, D. M., and Horrocks, L. A., 1988, Serum fatty acid proportions are altered during the year following acute Epstein-Barr Virus infection, Lipids, 23:981-988.

65. Williams, K. J., Tall, A. R., Bisgaier, C., and Brocia, R., 1987, Phospholipid liposomes acquire apolipoprotein E in atherogenic plasma and block cholesterol loading of cultured macrophages, J. Clin. Invest., 79:1466-1472.

66. Yeagle, P. L., 1988, Cholesterol and the cell membrane, in: "The Biology of Cholesterol," Yeagle, P. L., ed, p:121-145, CRC Press, Boca Raton.

67. Yust, I., Vardinon, N., Skornik, Y., Zakuth, V., and Shinitzky, M., 1988, Reduction of circulating HIV antigen in sero-positive patients after treatment with AL 721, IVth Intl. Conf. AIDS, Stockholm, Abst. No. 3530.

ULTRASTRUCTURE AND INTERACTIONS OF SOY LECITHIN DISPERSIONS

IN MILK SYSTEMS

Wolfgang Buchheim, Arnold Wiechen, Dieter Prokopek
and Andreas Funke

Federal Dairy Research Centre
Hermann-Weigmann-Str. 1
D-2300 Kiel, F.R. Germany

INTRODUCTION

Interactions between phospholipids and proteins are of decisive rele-
vance not only for biological systems but also for food systems. The formation of complexes between these two components may especially alter
colloidal properties of proteins because phospholipids modify the surface
charge thereby changing isoelectric point, degree of aggregation, solubility etc. As a result of these alterations physical properties of food
systems are affected. We report here about interactions between soy lecithin and milk proteins and their possible implications for compositional
and textural aspects in cheese manufacture [1,4,5].

MATERIALS AND METHODS

For these experiments a deoiled soy lecithin (20-23 % PC, 16-21 % PE,
12-18 % PI, ca. 15 % glycolipids; Lucas Meyer, Hamburg) was used and its
effects studied during the manufacture of a soft cheese. At first a 3 %
dispersion of lecithin in skim milk was made either by sonication or high
pressure homogenization (200 bar), partly heat sterilized (130°C, 15 min),
then acidified by a microbial starter culture and finally added to the
cheese milk with a final concentration of 500 mg lecithin per kg of milk.
The milk was subsequently processed into cheese [4,5]. Ultrastructure
was studied by transmission electron microscopy using the freeze-fracture
method [2]. Electron micrographs were used for particle size analysis [3].

RESULTS AND DISCUSSION

At the beginning of the dispersion process of lecithin in skim milk
larger, often globular particles are present (Fig.1) and small liposomal
particles develop at their periphery. After sonication or high pressure
homogenization a fine liposomal dispersion is present (Fig.2) which only
rarely shows association with colloidal milk proteins, i.e. casein micelles. Particle size analysis of these dispersions revealed similarities
as to volume frequency distribution (Fig.3) and distribution parameters
(Table 1). Liposome voluminosities of 3.1 and 4.2 ml/g indicate an
appreciable uptake of water (milk serum).

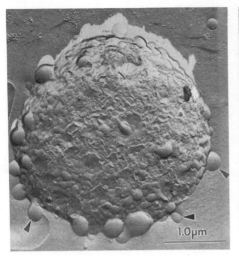

Fig.1. Larger lecithin particle in milk at the beginning of the dispersion process showing peripheral development of liposomes (arrows).

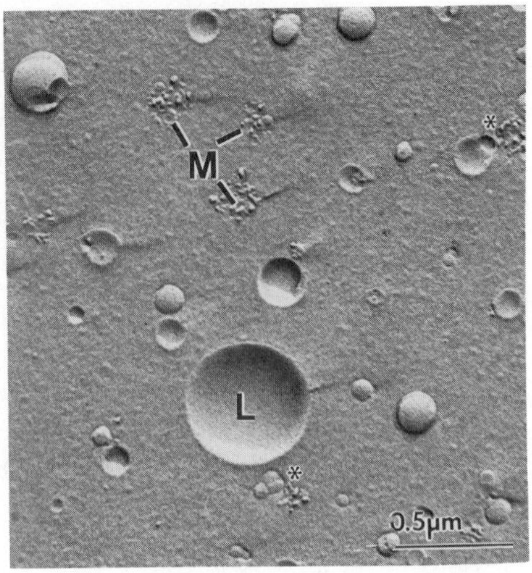

Fig.2. Sonicated lecithin dispersion in milk. Casein micelles (M) only rarely form complexes (asterisk) with liposomes (L)

Table 1. Distribution parameters of lecithin dispersions

	Method of dispersion	
	Sonication	Homogenization
volume fraction of liposomes*(ml/100 ml)	9.3	12.7
voluminosity of liposomes (ml/g)	3.1	4.2
surface to volume ratio S_v (m^2/ml)	32.2	29.6
volume-surface average diameter d_{32}(nm)	186	202

*3g lecithin per 100 g milk

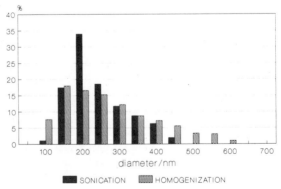

Fig.3. Volume frequency distributions of dispersed lecithin in milk

Depending on the heat treatment of the original lecithin/skim milk dispersion major differences are found in the amounts of lecithin retained in the cheese curd (i.e. coagulated protein mass) or getting lost with the whey (i.e. serum drained from the curd)[4,5]. The high retention of lecithin in the cheese curd in experiments with heat sterilized lecithin dispersions, i.e. 85 % versus only ca. 30 % with non-sterilized dispersions indicated a different degree of lecithin-protein interaction. Electron micrographs of sterilized dispersions (Fig.4) reveal a distinct increase in particle size, development of more irregular shapes and peculiar inner fine structures, and the formation of complexes with protein particles (casein micelles and/or heat denatured whey proteins).

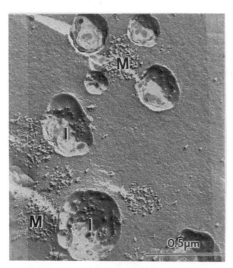

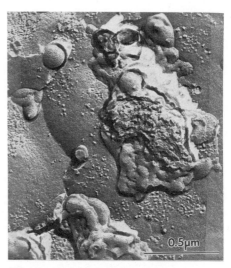

Fig.4. Heat-sterilized lecithin dispersion in milk. Lecithin particles show alterations in size, shape and internal structure(I), and bind to casein micelles (M).

Fig.5. Acidification of heat-sterilized lecithin dispersion results in partial fusion of lecithin particles.

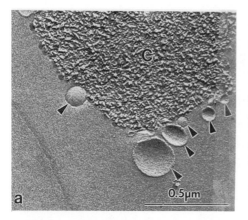

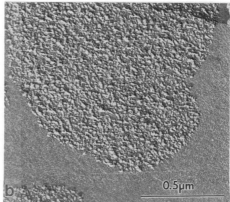

Fig.6. Soft cheese (a) with and (b) without lecithin. Liposomes (arrows) primarily adsorb to the surface of the casein strands (C).

During acidification of the lecithin/skim milk dispersion
the lecithin particles mostly fuse into large, irregularly shaped aggregates which show intimate association with acid-coagulated milk proteins
(Fig.5). During subsequent mixing with the cheese milk (i.e. a rise of
pH value to ca 6.0) and enzymic coagulation of the milk by rennet these
large lecithin aggregates partly transform again into liposomes. In the
soft cheese liposomes are preferentially located at the surface of the
casein matrix (Fig.6). This localization maybe the reason why lecithin -
treated cheeses showed a higher retention of moisture (ca. 6-11 %). The
peripheral accumulation of liposomal lecithin may inhibit further fusion
of casein strands (syneresis) during enzymic coagulation as a result of
surface charge.

SUMMARY

These studies have demonstrated that freeze-fracture electron microscopy is particularly suitable for following changes in dispersed lecithin
systems and interaction with proteins. The compositional and structural
complexity of food systems will, however, require more systematic experiments to obtain an optimal benefit from the functional properties of
lecithin.

REFERENCES

1. Bily, R.R. (1981): Addition of lecithin to increase yield of
   cheese. U.S. Patent 4277503.
2. Buchheim, W. (1982): Aspects of sample preparation for freeze-
   fracture/freeze-etch studies of proteins and lipids in food
   systems. Food Microstruct. 1: 189-208.
3. Buchheim, W., Falk, G. and Hinz, A. (1986): Ultrastructural and
   physicochemical properties of UHT-treated coffee cream. Food
   Microstruct. 5: 181-192.
4. Weber, N., Wiechen, A., Buchheim W. and Prokopek, D. (1985):
   Alterations of soybean lecithin during curd formation in
   cheese making. J. Agric. Food Chem. 33: 1093-1096.
5. Wiechen, A., Buchheim W. and Prokopek, D. (1985): Investigation
   on the distribution of soybean lecithin during cheese making
   by means of C14-labelling and electron microscopy.
   Milchwiss. 40: 402-406.

FRACTIONATION OF TOTAL LIPID EXTRACTS OF ANIMAL ORIGIN

Ladislas Colarow

Nestlé Research Centre, Nestec Ltd.

Vers-chez-les-Blanc, CH-1000 Lausanne 26, Switzerland

SUMMARY

Total lipid classes with distinctively different functional properties are usually represented by neutral lipids (NL), complex lipids (CL),
phospholipids (PL), glycolipids (GL), gangliosides (GS) and ceramidehexosides plus sulfatides (CS). The classes are required individually or in a
particular combination as micronutrients and liposomal agents for the formulation of advanced health products or cosmetic formulas. Their isolation
from total lipid extracts of animal origin is possible using three different procedures, which are described below (2). These procedures are suitable for industrial as well as analytical applications. Efficiency of the
whole process is demonstrated by fractionation of the total lipids extracted from bovine brain, cerebrospinal fluid, lungs, pancreas and delactosed
buttermilk. The resulting lipid fractions are evaluated by planar chromatography (HPTLC) whereas the composition of O-acyl (phosphoglycerides) and
N-acyl (sphingomyelin and glycolipids) moieties is analyzed by capillary
gas-chromatography (GC) after an acetyl chloride-catalyzed derivatization
(5). The methylation reaction is equally efficient for phospholipids and
their plasmalogens, thus producing both the fatty acid methyl esters and
dimethylacetals.

PROCEDURE 1: THE SEPARATION OF NL AND CL

One volume of total animal lipids is suspended by sonication in three
volumes of hexane at 40-50°C and passed through a large-diameter column
containing two volumes of activated silica gel. All complex lipids including phospholipids, ceramidehexosides, sulfatides and gangliosides, elute
by gravity in the form of micelles using 1-1.5 volume of hexane. Then the
neutral lipids are deadsorbed from the silica gel sorbent using 1-1.5 volume of a polar organic solvent (2).

PROCEDURE 2: THE SEPARATION OF PL AND GL

Pure complex lipids (Procedure 1) or alternatively, total animal lipids, are fractioned on a boric acid gel™ (2,7) fast-flow column which re-
tains, selectively, GS and CS while PL (alternatively, PL and non-polar
lipids) elute in a non-aqueous mobile phase. A mobile phase containing up

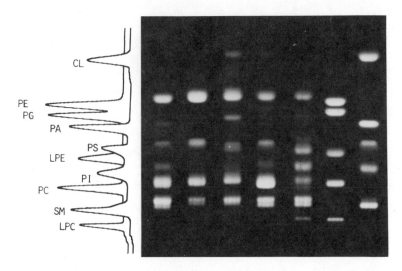

Fig.1.   HPTLC separation of the phospholipid fractions obtained by combining Procedures 1 and 2. From left to
right: bovine buttermilk, brain, cerebrospinal fluid,
lungs and pancreas. Phospholipid standards (extreme
right) and their fluorodensitometric peaks (extreme
left) are shown also. The lipids were analyzed on a
boric acid impregnated silica gel, plate using a
chloroform-MeOH-EtOH-triethylamine- 0.25 % KCl- 25 %
ammonium hydroxyde (40:20:20:35:6:1.5 v/v) mobile
phase containing 0.02 % NBD dihexadecylamine (Molecular Probes Inc.) as the fluorescent agent (3). The
chromatogram was photographed or scanned without any
additional detection procedure using a UV-light (366
nm) source and a 460-nm cut-off filter.
Abbreviations: phosphatidyl-choline (PC), -ethanol-
amine (PE), -serine (PS), -inositol (PI), -glycerol
(PG), lyso-PC (LPC), lyso-PE (LPE), phosphatidic acid
(PA), cardiolipin (CL) and sphingomyelin (SM).

to 20 % water (v/v) is used to dissociate the reversible complexes formed
between the dihydroxyboryl and glycolipid diol groups. Thus the aqueous
solvent system achieves a complete elution of all glycolipids without any
alteration of the stationary phase.   The column is reusable in ca 1-hour
cycles including the successive elution of PL and GL, followed by dehydra-
tion or regeneration of the gel with a non-aqueous solvent. One kg of the
gel is usually loaded with 200 g of complex or up to 300 g of total animal
lipids with recoveries representing ca 60 g of CS and GS as a single fraction. Composition of the phospholipid fraction is illustrated in Figure 1.
The glycolipid eluates can be lyophilized and then formulated as such, or
separated further into their CS (Figure 2) and GS (not shown) fractions as
described below.

PROCEDURE 3: THE SEPARATION OF GS AND CS

In the absence of PL, total GL can be easily separated into their GS
and CS fractions by different ion-exchange or solvent-partition methods.
The well known method of Folch et al. (4), originally described for the

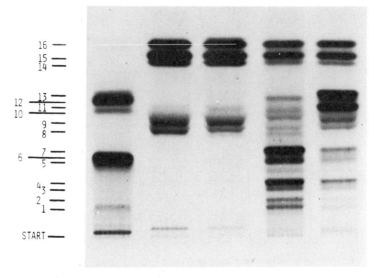

Fig.2. HPTLC separation of the ceramidehexoside plus sulfa-
tide fractions obtained by combining Procedures 1-3.
From left to right: bovine buttermilk, brain, cerebrospinal fluid, lungs and pancreas. The chromatogram
illustrates the presence of penta- (1-2), tetra- (3-
4), tri- (5-7), di- (11-13) and monohexosylceramides
(14-16) as well as sulfatides (8-10). Detection reagent: orcinol. Elution solvent: chloroform-methanol-
0.02 % CaCl$_2$ (65:25:4 v/v).

isolation and purification of total lipids from animal sources, is quite
efficient for the separation of total GL into their GS and CS fractions.
Total GL (see Procedure 2) are solvent partitioned in a chloroform-MeOH-
0.88 % KCl (8:4:3 v/v) system into the upper GS and the lower CS (Fig.2)
phases. This simple method produces excellent recoveries of all GS classes
except for those of hematosides which are characterized by the absence of
hexosamine moieties. Namely, trace amounts of hematoside species with 20:0
-24:0 fatty acid moieties are usually present as contaminants in the lower
CS phase.

RESULTS AND DISCUSSION

All animal PL (Figure 1) contain significant levels of sphingomyelin
(SM) and phosphatidylserine which, as lipid micronutrients, are absent in
vegetable sources. Also, SM is required in liposome formulations as a mem-
brane-rigidifying agent. The GL fractions (Figure 2) represent valuable
sources of monogalactosylceramides which, as hydroxy fatty acid carriers,
are present in human and absent in bovine or formula milks. Gangliosides
are used therapeutically in human peripheral neuropathies and, as demonstrated by Allen and Chonn (1), they also increase the resistance of liposomes in vivo against the reticuloendothelial system and thus prolongate
their circulation half-lives. The bovine brain PL fraction (Figure 1) is a
rich source of arachidonic (20:4) and docosahexaenoic (22:6) acids (Figure
3) which are especially required for formula milks such as those used for
premature babies or at an early stage of lactation.

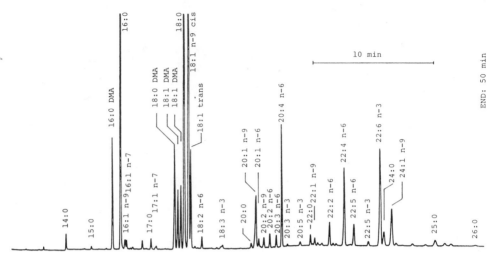

Fig.3. Composition of O-acyl and N-acyl moieties as obtained by capillary GC (8) after an acetyl chloride-catalyzed derivatization (5) of bovine brain PL as well as their plasmalogens. Both the PL-derived fatty acid methyl esters and plasmalogen-derived dimethylacetals (DMA) were identified using GC in combination with ion trap detector mass spectrometry.

In summary, the present procedures 1-3, used individually or in combination, represent an attractive approach for the separation of total animal lipids into their six principal classes. In particular, the boric acid gel procedure eliminates a serious gap in the lipid industrial and analytical methodology since pure PL and GL fractions are obtained in ca one-hour cycles. In comparison, it takes several days to isolate total GL using the widely accepted method of Saito and Hakomori (6) which is PL-destructive. In this context, the present procedures are not only rapid and feasible for industrial applications, but they also facilitate analytical possibilities in one of the most laborious areas of animal lipid research and methodology (2).

REFERENCES

1. Allen, T.M. and Chonn, A., 1987, Large unilamellar liposomes with low uptake into the reticuloendothelial system, FEBS Letters, 223: 42-46.
2. Colarow, L., 1988, Procédé de séparation des glycolipides d'un mélange lipidique et utilisation des fractions obtenues, in Patent pending, Registered No.4033'88-8, October 28th 1988, in Switzerland and other countries.
3. Colarow, L., 1989, Quantitation of phospholipid classes on thin-layer plates with a fluorescence reagent in the mobile phase, J.Planar Chromatography - Modern TLC, 2:19-23.
4. Folch, J., Lees, M. and Sloane-Stanley, G.H., 1957, A simple method for the isolation and purification of total lipids from animal sources, J.Biol.Chem., 226:497-509.
5. Lepage, G. and Roy, C.C., 1986, Direct transesterification of all lipid classes in a one-step reaction, J.Lipid Res., 27:114-120.
6. Saito, T. and Hakomori, S.I., 1971, Quantitative isolation of total glycolipids from animal cells, J.Lipid Res., 12:257-259.

7. Schott, H., Rudloff, E., Schmidt, P., Roychoudhury, R. and Kössel, H.,
    1973, A dihydroxyboryl-substituted methacrylic polymer for the
    column chromatographic separation of mononucleotides, oligonu-
    cleotides, and transfer ribonucleic acid, Biochemistry, 12:932-
    938.
8. Traitler, H. and Kolarovic, L., 1983, Application of silane-coupling
    agents for medium-polarity capillary columns, J.Chromatography,
    279:69-73.

SIGNIFICANT SURFACE ACTIVITY SHOWN BY THE MIXTURE OF PARTIALLY DEACYLATED

LIPIDS

Satoshi Fujita and Kazuaki Suzuki

ASAHI DENKA KOGYO K.K.
2-3-14 Nihonbashi-muromachi, Chuoku, Tokyo 103 Japan

INTRODUCTION AND SUMMARY

It is known that the existence of lecithin is essential for intestinal fat absorption[1 8 9 11 12 13]. Oil and fat, both dietary and biliary phospholipids (PLs), are partially hydrolyzed by lipases and phospholipase A-2 (PLA2) in the small intestine, and are changed into fatty acids (FAs), monoglycerides (MGs) and lysophospholipids (LPs). These deacylated lipids are solubilized into mixed bile salt micelles[2 6 14]. Because of the ubiquitous existence of PLA2, in all cells[3], the mixed solution of LPs and FAs may be very common. We examined the surface activity, in paticular, immersional and adhesional wettability, shown by an aqueous solution of soy lysophospholipid (SLP)/MG/FA, SLP/FA and SLP/MG, in vitro, and found that many lipid mixtures (compositions) showed significant surface activity when their MG and FA components consisted of unsaturated FA(ULFA) and/or medium chain FA(MCFA). A lipid mixture SLP/ medium chain MG (MCMG)/MCFA showed the highest surface activity. Wettability decreased rapidly when the moiety of saturated long chain FA in lipid mixtures increased. The higher ratio of MG and FA to SLP gave higher activity, and the addition of a solubilizer, such as bile salt, was necessary to dissolve them in water. The result of this study suggests that the surface activity of these mixed micelles affects intestinal digestion and absorption, and subsequently alters the level of the physiological phenomena which take place at the cell membrane.

EXPERIMENTAL

All SLPs were enzymatically (NOVO Lecithase 10-L) derived from soyPL. They consisted of mono and di-acylphospholipids (MDPLs). Soy lysophosphatidylcholine (LPC), from Nihon Shoji K.K.(Osaka, Japan), SLP containing 50%(Wt.), 66%(wt.), and 80%(wt.) of LPs(SLP50, SLP66, SLP80) were produced in our laboratory[7]. MGs, from Riken Vitamin K.K.(Tokyo, Japan), and FAs were distilled with a purity of 95-99%. Sodium taurocholate(STC), DIFCO Lab. 70% purity. Each component was dissolved in hexane-alcohol solution. After vacuum evaporation of the solvent, the lipid mixtures were dissolved in deionized water using a sonicator. Surface tension was measured by Wilhelmy's plate method[4] with a platinum plate ; immersional wettability was measured by the canvas disk method[10], and adhesional wettability (contact angle) was measured on bees wax.

RESULTS AND DISCUSSION

In the mammalian small intestine, the ratio of PLs (biliary and dietary) to triglycerides (TGs) depends on the ingested diet. A Japanese adult consumes 30-80g of TGs and 1-4g of PLs, and a Western adult consumes 60-150g of TGs and 1-8g of PLs per day. Quantitatively more important PL is endogenous PL (essentially PC) of hepatic origin (7-22g/day) which is secreted into duodenum[26]. The average molar ratio of TG/PL is unclear, however it may be in the range of 3/1-20/1.

Fig.1 shows the surface activity of lipid mixtures containing different FA components, in which the molar ratios of SLP/MG/FA are fixed at 1/3/6 and 1-3.7 molars of STC are added depending on their solubilities. It also shows the effect of different SLP components : pure soy LPC and soy MDPL (SLP66). Each mixture contained the same molar ratio of monolinolein, and differed only in FA components. With myristic acid it showed the least surface tension. A mixture which contained MCFA or ULFA showed significant wettability in both canvas disk wetting time and contact angle, whereas its activity decreased when it contained saturated long chain FA or short chain FA. There were almost no differences in surface activity between the two series containing soy LPC and SLP66. Fig.2 shows the canvas disk wetting time of typical lipid mixtures, i.e., a mixture of SLP/C10:OMG/C10:OFA/STC has the highest activity, which is comparable to that shown by Na-dioctylsulfosuccinate (Na-DOSS : Aerosol-OT type surfactant), the most powerful penetrating agent. Its contact angle on bees wax and surface tension, were lower than Na-DOSS. In general, the higher ratio of MG and FA to SLP or LPC the higher was the surface activity, and the addition of a solubilizer, such as bile salt, was necessary to dissolve these components in water. The more unsaturated the FA, the higher was the surface activity. Fig.3 shows the effect of different FAs in equimolar solutions of FA/LPL on surface tension (A), immersional wettability (B) and adhesional wettability (C), respectively. SLP/C14:OFA gave the minimal surface tension, and with C12:OFA it showed the maximal wettability, both canvas disk wetting time and contact angle. Polyenoic acid such as linolenic acid and eicosapentaenoic acid showed higher activities. MDPL/MG of which the FA moiety consisted of USFA and/or MCFA also showed the same phenomena as the former two types of mixtures. Similar effects were observed in mixtures with egg LPC.

It is recognized that the degrees of unsaturation and chain length of FAs play a decisive influence on the activity of their compositions (SLP/MG/FA and SLP/FA). SLP/C12:OMG/C12:OFA and SLP/C12:OFA have the highest activity, though this combination does not occur natually in vivo.

On the other hand, mixtures with saturated FA, long chain monoenoic acid such as C22:1 FA (erucic), trans C18:1 FA (elaidic) and stearic acid, showed lower activities.

CONCLUSION

In a comparison of the lipid mixtures (compositions) with some surfactants in several surface activities, as shown Table 1, it may be concluded that these lipid compositions should be useful as safe and effective biosurfactants for foods, drugs, cosmetics etc. Some of them showed not only excellent wettability but also good emulsifying ability and dispersibility. It is known that BS, PC, FA and MG in appropriate physiological ratio and concentration produce extremely small and stable emulsion of TG in vivo [5]. A composition of MDPL/MG/FA/STC is more effective for emulsification of TG than PL/MG/FA/STC. Thus, it is possible to prepare a composition which is suitable for a specific purpose.

MCFA and ULFA are readily absorbed via the intestine while absorbtion of saturated long chain FAs is more difficult. It would be useful, therefore, to consider the intestinal absorbability of lipids from the viewpoint of their surface activities. Our findings demonstrate that mixtures of partially deacylated lipids could act as penetrating agents in the intestinal absorption of nutrients. This may be very important since the surface activity of mixed micelles of LP/MG/FA and LP/FA may be ubiquitous in all living things. For example, they occur in seed germination, and subsequently alter the level of the physiological phenomena which take place at the cell membrane.

Table 1 Comparison of the surface activities of the three types of lipid compositions and commercial surfactants, 0.5% wt. aq. sol. at 25 ℃

Surfactant (trade name ®), lipid composition	Surface tension (dynes/cm)	Canvas disk Wetting time (second)	Contact angle on bees wax (degrees)
Sucrose mono stearate (Ryoto Ester S-1670)	34.8	438	64
POE(10mol)nonylphenol ether(ADK TOL NP-700)	31.2	4	39
Na-Dioctylsulfo-succinate (ADK COL EC-4500)	28.3	<0.5	35
Soy LPC	37.9	116	56
SLP 80	33.2	288	44
SLP 50	33.8	1140	57
SLP50/C18:2MG/C18:2FA /STC=1:2:3:4 (wt.)	28.4	18	35
SLP50/C10:0MG/C10:0FA /STC=1:2:3:4(wt.)	25.2	<0.5	22
SLPC/C18:2FA equimolar	27.8	65	33
SLPC/C10:0FA equimolar	26.2	9	32
SLP 80/C18:2FA equimolar	28.8	51	44
SLP 80/C10:0FA equimolar	27.3	37	35
SLP 50/C18:2MG=4:6(wt.)	28.2	37	42
SLP 50/C10:0MG=4:6(wt.)	27.6	7	37

(1) Ryoto Ester S-1670, MITSUBISHI KASEI SYOKUHIN K.K. JAPAN
(2) ADK COL EC-4500 (Aerosol-OT type surfactant), ASAHI DENKA KOGYO K.K. JAPAN
(3) ADK Tol NP-700 (Trition X-100 type surfactant), ASAHI DENKA KOGYO K.K. JAPAN

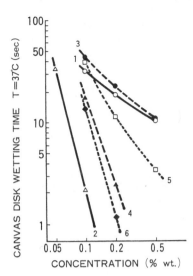

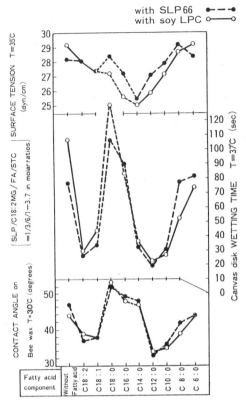

Fig.2 Wetting time vs. concentration curve for aq. sol. of lipid mixtures and of industrial surfactants.

1 : Soy LPC/C18:2MG/C18:2 FA/STC
  =1/2/3/3(wt.)
2 : Soy LPC/C10:0 MG/C10:0 FA/STC
  =1/2/3/2(wt.)
3 : SLP66/C18:2 MG/C18:2 FA/STC
  =1/2/3/4(wt.)
4 : SLP66/C10:0 MG/C10:0 FA/STC
  =1/2/3/4(wt.)
5 : POE (10mol) nonylphenolether
  (ADK TOL NP-700)
6 : Na-dioctylsulfosuccinate
  (ADK COL EC-4500)

Fig.1 Surface activity of lipid mixtures having different fatty acid compositions, 0.2% wt. aq. sol.

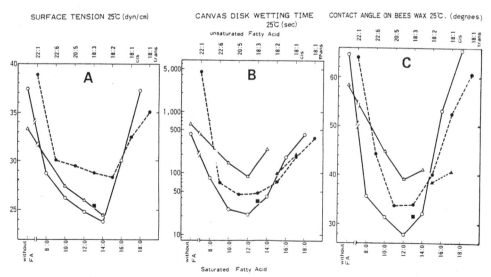

Fig.3 Surface activities of equimolar solutions of soy lysophospholipid/fatty acid. 3mM aq. sol.

soy LPC/sat.FA: —○—, soy LPC/unsat.FA: —●—, sat. FA : —△—, soy LP66/unsat. FA : --▲--, soy LPC/coconut FA : ■

REFERENCES

1. Beil, F. U., and Grundy, J., (1980), Studies on plasma lipoproteins during absorption of exogenous lecithin in man, J. lipid Res. 21 : 525-536.
2. Carey, M. C., Small, D. M., and Bliss, C. M., (1983), Lipid digestion and absorption, Ann. Rev. Physiol. 45 : 651-677.
3. de Haas, G.H., de Gier, J. van den Bosch, H., and Wirtz, K. W. A., (1986), Twenty-five years of lipid and membrane biochemistry in Utrecht, in " Lipids and Membranes : Past, Present and Future " Op den Kamp, J. A. F., et. al. eds., pp. 1-44, Elsevier, Amsterdam.
4. Harkims, W. D., and Anderson, T. F., (1937), A simple accurate film balance of the vertical type for biological and chemical work, Am. Chem. Soc. 59 : 2189-2197.
5. Linthorst, H. W., Bennett-Clark, S., and Holt, P. R., (1977), Triglyceride emulsification by amphipaths present in the intestinal lumen during digestion of fat, J. Colloid Interface Sci. 60 : 1-10.
6. Mead, J. F., Alfin-Slater, R. B., Howton, D. R., and Popjak, G., (1986), Digestion and absorption of Lipids, in " Lipids, Chemistry, Biochemistry and Nutrition". pp. 255-272, Plenum Press, New York.
7. Nakai, E., Suzuki, K., Satoh, S., and Katoh, M., (1989), Method of deacylation of phospholipid by phospholipase A-2, Japan Pat. laid open 89/16595.
8. O'Doherty P. J. A., Kakis, G., and Kuksis, A., (1973), Role of luminal lecithin in intestinal fat absorption, Lipids 8 : 249-255.
9. Rampone, A. J., and Lawrence, R. L., (1977), The effect of phosphatidylcholine and lysophosphatidylcholine on the absorption and mucosal metabolism of oleic acid and cholesterol in vitro, Biochim. Biophys. Acta 486 : 500-510.
10. Seyferth, H., and Morgan, O. M., (1938), The canvas disk wetting test, Am. Dyestuff Reptr. 27 : 525-532.
11. Tso, P., Balint, J. A., and Simmonds, W. J., (1977), Role of biliary lecithin in lymphatic transport of fat, Gastroenterology 73 : 1362-1367.
12. Tso, P., Lam, J., and Simmonds, W. J., (1978), The importace of the lysophosphatidylcholine and choline moity of bile phosphatidylcholine in lymphatic transport of fat, Biochim. Biophys. Acta 528 : 364-372.
13. Tso, P., Kendric, H., Balint, J. A., and Simmonds, W. J., (1981), Role of biliary phosphatidylcholine in the absorption and transport dietary triolein in the rat, Gastroenterology 80 : 60-65.
14. Yih-Fu Shian, (1987), Lipid digestion and absorption, in " Physiology of Gastrointestinal Tract" 2nd ed., Johnson, L. R., et al. eds., vol. 2, pp 1527-1556, Raven Press, New York.

QUANTIFICATION OF SOYBEAN PHOSPHOLIPID SOLUBILITY

USING AN EVAPORATIVE LIGHT SCATTERING MASS DETECTOR

P. Van der Meeren*, J. Vanderdeelen, M. Huys and L. Baert

State University of Ghent, Faculty of Agricultural Sciences
Department of Physical and Radiobiological Chemistry
Coupure Links 653, B-9000 Gent, Belgium

INTRODUCTION

Numerous HPLC procedures for the separation of phospholipids have been described during the last 10 years (2). One of the major problems has been their detection, especially if quantitative results are needed. Recently, the evaporative light scattering mass detector has been introduced (5,7). It measures the intensity of the light scattered by the particle cloud formed by the solute upon evaporation of the nebulized mobile phase. Using this detector, a fast and reproducible High Pressure Liquid Chromatographic method has been elaborated in order to separate all major soybean phospholipids. Since the response of the mass detector is not linearly related to the sample load, calibration curves were elaborated to enable quantification. Finally, the proposed method was used to investigate the solubility of phospholipids in organic solvents. The main objective of this research was to evaluate the possibility to fractionate phospholipids based upon their selective solubility. Moreover, it could be deduced which solvents were suitable to dissolve mixed phospholipid samples.

EXPERIMENTAL

Materials

HPLC-grade n-hexane and isopropanol were supplied by Alltech, whereas the remaining organic solvents, which were reagent grade, were obtained from UCB. The water was deionized, distilled and used freshly. Crude soybean lecithin was from Vamo Mills and commercial powdered soybean lecithin (Epikuron-100P) originated from Lucas Meyer. A highly purified phosphatidylcholine (PC) was from Nattermann. Triolein, phosphatidylinositol (PI) and phosphatidic acid (PA) were supplied by Serdary Research Laboratories, whereas phosphatidylethanolamine (PE) and phosphatidylglycerol (PG) were obtained from Sigma. All lipids were dissolved in chloroform and filtered through a 0.2 μm Dynagard filter.

---

* Research assistant of the Belgian National Fund for Scientific Research (N.F.W.O.)

Instrumental Set-up

A Waters model 590 isocratic HPLC pump, equipped with a solvent switcher (Waters) was used. The Waters Intelligent Sample Processor (WISP) could be programmed to inject up to 48 samples. The column consisted of 3 μm Spherisorb, packed in a 100 x 4.6 mm stainless steel column (Alltech). To avoid particulate contamination of the column, a 0.2 μm Uptight precolumn filter (Upchurch Scientific, Inc.) was inserted before the column. The phospholipids were detected by a mass detector (ACS), operating at an external carrier gas pressure of 2.0 bar, whereas the internal pressure amounted to 1.0 bar. The photomultiplier sensitivity of the detector was fixed at 3 and an evaporator set of 70 was selected. The peak areas were calculated by a Chromatopac C-R1A integrator (Intersmat).

Chromatographic Conditions

The initial part of the separation was performed by a mobile phase consisting of n-hexane, 2-propanol and water (58/39/3;v/v). After 5 minutes a second mobile phase, containing n-hexane, 2-propanol and water (55/44/5;v/v), was selected by the solvent switcher. After 15 minutes the first mobile phase was run again, so that the system was re-equilibrated after 25 minutes. The flow rate of the mobile phase amounted to 1.8 ml per minute throughout the whole experiment. At the end of each day, the column was flushed with approximately 50 ml of n-hexane. Degassing of the solvents was achieved by flushing with helium.

RESULTS AND DISCUSSION

Optimization of Mobile Phase

The method proposed is based upon the procedure described by Nasner and Kraus (3), where n-hexane, 2-propanol and water (8/8/1) were used to separate soybean phospholipids on a Lichrosorb stationary phase. In order to obtain a good resolution of the major components, the water content of the mobile phase seemed to be very critical, whereas the hexane/2-propanol ratio was of minor importance; in the experiments described below a ratio of 55/44 was always maintained. Thus, using a mobile phase with only 2% of water, all phospholipids remained fixed on the stationary phase. On the other hand, a mobile phase containing 3% of water caused PE to elute after 11 minutes, whereas PI eluted as a very broad peak. To enable elution of PA and PC, at least 4% of water was required. On the other hand, a water content of more than 5% resulted in the elution of all phospholipids within less than 5 minutes, so that baseline separation couldn't be obtained anymore. Thus, a stepwise gradient was selected; during the first step the mobile phase consisting of hexane, 2-propanol and water (58/39/3) yielded a good resolution of neutral lipids (NL) and PE, whereas a 55/44/5 mixture of these solvents resulted in a complete separation of PI, PA and PC (Fig. 1; next page).

Reproducibility

The reproducibility of the method proposed was investigated using both a crude soybean lecithin and a highly purified PC, of which respectively 213 and 44 μg were injected. These samples were analyzed at three different days, each time in 4 successive runs. The coefficients of variation of the resulting peak areas are summarized in Table 1. It can be seen that the coefficient of variation of four successive analyses

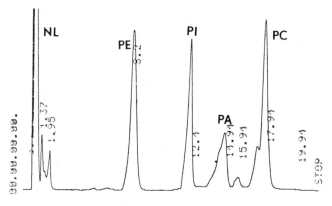

Figure 1. HPLC separation of 213 μg of crude soybean lecithin

amounts to about 5% for all compounds, except from PE, which displayed only 2% of variation. It was shown that these small deviations were mainly due to fluctuations on the injected volume. In the last column of Table 1, the coefficient of variation of the mean values determined at three different days is indicated; the coefficient of variation was about 5% for the sharpest peaks such as PE and PC, whereas higher values were obtained for broader peaks. The large influence of the carrier gas pressure partly explains the coefficient of variation. Moreover, slight changes in peak shape can affect the detector response to a large degree. Besides, baseline instability will especially influence the peak area of the smaller and broader PA peak. Finally, some minor components sometimes co-eluted with PA, thus resulting in an overestimation of its peak area.

TABLE 1

REPRODUCIBILITY OF THE PEAK AREAS RESULTING FROM 213 μg OF CRUDE SOYBEAN LECITHIN AND FROM 44 μg OF PURIFIED PHOSPHATIDYLCHOLINE (PPC)

| | Coefficient of variation (%) | | | |
	Day 1	Day 2	Day 3	Days 1-3
NL	7.5	1.9	2.8	0.8
PE	2.0	1.2	2.1	5.4
PI	4.5	10.1	2.7	9.7
PA	4.7	6.5	2.1	17.7
PC	4.4	5.7	4.4	4.7
PPC	3.1	1.9	7.3	5.7

Calibration curves

To enable quantitative estimations of the separated phospholipids,
calibration curves were needed. Hence, the peak areas originating from
1 up to 50 μg of pure phospholipids were determined. As stated by
Oppenheimer and Mourey (4) the calibration curves were sigmoidal.
Despite the universal character of the mass detector, different curves
were obtained for the various phospholipids. This was caused by
differences in peak shape: the detector response was inversely
proportional to the peak width.

Solubility

The solubility of the soybean phospholipids was tested by analyzing
5 μl of the filtered supernatant obtained after shaking 2 g of powdered
soybean lecithin in 50 ml of solvent for at least 48 hours. It was shown
that the apolar solvents hexane, diethyl ether, dichloromethane,
chloroform and isoamyl acetate were very suitable solvents for the most
abundant soybean phospholipids such as PA, PI, PE and PC; the polarity of
these solvents, as defined by Snyder (6), ranged from 0.1 to 4.1. On the
other hand, alcohols, characterized by a polarity between 3.9 and 5.1,
yielded quite different results. From Fig. 2, it follows that the
solubility of PC, PE and PG decreased with increasing length of the alkyl
group of the alcohol, whereas PA and PI dissolved only in methanol and
seemed to be insoluble in higher alcohols. On the other hand, the
solubility of neutral lipids was independent of the structure of the
alcohol. Acetone and acetonitrile, whose polarity amounts respectively

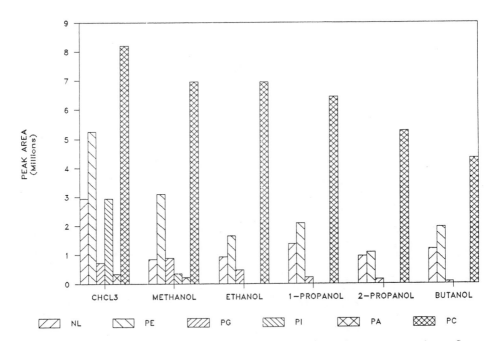

Figure 2.   Solubility of powdered soybean lecithin in a series of
            alcohols as compared to chloroform.

to 5.1 and 5.8, were very poor phospholipid solvents: only a few percent
of the added amount was dissolved. Besides neutral lipids, the filtered
supernatant contained primarily PC. According to Baer and Mahadevan (1),
the solubility would be caused by molecular species containing short
fatty acid chains. Finally, a completely different behavior was observed
for ethylene glycol as well as for water: by shaking, a homogeneous
dispersion was formed. Using quasielastic laser light scattering, a
z-average particle diameter of 200 to 400 nm was obtained, indicating the
formation of multilamellar vesicles in these highly polar solvents.

REFERENCES

1. Baer, E., and Mahadevan, V., 1959, Synthesis of L-α-lecithins
      containing shorter chain fatty acids. Water-soluble
      glycerolphosphatides. I. J. Am. Chem. Soc. 81:2494-2498.
2. Christie, W.W., 1987, "High-performance Liquid Chromatography and
      Lipids", Pergamon Press, Oxford.
3. Nasner, A., and Kraus, L., 1981, Trennung einiger Bestandteile des
      Lecithins mit Hilfe der Hochleistungs-Flüssigkeits-
      Chromatographie. II. J. Chromatogr. 216:389-394.
4. Oppenheimer, L.E., and Mourey, T.H., 1985, Examination of the
      concentration response of evaporative light-scattering mass
      detectors. J. Chromatogr. 323:297-306.
5. Robinson, J.L., and Macrae, R., 1984, Comparison of detection systems
      for the high-performance liquid chromatographic analysis of
      complex triglyceride mixtures. J. Chromatogr. 303:386-390.
6. Synder, L.R., 1978, Classification of the solvent properties of
      common liquids. J. Chromatogr. Sci. 16:223-234.
7. Stolywho, A., Colin, H., Martin, M., and Guiochon, G., 1984, Study
      of the qualitative and quantitative properties of the light-
      scattering detector. J. Chromatogr. 288:253-275.

BIODISPOSITION AND HISTOLOGICAL EVALUATION OF TOPICALLY APPLIED RETINOIC ACID IN LIPOSOMAL, CREAM AND GEL DOSAGE FORMS

W.C. Foong†, B.B. Harsanyi† and M. Mezei*

College of Pharmacy* and Faculty of Dentistry†

Dalhousie University, Halifax, N.S. Canada B3H 3J5

In search for improved dermatological products attempts are being made to design new vehicles or utilize drug carriers to optimize topical therapy. Several reports have indicated that the liposomal form provided higher drug (triamcinolone, progesterone, econazole, minoxidil and local anesthetics) concentration in the skin but similar or lower concentrations in the internal organs, than the conventional dosage forms, i.e., ointment, cream, gel or lotion (3-5). Biocompatibility studies indicated that liposomes are non-toxic and are safe vehicles (2). Topical retinoids have an increasing role in topical therapy. The existing products have many disadvantages, e.g. high degree of irritation and the chance of systemic absorption which may lead to serious side-effects. The objective of the present study was to develop a liposomal retinoid product and to evaluate the *in vivo* biodisposition and histological effects of retinoic acid applied topically in liposomal, gel and cream form on guinea pig skin.

MATERIALS AND METHODS

Biodisposition Studies

Liposomal Retinoic Acid composed of Soy Phospatide, cholesterol and retinoic acid in a molar ratio of (1.0:0.95:0.03) (4.0:2.0: 0.05 % w/w) was prepared using the method described by Mezei (5). In the cream formulation, retinoic acid was incorporated into Dermabase®. Both liposomal and cream forms contained retinoic acid with 20μCi/g of product.

The dorsal hair of albino guinea pigs' (300-350g) backs was clipped 24 hours before treatment. The animals were randomly divided into two groups; one group was treated with the liposomal and the other with the cream preparation. The treatment consisted of applying 100 mg preparation (either liposomal or cream formulation) on 10 cm^2 area for five days in a twice a day dosage regimen. The animals were sacrificed by carbon dioxide, 4 hours after the last dose. Drug disposition in the skin, internal organs and in the urine was determined

Phospholipids
Edited by I. Hanin and G. Pepeu
Plenum Press, New York, 1990

using radioactive tracer technique. The skin samples were horizontally sliced with a keratotome, the first 0.2 mm slice was designated as the epidermis, the next 0.5 mm as the dermis and the remaining as subcutaneous tissue.

Statistical evaluation was done by Student's t test.

Histological Studies

The liposomal products were prepared as described above but with non-radioactive ingredients, and they contained 0.01% tretinoin or tretinoin palmitate. The cream and gel formulations were commercial products: Vitamin A Acid Cream and Vitamin A Acid Gel (Rorer Canada Inc.) which contained 0.01% Tretinoin USP.

The backs of four guinea pigs were shaved 24 hours before treatment and six 1 cm^2 treatment sites were outlined on each animal. The treatments were: 1. Untreated control; 2. Empty Liposomes; 3. Liposome-entrapped tretinoin; 4. Tretinoin cream; 5. Tretinoin gel; and 6. Liposome-entrapped tretinoin palmitate. The treatment sites were randomized. Each animal was treated topically (10 mg) once a day for seven days and sacrificed 4 hours after the final dose. Treatment and control areas were excised, fixed in 10% formalin and processed for histology. Epithelial and dermal thickness were measured on 5-micron thick sections using a calibrated eye piece on the Zeiss microscope. Statistical evaluation was by Analysis of Variance.

RESULTS AND DISCUSSION

Biodisposition

Results indicated a three and two times higher drug concentration in the epidermis and dermis, respectively, of guinea pigs treated with liposomal retinoic acid as compared to those treated with the cream forms. In the plasma and urine, the pattern was reversed; the cream-treated animals had approximately two fold higher concentrations in plasma and urine than the liposome-treated group (Table 1).

Table 1. The effect of liposomal encapsulation on the bio-disposition of tretinoin in guinea pig after five days of treatment.

Tissue	Cream-treated µg per gram tissue	Liposome-treated µg per gram tissue	
Epidermis	66.941 ± 10.342	175.078 ± 16.927	$p < 0.01$
Dermis	10.057 ± 3.779	28.451 ± 5.931	$p < 0.01$
Subcutaneous Tissue	2.575 ± 1.213	1.122 ± 0.590	$p < 0.01$
Plasma	0.050 ± 0.005	0.028 ± 0.002	$p < 0.01$
Urine (5 day excretion)	23.876 ± 4.598	9.846 ± 3.209	$p < 0.01$

Histological Evaluation

Microscopic appearances of samples obtained from the untreated control site and from the site treated with empty liposomes (groups 1 and 2) were essentially the same: they were covered by stratified squamous epithelium; and showed occasional inflammatory cells in the lamina propria; numerous hair follicles and sebaceous glands in the dermis; and a well defined muscle layer, the Panniculus Carnosus.

Liposome-tretinoin treatment (group 3), however, caused hypergranulosis, increase of mitoses in the basal and parabasal layers and epithelial hyperplasia, with the epithelial thickness being significantly greater (p < 0.01) than in the untreated control site (Fig 1). With regards to other features, particularly inflammation, the appearances were the same as in groups 1 and 2.

Tretinoin cream treatment (group 4) showed hypergranulosis and epithelial hyperplasia, with the epithelial thickness being significantly greater (p < 0.01) than untreated control, but not significantly greater than group 3. Microscopically, tretinoin cream showed a scab of fibrin and inflammatory cells on the epithelial surface; migration of inflammatory cells through the epithelium; foci of mild to moderate inflammatory infiltrate limited to the upper dermis; and loosening of the stratum spinosum with accumulation of PAS positive material. The microscopic features were absent in groups 1, 2 and 3.

Tretinoin gel (group 5) resulted in severe inflammatory changes. These included: ulceration with complete loss of epithelium in some areas and regenerating epithelium in others; blisters with inflammatory cells; pools of PAS-positive material and collections of inflammatory cells within the epithelium; a psoriasiform dermatitis with deep, proliferating rete ridges and markedly hyperplastic epithelium at the edges of the ulcers. Hair follicles and sebaceous glands had disappeared completely in the treatment areas and were replaced by a proliferation of epitheliod and inflammatory cells. In contrast, groups 1 through 4, showed and intact dermis with normal adnexae. The inflammation in group 5 involved the muscle layer and usually extended beyond the Panniculus Carnosus.

Liposome-tretinoin palmitate (group 6) showed similar appearances to liposome-tretinoin, but the epithelial changes were somewhat less marked and did not reach statistical significance with the small number of samples used (Fig. 1).

The epidermal hyperplasia observed in this study (Fig. 1) is similar to that reported by Schiltz et al. (1986); their values were higher (200 vs. 120 microns range), but are pro-portional to the concentrations used, 0.05%, by Schiltz et al. (7) vs. 0.01% in our study. Severe psoriasiform dermatitis has been observed before (6) as the accumulation of PAS positive material in the epidermis, as well as blister formation (1). These side-effects were most striking following treatment with tretinoin gel; present partially and to a mild degree following treatment with tretinoin cream; and absent in the case of treatment with liposomal formulations. Thus liposomal encapsulation resulted in the absence of undesirable side effects of topical tretinoin treatment.

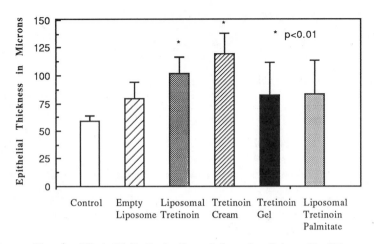

Fig. 1. Effect Of Tretinoin Formulations On Guinea Pig Skin

CONCLUSION

Liposomal encapsulation of retinoids can provide selective delivery, i.e. higher drug concentrations at the target sites, epidermis and dermis and lower concentrations in non-target organs. This may lead to increased efficacy and decreased toxicity of topically applied retinoic acid. In addition, liposomal retinoid dosage forms produced no adverse effects in contrast to the cream or gel dosage forms, suggesting that the liposomal form of retinoids is probably safer, and could result in better patient compliance than the presently available products.

REFERENCES

1. Elias, P.M. (1986): Epidermal effects of retinoid: Supra-molecular observations and clinical implications. J. Am. Acad. Dermatol. 15:797-809.
2. Foong, W.C., Harsanyi, B.B. and Mezei, M. (1989): Effect of liposomes on hamster oral mucosa. J. Biomed. Mater. Res. (in press).
3. Gesztes, A. and Mezei, M. (1988): Topical anaesthesia of the skin by liposome-encapsulated tetracaine. Anesth. Analg. 67:1079-81.
4. Mezei, M. (1988a): Liposomes in the topical application of drugs, in Liposomes as Drug Carriers, G. Gregoriadis, ed., New York, John Wiley & Son, pp 663-677.
5. Mezei, M. (1988b): Multiphase liposomal drug delivery system, U.S. Patent No. 4,761,288.
6. Orfanos, C.E., Ehlert, R. and Gollnick, H. (1987): The Retinoids. A review of their clinical pharmacology and therapeutic use. Drugs 34:459-503.
7. Schiltz, J.R., Lanigan, J., Nabial, W., Petty, B. and Birnbaum, J.E. (1986): Retinoic acid induces cyclic changes in epidermal thickness and dermal collagen and glyco-saminoglycan biosynthesis rates. J. Invest. Dermatol. 87(5):663-667.

RELATIONSHIP BETWEEN IODINE VALUE AND PHASE TRANSITION
TEMPERATURE IN NATURAL HYDROGENATED PHOSPHOLIPIDS

A. Moufti, I. Genin, G. Madelmont, M. Deyme,
M.Schneider* and F. Puisieux

Laboratoire de Pharmacie Galénique et Biopharmacie, UA CNRS 1218
Centre d'études pharmaceutiques, Université de Paris XI
92290 Chatenay-Malabry, France
*Lucas Meyer GmbH and Co., Ausschläger Elbdeich, 62
 2000 Hamburg 28, RFA

INTRODUCTION

The ability of phospholipids which make up liposomal bilayers to retain entrapped drug in the aqueous spaces might be altered by their phase transition temperature (Tc), since it is well known that drug leakage is markedly reduced if liposomes are kept at a storage temperature below the Tc of these phospholipids.

In addition the physical stability and integrity of these liposomes could be maintained over longer periods of time.

Natural phospholipids exhibit Tc values which are too low for the purpose of drug delivery system design. However, partial or total hydrogenation of these phospholipids has been proposed in order to increase their Tc, in an attempt to improve the thermal properties of these compounds.

It has been previously shown that decreasing the number of double bonds in the acyl chains of pure synthetic phospholipids increased the Tc value (1). For example, as a result of a double bond disappearance, the Tc of dioleoylphosphatidylcholine (C18:1) is -22°C, while the Tc of distearylphosphatidylcholine (C18:0) is 55°C.

Most phospholipid manufactures provide an iodine value (Iv) to designate the unsaturation degree of their products; thus, this value is an expression of unsaturation. In fact, this value can also represent the reciprocal of the degree of hydrogenation.

In this work an attempt was made to correlate the Iv and Tc of various partially or completely hydrogenated hydrated soya phospholipids.

MATERIALS AND METHODS

Partially and completely hydrogenated soya phosphatidylcholines (Epikuron 200 HR) with known Iv values were kindly supplied by LUCAS MEYER.

DIFFERENTIAL SCANING CALORIMETRY (DSC) SAMPLE PREPARATION

Accurately weighed phospholipids samples (3 mg) were placed in a cup, to which 3 µl of tridistilled water were added. The cup was rapidly sealed and incubated at 40°C for 2 hours. The samples were subjected to heating cycles in the range of 5°C to 65°C using a DuPont de Nemours thermal analyser equipped with a thermogram recorder and DSC calorimeter type 990-910; the heating rate was 5°C. min^{-1}.

RESULTS AND DISCUSSION

The various Tc values yielded by DSC analysis and the corresponding Iv values are presented in Table 1 .

Table 1. Iv data and corresponding Tc values in the different hydrogenated soya phospholipids.

	Iodine values Iv	Tc °C
A	2.0	50.50
B	6.1	48.25
C	25.0	36.75
D	35.0	30.00
E	45.4	21.00

Analysis of the correlation by means the linear " least square regression" procedure revealed the existence of an inverse linear relationship between the Tc and Iv of soya phosphatidylcholine compounds having various hydrogenation levels (Fig 1).

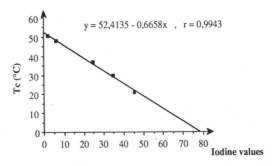

Figure 1. Tc plot as a function of Iv for various hydrogenated soya phosphatidylcholines.

It should be added that other parameters such as chain length, branched chain and the nature of the head group might also alter the Tc (1).

The Tc of partially hydrogenated soya phosphatidylcholine was hardly detectable at low degrees of hydrogenation (thermograms D and E) but it was obvious in the almost totally hydrogenated sample (thermograms A, B and C) (Fig 2).

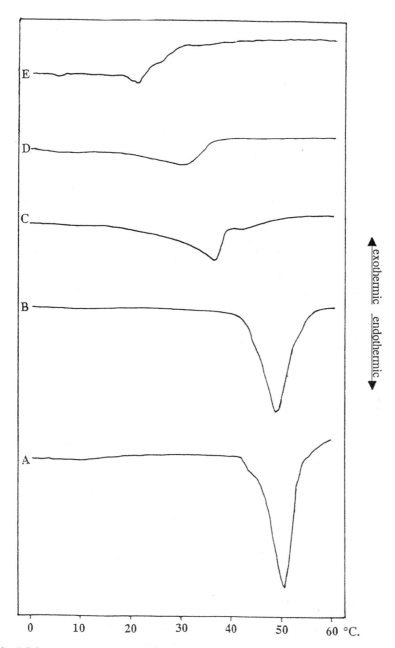

Figure 2. DSC curves for fully hydrated partially (B, C, D and E) or totally (A) hydrogenated soya phosphatidylcholines, 50% (W/W) water. Heating rate 5°C. min-1.

Thus, the linear relationship between Tc and Iv could be considered as a useful tool for selection of an appropriate phospholipid with given Iv and Tc values.

CONCLUSION

Partial or total hydrogenation of natural phospholipids has been revealed to be a useful tool in the development of potential drug delivery systems.

It would be possible to estimate the Tc if the Iv was known. It should, however, be emphasized that this value (Tc) is reliable only in phospholipids having the same acyl chains, and which are from the same origin but with different degrees of unsaturation.

ACKNOWLEDGEMENTS

We would like to thank Pr. S. Benita for his helpful discussions and English editing and Mr. N. Ammoury for his technical assistance.

REFERENCES

1- Szoka, F., and Papahadjopoulos, D. (1980): Comparative properties and methods of preparation of lipid vesicles (Liposomes). Ann. Rev. Bioeng. 9: 467-508.

2- Ter-Minassian-Saraga, L. and Madelmont, G. (1985): Differential scanning calorimetry studies of hydration forces with phospholipid multilamellar systems. J. Coll. Interf. Sc. 85: 375-388.

HEXADECYLPHOSPHOCHOLINE (D-18506): A NEW PHOSPHOLIPID WITH HIGHLY

SELECTIVE ANTITUMOR ACTIVITY

W. Schumacher, J. Stekar, P. Hilgard, J. Engel
ASTA Pharma AG, D-6000 Frankfurt 1
H. Eibl, Max-Planck-Institut, D-3400 Göttingen
C. Unger, University Hospital, D-3400 Göttingen

Ether lipids have several biological functions as important components of mammalian cells. Considerable interest has recently focussed on the "platelet activating factor" (PAF; 1-alkyl-2-acetyl-sn-glycero-3-phosphocholine), which seems to be of pathogenic importance in the mediation of allergic and inflammatory reactions. A number of analogs of PAF whithout platelet aggregating activity are cytotoxic for a variety of cell types[3]. Racemic 1-octadecyl-2-methyl-glycero-3-phosphocholine (Et-18-OCH3; INN: Edelfosine) was the prototype in this class of new anti-cancer agents to undergo preclinical evaluation.

Recently alkylphosphocholines were investigated for their inhibitory effect on tumor cells. Hexadecylphosphocholine (D-18506; INN: Miltefosine) was found to have highly selective antineoplastic activity in vitro and in vivo[2].

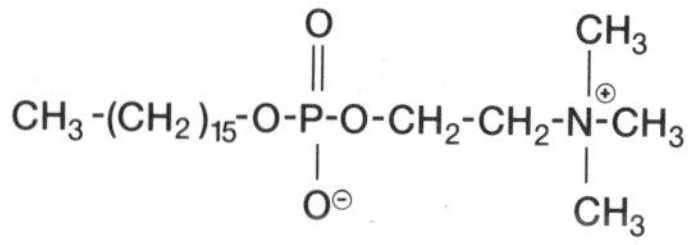

$$CH_3\text{-}(CH_2)_{15}\text{-}O\text{-}\overset{\displaystyle O}{\underset{\displaystyle O^{\ominus}}{P}}\text{-}O\text{-}CH_2\text{-}CH_2\text{-}\overset{\displaystyle CH_3}{\underset{\displaystyle CH_3}{N^{\oplus}}}\text{-}CH_3$$

Figure 1. Chemical Structure of Hexadecylphosphocholine

D-18506 was tested on the methyl-nitroso-urea (MNU)-induced mammary tumor in the Sprague Dawley (SD) rat[1]. In this experiment the drug reduced the total tumor mass and the number of tumors below the level at the beginning of therapy.

Phospholipids
Edited by I. Hanin and G. Pepeu
Plenum Press, New York, 1990

We have tested D-18506 in autochthonous DMBA (dimethylbenzanthracene)-induced mammary carcinoma of the SD-rat. When the animals received D-18506 daily p.o. for 5 weeks (ten rats 230-270 g body weight per dose group) the maximally tolerated dose was 46,4 mg/kg/day.

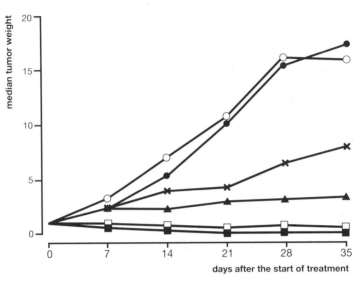

Growth curves of DMBA-induced mammary tumors in rats under daily treatment with D-18506 for 5 weeks. Test groups each consisted of 10 animals.

○——○ control   ●——● 2.15 mg/kg/day   x——x 4.64 mg/kg/day
▲——▲ 10 mg/kg/day   □——□ 21.5 mg/kg/day   ■——■ 46.4 mg/kg/day

Figure 2

As Fig. 2 shows, tumor growth was considerably suppressed and the therapeutic effect correlated clearly with the daily dose administered. Already 4,64 mg/kg resulted in a marked inhibition, while complete growth retardation was seen at doses between 10 and 21,5 mg and 46,4 mg consistently induced a regression of all tumors to below the level of palpability.

To further evaluate a possible difference between the antineoplastic efficacy of D-18506 and that of other chemotherapeutic agents whose mechanism is well known, we compared D-18506 with cyclophosphamide (CP; Endoxan) in the DMBA-tumor.

As shown in Fig. 3, D-18506 reduced the total tumor mass below the level at the start of the experiment. CP, however, even when applied at the maximally tolerated single dose inhibited tumor growth only to about 50 % of the control.

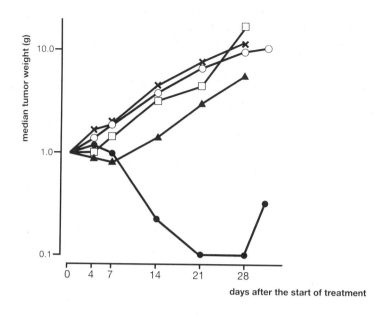

Comparative activity of D-18506 and cyclophosphamide against DMBA-mammary carcinoma of the Sprague-Dawley-rat. When the tumors attained a mass of 0.8-1.2g, the animals received a single i.v. dose of cyclophosphamide. Treatment with D-18506 started at the same tumor mass but was conducted for 28 days (31.6 mg/kg daily p.o., 5 times per week).
Tumor weight: mean values; group size: 9 animals.

O ——— O **Control animals (vehicle alone)**
● ——— ● **D 18506, 31.6 mg/kg/day**
x ——— x **Cyclophosphamide, 68.1 mg/kg/ i.v.**
□ ——— □ **Cyclophosphamide, 100 mg/kg i.v.**
▲ ——— ▲ **Cyclophosphamide, 147 mg/kg i.v.**

Figure 3

Preliminary clinical results regarding D-18506 have been reported for the topical application of the drug. In a pilot trial patients with ulcerating recurrences of a mammary carcinoma received local treatment of their metastases[4]. We have recently initiated a phase I trial, using a capsule formulation of D-18506, in order to establish the maximum tolerated dose for oral administration; this will be followed by disease-oriented phase II trials.

REFERENCES

1. Berger, M., Muschiol, C., Schmähl, D., Unger, C. and Eibl, H., 1985, Chemotherapeutische Studien zur Struktur-Wirkungs-Beziehung zytotoxischer Alkylphospholipide an chemisch induzierten Mammacarcinomen der Ratte, in: "Aktuelle Onkologie 34 - Die Zellmembran als Angriffspunkt der Tumor-

therapie", Unger, C., Eibl, H., Nagel, GA. eds., W. Zuckschwerdt, Munich, p. 27-36.

2. Fleer, E.A.M., Unger, C., Kim, D.J., Eibl, H., 1987,
   Metabolism of Ether Phospholipids ans Analogs in Neoplastic Cells,
   Lipids 22: 856-861.

3. Hoffmann D.R., Hajdu, J., Snyder, F., 1984,
   Cytotoxicity of platelet activating factor and related alkyl-phospholipid analogs in human leukemia cells, polymorphonuclear neutrophils, and skin fibroblasts, Blood 63: 545-552.

4. Unger, C., Eibl, H., Nagel, GA., von Heyden, HW., Breiser, A., Engel, J., Stekar, J., Peukert, M., Berger, M., 1989, Hexadecylphosphocholine in the Topical Treatment of Skin Metastases: A Phase I Trial,
   Contrib. Oncol. 37: 219-223.

LIPOSOMES AS CARRIERS FOR COSMETICS - A FREEZE-FRACTURE
ELECTRON MICROSCOPY STUDY

Brigitte Sternberg and Michael Schneider

Electronmicroscopical Lab., Medical School.
Friedrich-Schiller-University, Jena, G.D.R.
Lucas Meyer Gmbh & Co., Hamburg, F.R.G.

INTRODUCTION

Liposomes first described in 1965 (1) and initially used as excellent models for studying membranes have been considered more frequently as vehicles for medical drugs during the last 10 years (2). In fall 1986 the first commercial product appeared on the market with liposomes as mini-containers; however for cosmetically active principles and not for drugs.

In liposomes, amphiphilic phospholipid molecules are arranged in a closed spherical bilayer. Lipophilic ingredients are entrapped in the hydrophobic region, while hydrophilic substances are held in the aqueous internal space of the artificial microcapsule.

Many of the latest preparations from very well known cosmetic houses are based on liposome dispersions. This "new generation" of skin care products claims to remove wrinkles or improve the texture of the skin by combating the aging process, or speeding up the renewal of cells.

In this paper we present electron micrographs of some commercially available formulations and of liposomes made of lipids from the Lucas Meyer Company.

FREEZE-FRACTURE ELECTRON MICROSCOPY

Freeze fracture electron microscopy is a useful technique:
- to investigate liposomes in their natural shape, size and to a certain extent concentration; it is time-dependent, on the age of the dispersions;
- to classify liposomes as multi- or unilamellar vesicles;
- to examine the hydrophobic interior of liposome membranes; and
- to distinguish between liposomes and non-liposomal objects.

Liposome containing creams or dispersions were quenched for freeze-fracture electron microscopy after incubation for 20 minutes at required temperatures (mostly room temperature) using the sandwich technique (cooling rate 10° K/sec) and liquid propane as described by Sternberg

Phospholipids
Edited by I. Hanin and G. Pepeu
Plenum Press, New York, 1990

291

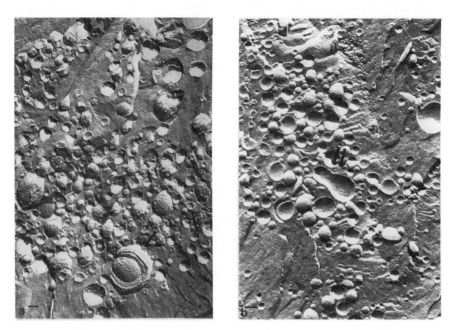

Figures 1a and 1b: Liposomes as carriers for cosmetics.

et al. (5). The specimens were fractured and shadowed in a Balzers BAF
400D freeze-fracture device at -150° C. The cleaned replicas were
examined in a Jeol 100B or Tesla BS 500 electron microscope.

RESULTS AND CONCLUSIONS

Freeze-fracture electron micrographs of some commercially available
formulations show mostly circular shaped liposomes in high concentrations. As an example CAPTURE-liposomes are shown in Figs. 1a and 1b.
These liposomes are mainly SMV (Small Multilamellar Vesicles) and SUV
(Small Unilamellar Vesicles) with diameters around 100 nm. NIOSOMES and
LIPODERMIN-liposomes are clearly larger, whereas EXTRAIT LIPOSOMAL shows
the smallest liposomes (about 50 nm) of all commercially available
creams studied.

The fracture faces of membranes of most of the liposome formulations
are smooth. Although the fracture faces of CAPTURE-liposomes (Figs. 1a
and 1b) and NIOSOMES are not as smooth, they show no special lipid
structures as ripples or other long range ordered structures.

There are only small changes of the investigated parameters after
storage of the cosmetic-liposomes in a refrigerator at 4°C for a year.
Sometimes some fused liposomes are noticeable (Fig. 1b).

Freeze-fracture electron micrographs of liposomes made of lipids from
the Lucas Meyer Company show similar results: Those liposomes, prepared
from soybean lipids, are circular shaped and mostly small multi- and
unilamellar with diameters of about 100 nm. Their fracture faces are
smooth, and there are also small changes of the investigated parameters

only after storage for one year (Figs. 2a and 2b). Thus EMULMETIK 160 and EMULMETIK LS, extracted from soybeans and produced by the Lucas Meyer Company, are suitable lecithin mixtures for preparing cosmetic type liposomes.

Liposomes of some commercially available formulations as well as liposomes made of lipids from the Lucas Meyer Company are clearly different (not shown) from non-liposomal objects as oil droplets or non-encapsulated substances such as vitamins or hyaluronic acid.

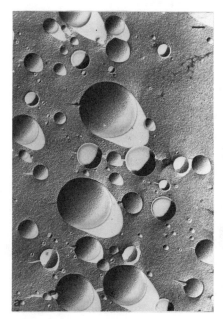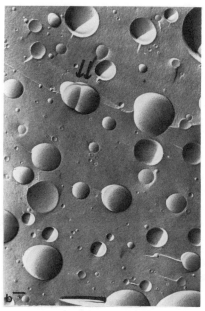

Figures 2a and 2b: Liposomes as carriers for cosmetics.

ADVANTAGES OF LIPOSOMES AS COSMETIC CARRIERS

Liposomes containing cosmetics as a new tool in skin care products have certain advantages:
- Thanks to their biochemical nature and their structural identity with cellular membranes, liposomes are able to interact with cells. Different interaction types have been identified, depending on the liposome composition and the type of cells encountered (4). Specifically, these are: endocytosis, transfer or exchange of phospholipids, and fusion.
- Compared to other vehicles applied on the skin, utilization of the liposomal form allows one to increase five fold the effective concentration in the epidermis and three fold in the dermis (3). The liposomes containing cosmetics seem to be sufficiently small and active even to reach the dermis.
- Liposomes are micro-depots for cosmetics. The ability of liposomes to release their contents slowly makes them excellent candidates for time-release cosmetic-delivery systems.
- Thanks to their physical stability (as shown in Figs. 1b and 2b) liposomes containing cosmetics may be stored for a long time without modification of their properties.

REFERENCES

1. Bangham, A.D., Standish, M.M., and Watkins, J.C., 1965, Diffusion of univalent ions across the lamellae of swollen phospholipids. J. Mol. Biol. 13:238-252.
2. Gregoriadis, G., 1988, Liposomes as Drug Carriers, John Wiley, New York.
3. Mezei, M., 1980, Liposomes - a selective drug delivery system for the topical route of administration. Life Sci. 26:1473-1477.
4. Ostro, M., 1983, Liposomes, Marcel Dekker Inc., New York.
5. Sternberg, B., Gale, P., and Watts, A., 1989, The effect of temperature and protein content on the dispersive properties of bR from H. halabium in reconstituted DMPC complexes free of endogenous purple membrane lipids: A freeze-fracture electron microscopy study, BBA 980:117-126.

CHARACTERIZATION OF LECITHINS AND PHOSPHOLIPIDS BY

HPLC WITH LIGHT SCATTERING DETECTION

Bengt Herslöf, Urban Olsson and Per Tingvall

LipidTeknik
P.O.Box 152 00, S-104 65 Stockholm
Sweden

INTRODUCTION

Plant lecithins are obtained in the degumming process in the production of vegetable oils. They are normally by-products and the crude material consists of a complex mixture of lipids and other components such as pigments and carbohydrates [2]. The use of lecithins in food is mainly for functional resons, e.g. as emulsifiers. Multicomponent natural products such as the lecithins are also multifunctional and applications are often found through tedious testing procedures. In specific applications it may even be necessary to test production batches before delivery due to batch-to-batch variations which could significantly change the properties of the product in question.

In order to understand the properties of the multicomponent lecithins it is important to be able to characterize the constituents properly. Chromatography is the preferred methodology due to the polarity range and non-volatility of the lipids in lecithins. Thin layer chromatography has until recently been the only universal technique applicable to lecithins. However, in most cases TLC is used qualitatively because of the general difficulties to quantitate such separations.

HPLC has been successfully used for the separation of all kinds of lipids, classes as well as molecular species [1]. The most commonly employed detectors for this kind of work are refractometers and UV-detectors. However, it has been difficult to work with gradient elution since RI-detectors are not amenable to changes in solvent composition. Furthermore, UV-detectors are not compatible with strong baseline shifts at the required low wavelengths at which lipids are detected most effectively.

In recent years it has been possible to attack detection problems of the above mentioned type by the introduction of new types of commercially available evaporative detectors such as flame ionization detectors and light scattering detectors. In this work two types of commercially available light scattering detectors have been used.

Phospholipids
Edited by I. Hanin and G. Pepeu
Plenum Press, New York, 1990

METHODS AND MATERIALS

Straight Phase HPLC

Gradient system CM 4000 (Milton Roy, U.S.A.). Lichrospher 100 Diol, 5 um, 250x4.0 mm (E. Merck, West Germany). Gradient: Hexane/isopropanol/ acetic acid (75/25/1.5 w/w/w) to water/isopropanol/acetic acid (21/78/1.5 w/w/w). ACS 750/14 Mass Detector (Applied Chromatography Systems, U.K.).

Reversed Phase HPLC

Gradient system LC 6000 (Shimadzu, Japan). Lichrospher RP 18, 4 um, 250x4.0 mm (E. Merck). Gradient: Methanol/water/acetonitrile (80/10/10 w/w/w) to methanol/water/acetonitrile (90/5/5 w/w/w). Cunow DDL 11 Light Scattering Detector (Cunow, France).

Gas Chromatography

Capillary GC 3500 (Varian, U.S.A.) with on-column injection system. DB-wax 30 m (0.32 mm). FID.

Materials

Soybean lecithin (Lucas Meyer, West Germany). Sunflower lecithin (Vamo Mills, Belgium). Rapeseed lecithin (Exab, Sweden).

DETECTION PRINCIPLE

The advent of commercially available evaporative detectors working on the light scattering principle has made it possible to utilize the full separation potentials of HPLC. In these detectors the effluent from the column is evaporated in a stream of air or nitrogen in a nebulizer, leaving the non-volatile solutes as droplets or particles. These are carried by the flow through a light beam, i.e. the light scattering detector. The amount of scattered light is related to the amount and nature of the non-volatile material. Thus, even strong gradients do not change the baseline level.

PHOSPHOLIPID APPLICATION

Phospholipid classes are traditionally separated by TLC adsorption chromatography. By transformation to straight phase HPLC we have investigated the total composition of three commercial lecithins (Table 1). The presence of neutral lipids (mainly triglycerides) and carbohydrates are factors beside the proportions of the individual phospholipids that affect the overall properties of the lecithins
(Fig 1 a).

LipidTeknik has developed large scale chromatographic procedures to isolate pure phospholipids from lecithins. The fatty acid compositions of the phosphatidylcholines (CPL PC) are significantly different from each other (Table 2). The PC:s were further separated by reversed phase HPLC (Fig 1 b). The peaks, representing molecular species, were collected and submitted to fatty acid determination by GC. The result (Table 3) reveals the specific fatty acid combinations, a characteristic and consistent feature of the individual plant.

Table 1   Lipid class compositions of lecithins (weight %)

	Rapeseed	Soybean	Sunflower
Neutral lipids[a]	39.9	42.1	43.1
Phosphatidic acid	6.3	3.6	2.8
Phosphatidylethanolamine	12.1	16.9	10.1
Phosphatidylserine	1.7		1.3
Phosphatidylcholine	21.9	20.5	24.6
Lysophosphatidylethanolamine	0.1		
Phosphatidylinositol	11.2	8.7	13.3
Lysophosphatidylcholine	0.8	0.9	
Carbohydrates[b]	6.0	7.1	5.2
Total	100.0	100.0	100.0

a) Mainly triglycerides   b) *e.g.* sucrose, raffinose, stachyose

Table 2   Fatty acid composition (weight %) of CPL[c] Phosphatidylcholines

Fatty acid	Rapeseed	Soybean	Sunflower
16:0	6.4	15.2	10.8
18:0	0.4	2.8	3.9
20:0			0.1
22:0			0.4
16:1 n-7	0.3		
18:1 n-9	51.8	6.1	11.4
18:1 n-7	4.0	1.3	0.7
18:2 n-6	31.0	67.7	71.5
18:3 n-3	4.4	6.4	0.3
Minors[d]	1.9	0.6	1.1
Total	100.0	100.0	100.0

c) Chromatographically Purified Lipids   d) <0.4% each

Table 3   Distribution of molecular species (weight %) in CPL[c] Phosphatidylcholines

Carbon No	Species[e]	Rapeseed	Soybean	Sunflower
36:5	18:2/18:3	1.3	6.1	
36:4	18:2/18:2	8.3	47.5	57.3
36:4	18:1/18:3	4.2	0.2	
34:3	16:0/18:3		0.6	
36:3	18:1/18:2	35.1	10.5	16.7
34:2	16:0/18:2	6.2	30.0	19.5
36:2	18:1/18:1	36.1	0.2	1.5
34:1	16:0/18:1	7.9	1.3	0.6
36:2	18:0/18:2		3.0	3.9
Total		99.1	97.4	99.5

e) No stereospecific order

CONCLUDING REMARKS

The evaporative detectors are useful tools in the characterization of non-volatile materials such as lecithins and phospholipids. Using this new medium we have shown that lecithins exhibit similar patterns of lipid class compositions. However, the phosphatidylcholines from the three sources are different from each other in molecular species distributions. For example, CPL PC from rapeseed contains more than 50% of dioleoylphosphatidylcholine, which is very much different from that of CPL PC from soybean. Thus, HPLC with light scattering detection is a useful new technique with considerable potential for use in the content analysis and the characterization of different phospholipids.

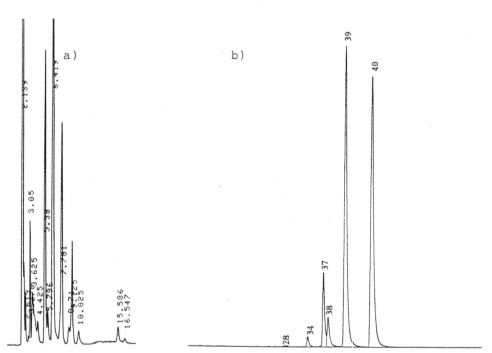

Fig 1   a) Straight phase HPLC of rapeseed lecithin. 0-4.425 neutral lipids; 5.38 PE; 6.419 PC; 7.701 PI; >8.71 carbohydrates.
b) Reversed phase HPLC of CPL PC from rapeseed. 34 (18:2/18:3), 37 (18:2/18:2), 38 (18:1/18:3), 39 (16:0/18:2,15%;18:1/18:2,85%), 40 (16:0/18:1,18%;18:1/18:1,82%)

REFERENCES

1. Szuhaj, B.F. and List G.R., ed. (1985): "Lecithins", American Oil Chemists' Society, U.S.A.

2. Christie, W.W. (1987): "HPLC and Lipids", Pergamon Press, Oxford, U.K.

THE INTERNATIONAL LECITHIN STUDY GROUP (I.L.S.G.):

History, Aims and Structure

Tom R. Watkins

Jordan Heart Research Group
Montclair, NJ 07042 USA

Introduction

I take this opportunity to thank the chairs, Drs. Hanin and Pepeu, the sponsor Lucas Meyer and each of you participants for your interest and enthusiasm today. My charge as the Moderator of the I.L.S.G. has been to introduce you to the Group, trace its brief history, its scope and aims and to sketch its proposed structure. Most drivers do not enjoy driving the car while looking in the rear-view mirror. Before looking ahead, we shall look back briefly to the origin of the I.L.S.G.

I HISTORY

At the conclusion of the IVth International Colloquium on Lecithin held in Chicago, IL, in September, 1986, Dr. Michael Schneider issued an invitation to those present interested in discussing some common concerns related to lecithin and phospholipid nomenclature, quality criteria, and analysis to meet later to do so. Scientists from Industry and Academia agreed to meet. Original participants included: I. Berry, Pharmacaps, Elizabeth, NJ, USA; H. Graham, Lipton, Englewood Cliffs, NJ, USA; B. Gunther & H. Koof, Nattermann, Koln, W. Germany; M. Murari, Liposome Technology, Inc., Menlo Park, CA, USA; W. van Nieuwenhuyzen, Loders Croklaan, Wormerveer, Holland; F. Pregnolato & G. Toffano, Fidia, Padova, Italy; M. Schneider & R. Ziegelitz, Lucas Meyer, Hamburg, W. Germany; B. Szuhaj, Central Soya, Ft. Wayne, IN, USA; and me.

Improved chemical technique in separation science has led to newer technologies, new applications of production and use of lecithin and phospholipid. Reports of scientific studies of highly purified lecithins, or phospholipids in some cases, have appeared in the literature; the material has been described with varying degrees of precision. Some material was identified simply as **lecithin**, yet treated as if it was more than 95% phosphatidylcholine. On the other side, some material was reported as **phospholipid**, e.g., phosphatidylcholine, which was possibly a blend containing from 55 - 80% phosphatidylcholine, far from the presumed 95% enrichment implied. Many of the attendees of the IVth International Colloquium had observed this problem and hoped to discuss it.

In the marketplace confusion also existed. Lecithin would bear a label stating 'soybean lecithin', or '95% soy phosphatides', or 'PC-55' (meaning 55% phosphatidylcholine), or 'Triple PC' (also implying 55% phosphatidylcholine). The use of so many different terms for remarkably similar products has generated confusion in the consumer's mind. Food codes do exist -- many outdated -- which set the standards used to identify lecithin samples. Yet these terms have not been used clearly in labeling food supplements.

With the newer separation and isolation techniques attention has shifted to fractionated lecithins highly enriched in phosphatidylcholine, phosphatidylserine, or other phospholipids. These phospholipids, in general, have not been clearly defined in food, cosmetic or pharmaceutical codes. This has led to further potential confusion, especially on two levels: 1) the scientific literature, as in the case of the lecithins; 2) the marketplace at several levels of disclosure and labelling.

Every label disclosure amounts to a claim about a product's identity. Though widely accepted tests exist in food and formulary law about appropriate assays to identify food-grade lecithin (e.g., acetone insolubles, phosphorus, toluene insolubles, peroxide value), beyond food, few useful standards exist honored throughout the lipid community.

Many questions have been raised. Which test would accurately and economically identify the **phospholipid** content of a fractionated lecithin, or synthetic sample? Does peroxide value adequately describe the oxygen exposure and deterioration of a sample to be used in liposomes? What other **quality criteria** of phospholipids will be useful, besides phospholipid composition and oxidative stress? How would this kind of information best be disclosed on a label? These illustrate some of the questions pondered.

The people who formed the nucleus of the I.L.S.G. reasoned that if the **terminology** could be made more precise and appropriate **criteria of quality** accepted throughout the industry and academia, then many benefits would be expected. First, the more precise terminology could help to protect the producer, vendor and consumer. Second, it could help to reduce misleading labelling, or outright fraud in the marketplace. Third, it would allow researchers trying to confirm results from another laboratory a better chance of setting conditions as reported to be able to confirm or deny the original data. This would be expected to expedite progress from laboratories around the globe as reported in colloquia and journals. Fourth, it could also lead to simpler, more uniform testing to verify label statements, after acceptable, analytical methods have been evaluated and recognized.

The Group has met three times since the IVth Colloquium: New York, Venice and Boston. The following is a brief summary of resolutions proposed as an outcome of deliberations during these meetings. Descriptions of **lecithin** and **phospholipid**, used only **as excipients, not pharmaceuticals**, have been drafted.

Lecithin

> **Lecithin** is a mixture of glycerophospholipids obtained from animal, vegetable or microbial sources, containing varying amounts of substances, such as sphingosylphospholipids, triglycerides, fatty acids and glycolipids.

Where the market practice requires further description of a product, such as in Europe, the vendor will so specify as the local law requires. A typical label format would be: (hydrogenation); and major phospholipids.

In the United States lecithin has GRAS status (i.e., it is Generally Recognized as Safe). Since other qualifiers like **modified** may suggest or connote another substance to the consumer, such descriptors would be avoided in U.S. labelling.

Phospholipid

For **phospholipid** the Group worked to make a broad description that would not be product - or application - specific. This should simplify further revision later, as new products and applications enter the market. The proposed statement reads:

> **Phospholipids** are generally purified or of synthetic origin. These products must be of specified high quality, identified by I.U.P.A.C. nomenclature.

Our goal here has been to use standard chemical terms.

How might these statements be used outside of the Group? The Group decided to contact editors of scientific journals and of chemical supply catalogs, and officers of societies, such as the D.G.F and A.O.C.S., in order that they appeal for use of precise terms in papers that they print and in product statements. At this time the editors of **Lipids** and **Chemistry and Physics of Lipids** have agreed to make such an appeal in an editorial for the I.L.S.G. Many others will be contacted.

To begin to solve the problem of industry standards being imprecise, narrow, or non-existent, for phospholipids in the States and the European community, the Group has resolved to contact the appropriate pharmacopeias, draft proposals and submit them. The Group has decided to address the U.S.P. **Forum** first.

The U.S.P. staff is well aware of problems with phospholipid terminology. They have told me that they will welcome from us a proposed monograph, or series for each major phospholipid. We are about to begin drafting these monographs; some updated lecithin statements may also be proposed.

If the Group accomplishes even some of these goals, then we will be helping to protect each other and the consumer. The researcher will benefit, too, from this work.

Another major project has been the assembling of a 'Master Monograph' listing of assays currently used to establish the identity of a sample. With these together, the Group plans to rate the selectivity, sensitivity and economy of these assays. Thereafter, we shall select appropriate **quality criteria** for lecithin (where needed) and phospholipid to be used for a particular application. This, too, will require close study of assays now accepted or proposed. We have collated most of these assays now in use.

This precis provides a sketch of what we have been working upon recently. Other challenges also deserve our attention.

II. OTHER ACTIVITIES

The I.L.S.G. can also serve a public relations function for the Industry and the rest of the scientific community. General educational articles about lecithin, as well as phospholipid, may be written for the lay reader to be published in health magazines, such as <u>American Health</u> and <u>Prevention</u>. Not chemical treatises, these would be written to inform and clarify terms, giving an overview of selected health benefits of various product.

Single-page statements highlighting features of these materials also ought to be published in health magazines that might be distributed in doctors' offices. These would serve as non-branded advertisements designed to inform.

A newsletter should be issued by the Group, perhaps twice a year. This would be aimed at informing professionals in the medical and university communities about recent research and beneficial uses of phospholipids. Meetings would also be advertised in such a newsletter.

A toll-free consumer telephone 'hotline' number could be maintained with an agency such as the National Health

Referral Service. This would reach an audience needing information not reached by other media. A one-page flyer could also be mailed to such enquirers.

Eventually, the Group should also be sponsoring special seminars, symposia and colloquia. Seminars could be offered 'piggy-backed' with meetings of other scientific societies, such as the American Chemical Society, American Oil Chemists' Society, and European Biochemical Society. These could be one-day lectureships about lecithin and phospholipid similar in style to those organized by Dr. B. Szuhaj with the sponsorship of the A.O.C.S.

III.   STRUCTURE

An appropriate vehicle to accomplish these and other aims would be a not-for-profit corporation which gives an assembly such as the I.L.S.G. recognized legal identity. Incorporation would give it birth. The delivery in the birthing room would be accomplished by the drafting of a constitution, plus <u>articles</u> defining the <u>aims</u> of the Group. To this would be added <u>by-laws</u> which would indicate the <u>requirements</u> of membership, <u>order</u> of meetings, and other details about the conduct of business and collection of dues.

Responsibilities important to the group in achieving its aim would be met by several committees with particular assignments. The Public Policy Committee would develop a series of contacts with congressional folks, the F.D.A., the U.S.P., etc., to address issues of common concern as they arise.

A Consumer Education Committee would offer material in appropriate format, with copy written at a level understandable by the average consumer, with appealing graphics support. Initially this committee would have to identify themes that would begin to meet media aims. Final products such as fact sheets, e.g. "Phospholipid Updates", should be put into their agenda.

In conjunction with the Group, a Science and Industrial Education Committee would exist to identify particular needs and interest of the Industry.

I am issuing a challenge to plant seeds today for tomorrow's harvest; to carry the message out through the various media. Will you join me in the challenge, or are you willing to wait until others less qualified speak in your behalf, spreading misinformation and passing regulations or laws that cause the industry and the consumer unnecessary grief?

PARTICIPANTS

ARNDT, Dieter   Academy of Sciences of German Democratic
                                     Republic
                              Berlin-Buch/GDR

BAERT, L.   State University of Ghent
                              Gent/Belgium

BARRATT, G.   Université de Paris XI
                              Chatenay-Malabry/France

BELLINI, Fabrizio   Fidia Research Laboratories
                              Abano Terme/Italy

BOARATO, Elena   Fidia Research Laboratories
                              Abano Terme/Italy

BRUNI, Alessandro   University of Padova
                              Padova/Italy

BUCHHEIM, Wolfgang   Federal Dairy Research Centre
                              Kiel/F.R.G.

CADERNI, G.   University of Florence
                              Florence/Italy

COLAROW, Ladislas   Nestlé Research Centre, Nestec Ltd.
                              Lausanne/Switzerland

COUVREUR, P.   Université de Paris XI
                              Chatenay-Malabry/France

DAVIS, Stanley S.   University of Nottingham
                              Nottinghan/United Kingdom

DELATTRE, J.   Université de Paris V
                              Paris/France

DEVISSAGUET, J.Ph.   Université de Paris XI
                              Chatenay-Malabry/France

DEYME, M.   Université de Paris XI
                              Chatenay Malabry/France

DiPATRE, P.L.   University of Florence
                              Florence/Italy

EIBL, H.	Max-Planck-Institut Göttingen
ENGEL, J.	ASTA Pharma Frankfurt
FELDHEIM, Walter	University of Kiel Kiel/F.R.G.
FOONG, W.C.	Dalhousie University Halifax/Canada
FUJITA, Satoshi	Asahi Denka Kogyo K.K. Tokyo/Japan
FUNKE, Andreas	Federal Dairy Research Centre Kiel/F.R.G.
GENIN, I.	Université de Paris XI Chatenay Malabry/France
GREGORIADIS, Gregory	University of London London
GUIDOCIN, Diego	Fidia Research Laboratories Abano Terme/Italy
HANIN, Israel	Loyola University of Chicago Maywood, IL/U.S.A.
HARSANYI, B.B.	Dalhousie University Halifax/Canada
HAWTHORNE, John N.	Nottingham University Medical School Nottingham/U.K.
HERMETTER, A.	Graz University of Technology Graz/Austria
HERSLØF, Bengt	LipidTeknik Stockholm/Sweden
HILGARD, P.	ASTA Pharma Frankfurt
HORROCKS, Lloyd A.	The Ohio State University Columbus, OH/U.S.A.
HUBER, Wolfgang	HK Biomedical, Inc. Berkeley, CA/U.S.A.
HUYS, M.	State University of Ghent Gent/Belgium
KIDD, Parris M.	HK Biomedical, Inc. Berkeley, CA/U.S.A.
LACHMANN, Burkhard	Erasmus University Rotterdam/The Netherlands
LAGANIERE, Serge	Loyola University of Chicago Maywood, IL/U.S.A.

MADELMONT, G.	Université de Paris XI Chatenay Malabry/France
MEZEI, M.	Dalhousie University Halifax/Canada
MIETTO, Lucia	Fidia Research Laboratories Abano Terme/Italy
MILAN, Fabrizio	Fidia Research Laboratories Abano Terme/Italy
MOUFTI, A.	Université de Paris XI Chatenay Malabry/France
NUNZI, Maria	Fidia Research Laboratories Abano Terme/Italy
OLSSON, Urban	LipidTeknik Stockholm/Sweden
PALTAUF, F.	Graz University of Technology Graz/Austria
PEPEU, G.	University of Florence Florence/Italy
PROKOPEK, Dieter	Federal Dairy Research Centre Kiel/F.R.G.
PUISIEUX, F.	Université de Paris XI Chatenay-Malabry/France
REDZINIAK, Gérard	Parfums Christian Dior St. Jean de Braye/France
ROBLOT-TREUPEL, Roblot	Université de Paris XI Chatenay-Malabry/France
SCHNEIDER, M.	Lucas Meyer GmbH and Co. Hamburg/F.R.G.
SCHUMACHER, W.	ASTA Pharma Frankfurt
STEKAR, J.	ASTA Pharma Frankfurt
STERNBERG, B.	Friedrich Schiller University Jena/G.D.R.
STRICKER, Herbert	Biopharmazie der Universität Heidelberg/FRG
SUZUKI, Kazuaki	Asahi Denka Kogyo K.K. Tokyo/Japan
TINGVALL, Per	LipidTeknik Stockholm/Sweden
TOFFANO, Gino	Fidia Research Laboratories Abano Terme/Italy

UNGER, C.	University Hospital Göttingen
VANDERDELLEN, J.	State University of Ghent Gent/Belgium
VAN DER MEEREN, P.	State University of Ghent Gent/Belgium
VANNUCCHI, M.G.	University of Florence Florence/Italy
VIOLA, Giampietro	Fidia Research Laboratories Abano Terme/Italy
YESAIR, David W.	BioMolecular Products, Inc. Byfield, MA/U.S.A.
WATKINS, Tom R.	Jordan Heart Research Group Montclair, NJ/U.S.A.
WIECHEN, Arnold	Federal Dairy Research Centre Kiel/F.R.G.
ZANOTTI, Adriano	Fidia Research Laboratories Abano Terme/Italy
ZEISEL, Steven	Boston University School of Medicine Boston, MA/U.S.A.
ZUBENKO, George	University of Pittsburgh School of Medicine Pittsburgh, PA/U.S.A

LEFT TO RIGHT
FRONT ROW: W. Feldheim, A. Bruni, H. Stricker, B. Lachmann, J.N. Hawthorne,
            L.A. Horrocks, S. Laganiere
BACK ROW: I. Hanin, G. Caderni, T.R. Watkins, G.S. Zubenko, F. Paltauf, G. Gregoriadis,
            U. Pfeiffer, G. Barratt, D. Arndt, D.W. Yesair, P.M. Kidd, S.H. Zeisel

MISSING ON THE PHOTOGRAPH: S.S. Davis, M.G. Nunzi, G. Pepeu, F. Puisieux, G. Redziniak

INDEX

(PL=Phosphospholipid)

Alzheimer's disease (continued)
    and platelet membrane (continued)
        fluidity increased, 205-208
            in Down's patient, 207
    and reticulum,endoplasmic
        abnormality inherited, 208
    and tangle,neurofibrillary, 208
    and tetrahydroaminoacridine, 44
Aminoglycoside, 100
p-Aminosalicylic acid, 95
Amphotericin B, 76,77,115,128,129,
        146,147,181
    in liposome, 128-129
Ampicillin in liposome, 142
Anthracycline, 179
    analogs, 91
Antibody,monoclonal,in liposome,
        155
Antibiotics in liposome, 128-129
    and target inaccessible, 123
Animal experimentation, 15
    methods, 15
Antimetabolite, 179
Antimony in liposome, 141
Ara-C in liposome, 149,156,157
Arachidonic acid, 51-58,243
Archaebacteria,thermoacidophilic, 1
    and tetraether glyceroglycolipids,
        1
ARDS, see Respiratory distress
        syndrome,acute
Asialoglycoprotein, 127
Aspartate aminotransferase, 228
Asthma and surfactant,pulmonary,192
Atherosclerosis, 26
Autacoid, 59

Beef tallow, 169
Betaine, 220-223
Betamimetic,bronchodilatory, 192
Bile salt, 94,96
Bleomycin, 178,181
Blood
    and liposome behavior, 124-126
    pressure, 190
    substitute, 73
Borontrifluoride etherate, 10
Brain, 3
    and aging 43,208,213-218
        and PL, 22
        and PL therapy for, 213
        and polyunsaturated fatty acids,
            51-53
    and cell,cholinergic,decrease,214
    immunocytochemistry, 214
    and lipid composition changes, 213
    and neuron loss, 43,208,215
    and phosphatydilserine, 213-218
    and PL as therapy, 213
Butanol, 276

Calcium, 6,14,23,197,232-234
    gating theory, 233
    and muscle contraction, 233
Calorimetry,differential scanning,
        284
Cancer chemotherapy, 179-180
    and liposome, 128,179-180
Candidiasis, 128-129
5,6-Carboxyfluorescein
    release from liposome, 107-111
        dose-response relationship,
            109-111
Carcinogenesis in colon
    and dietary fat, 167-174
Cardiolipin, 178
Cardiovascular disease, 197
    and food, 197
    lipid hypothesis of, 197
    and lipid peroxidation, 197
Casein, 64
Castor oil, 94
Catecholamine, 234
Cell
    activation,immune, 242-243
    proliferation
        and carcinogenesis in colon,
            169
        and dietary fat, 169
Ceramidehexoside, 261,263
Ceroid, 197
Cheese,soft and lecithin, 257-260
Chemoembolization, 135 see
        Microcapsule, Microsphere
Chemotherapy and liposome
    in cancer, 128
    in infection,microbial, 128-129
Chloroform, 276
Cholesterol, 59,116,125,242
    excess in plasma membrane, 245
        and human diseases, 245
    in extrahepatic cells, 246
    increase with aging, 244-247
    and lecithin, 198-202
    and PL, 197-204
    reverse transport hypothesis,247
Choline, 167-168
    deficiency, 219-231
        and acetyltransferase, 214
        and carcinogenesis, 219, 223-
            226
        dietary, 198,199
        and dysfunction,hepatic, 219,
            220
        and fat metabolism, 168, 171-
            174
        and hepatocarcinoma,murine,
            223-224
        in human, 226-229
            induced,experimentally, 226

Choline (continued)
  deficiency (continued)
    and infertility, 219
    kinase, 223
    and liver infiltration,fatty,
      219
    and muscle function test, 228-
      229
    and neurotransmission,cholin-
      ergic, 228
    and phosphatidylcholine biosyn-
      thesis, 223
    and renal function, 219
    metabolism, 221-223
    in milk during lactation, 221
    in neonate, 220-221
    as nutrient,essential,for humans,
      219
    pathway,metabolic, 220
Choline glycerophospholipids, 52,55
Chromaffin cell,medullary, 233
Chromatography
  charring reaction, 18
  column - , 15-16
  fluorescence - , 18
  gas - , 296
  liquid -, 19-23,26,273
  planar, 16-19
  thin layer, 16-19
Chylomicron, 246
Cisplatin, 178,179
Cis-platinum, 153
  in liposome, 153
Coemulsifier
  PL and triglyceride, 107
Coenzyme Q, 75
Colon cancer and dietary fat, 164-
    174
  and toxicity to intestinal mucosa
    169-174
Corn oil,dietary, 169-174,200
  and cell proliferation in colon,
    169
Corticoid, 192
  bronchodilatory, 192
Cosmetics in liposome, 291-294
Cotton seed oil,emulsified, 70
Cremophor, 74
Cromolyn, 100
Cyclooxygenase, 51,52
Cyclophosphamide, 178,288,289
Cyclosporin A, 95
Cytidylyltransferase, 223
Cytochalasin B, 64
Cytokines, 180, 242
  interferon, 180
  interleukin, 180
Cytosine arabinoside, 179

Daunomycin, 90,92
Daunosamine, 90
Death due to cardiovascular disease,
    197
Decay-accelerating factor, 242
Dementia, see Alzheimer's disease
Dexamethasone palmitate, 75
1,2-Diacylglycerol, 6,7,15,23,59,
    168,223-226,232,233,249
  and carcinogenesis, 224,226
  as second messenger, 234
  synthesis in liver, 224
Diacylglycerol phosphotransferase,
    223
DiacylPL, 64
  internalization,cellular, 64
  by endocytosis, 64
  by translocation, 64
Diazepam, 73,76
Dicyclohexylcarboiimide, 9
Diethylaminoethylcellulose, 15
Differential scanning calorimetry,
    284
Dihomoprostaglandin, 52
Dihomothromboxane B_2, 51
Diisopropylphenol, 74
Dimethylbenzanthracene
  and mammary tumor in rat, 288,289
Dimiristoylphosphatidylcholine, 116
Dipalmitoylphosphatidylcholine, 1
Dipalmitoylphosphatidylserine, 116,
    182
1,6-Diphenyl-1,3,5-hexatrien fluor-
    escence, 205
  for platelet membrane fluidity
    205
Diphtheria toxoid, 130
Diradylglyceroacetate, 54
Distearoylphosphatidylcholine, 125
DMSO-sodium, 9
DNA as target of anticancer agent,
    178
Docosahexaenoic acid, 52
Down's syndrome patient with Alz-
    heimer's disease, 207
Doxorubicin(adriamycin), 90-92,115,
    155-156,177-179,187
  in liposome, 156
  toxicity, 92,144-145,178
    and cardiomyopathy, 145
Draize test for eye irritation in
    rabbit, 107
  for detergent, 107
  for surfactant, 107
Drug
  absorption, 93
  carrier, 73-76,134-135
  chemoembolization, 134-135
  delivery systems, 73-76,83-84

Drug (continued)
  delivery systems (continued)
    benefits, 83
    classification of, 84
    emulsion
      advantages,listed, 74
      disadvantages,listed, 74
    fate, 134
    lipophilic, 86-101
    and liposome, 86
    targeting, 73-76
    withdrawal and PL, 241-255

Econazole in liposome, 279
Edelfosine, 287
Edema,pulmonary, 189
Egg yolk, 3
  lecithin, 4,5,70,80
    PL composition of, 4
  phosphatidylcholine,
    fatty acid composition of, 4
Eicosanoid, 62,177
Eicosapentaenoic acid, 268
Elaidic acid, 268
Electrolyte, 71
Electron microscopy
  freeze-fracture technique, 291-292
Elution,isocratic, 23,26
Emulsion
  colloidal, 76
  evaluation, 79
    by gamma-scintigraphy, 79
    by macrophage, 79
  flocculation and electrolyte, 77
  liver uptake of, 80
  in medicine, 69
  microemulsion, 69
  mixture with drug, 76
    drug release from, 76
  and oleic acid, 72
  particle size, 78
  perfluochemical as blood sub-
    stitute, 73
  evaluation of, 76-79
  stability test, 79
Endocytosis, 64
  of liposome, 138
Endotoxin, 180
Endoxan, 288
Enterocyte
  and PL uptake, 62
Epstein-Barr virus, 250
Erucic acid, 268
Estradiol, 95
Ethanol, 276
  and membrane hyperfluidization,
    251
  and PL,dietary,beneficial, 251
Ethanolglycerophospholipid, 53
Ethanolamine, 9

Ethanolamineglycerophospholipid,
  51,55
  acyl group composition, 52
Ethanolamine plasmalogen, 2,51,55
Ether lipid, 178
  structure, 2
Ethynyl estradiol-3-cyclopentyl ether,
  95
Evaporative light scattering mass
  detector, 273-277
Eye irritation test in rabbit,
  see Draize test

Factor VIII in liposome, 150
Factor X and coagulation cascade,45
Fat
  dietary, 167-174
    absorption, 168
    and cell proliferation,intestin-
      al, 169
    and choline, 167-168
    and colon cancer, 167
    and methionine, 167
    and phosphatidylcholine, 167-168
    and toxicity to intestinal muco-
      sa, 169-174
    transport of, 168
  emulsion , 70,73-76
    for drug delivery, 73-76
    for drug targeting, 73-76
Fatty acids
  free, 15,223,268
  unsaturated in brain, 51-53
Fenretinamide, 98-100
Field desorption mass spectrometry,
  29
Fluorescence, see Fluorophotometry
Fluorophotometry, 109
5-Fluorouracil, 179
Folate metabolism, 221-223
Folch extraction, 15
Freeze-fracture electron microscopy,
  291-292

Ganglioside, 261,263
Gaucher's disease, 129
Gentamycin, 95
Glycerol, 9
Glycerophosphocholine, 6
Glycosylphosphatidylinositol, 235
Glycerophospholipid
  of brain, 22
    of mouse, 53-57
  properties, 4-7
  and reverse phase high-performance
    liquid chromatography, 23
  species,molecular, 23
  structure, 2
  synthesis, 9
Glycerophosphoric acid diester, 6
Glycolipid, 124